T0235727

Cambridge Imperial and Post-Colonial Studies Series

General Editors: Megan Vaughan, Kings' College, Cambridge and Richard Drayton, King's College London

This informative series covers the broad span of modern imperial history while also exploring the recent developments in former colonial states where residues of empire can still be found. The books provide in-depth examinations of empires as competing and complementary power structures encouraging the reader to reconsider their understanding of international and world history during recent centuries.

Titles include:

Tony Ballantyne
ORIENTALISM AND RACE
Aryanism in the British Empire

Peter F. Bang and C.A. Bayly (*editors*)
TRIBUTARY EMPIRES IN GLOBAL HISTORY

Gregory A. Barton
INFORMAL EMPIRE AND THE RISE OF ONE WORLD CULTURE

James Beattie
EMPIRE AND ENVIRONMENTAL ANXIETY, 1800–1920
Health, Aesthetics and Conservation in South Asia and Australasia

Rachel Berger
AYURVEDA MADE MODERN
Political Histories of Indigenous Medicine in North India, 1900–1955

Robert J. Blyth
THE EMPIRE OF THE RAJ
Eastern Africa and the Middle East, 1858–1947

Rachel Bright
CHINESE LABOUR IN SOUTH AFRICA, 1902–10
Race, Violence, and Global Spectacle

Larry Butler and Sarah Stockwell
THE WIND OF CHANGE
Harold Macmillan and British Decolonization

Kit Candlin
THE LAST CARIBBEAN FRONTIER, 1795–1815

Nandini Chatterjee
THE MAKING OF INDIAN SECULARISM
Empire, Law and Christianity, 1830–1960

Esme Cleall
MISSIONARY DISCOURSE
Negotiating Difference in the British Empire, c.1840–95

T.J. Cribb (*editor*)
IMAGINED COMMONWEALTH
Cambridge Essays on Commonwealth and International Literature in English

Bronwen Everill
ABOLITION AND EMPIRE IN SIERRA LEONE AND LIBERIA

Ulrike Hillemann
ASIAN EMPIRE AND BRITISH KNOWLEDGE
China and the Networks of British Imperial Expansion

B.D. Hopkins
THE MAKING OF MODERN AFGHANISTAN

Ronald Hyam
BRITAIN'S IMPERIAL CENTURY, 1815–1914: A STUDY OF EMPIRE AND EXPANSION
Third Edition

Iftekhar Iqbal
THE BENGAL DELTA
Ecology, State and Social Change, 1843–1943

Cambridge Imperial and Post-Colonial Studies Series
Series Standing Order ISBN 978–0–333–91908–8 (Hardback)
978–0–333–91909–5 (Paperback)
(*outside North America only*)

You can receive future titles in this series as they are published by placing a standing order. Please contact your bookseller or, in case of difficulty, write to us at the address below with your name and address, the title of the series and the ISBN quoted above.

Customer Services Department, Macmillan Distribution Ltd, Houndmills, Basingstoke, Hampshire RG21 6XS, England

Insanity, Race and Colonialism

Managing Mental Disorder in the Post-Emancipation British Caribbean 1838–1914

Leonard Smith

Honorary Senior Research Fellow, University of Birmingham, UK

First published 2014 by
PALGRAVE MACMILLAN

Palgrave Macmillan in the UK is an imprint of Macmillan Publishers Limited, registered in England, company number 785998, of Houndmills, Basingstoke, Hampshire RG21 6XS.

Palgrave Macmillan in the US is a division of St Martin's Press LLC, 175 Fifth Avenue, New York, NY 10010.

Palgrave Macmillan is the global academic imprint of the above companies and has companies and representatives throughout the world.

Palgrave® and Macmillan® are registered trademarks in the United States, the United Kingdom, Europe and other countries.

ISBN 978-1-349-43998-0   ISBN 978-1-137-31805-3 (eBook)
DOI 10.1057/9781137318053

Contents

Illustrations

Figures

Tables

Acknowledgements

The completion of this book would not have been possible without assistance, advice and support from a number of people and organisations.

I must firstly acknowledge the contribution of people in the Caribbean. Valuable advice and guidance was received from Professor Pedro Welch. The always-stimulating dialogue with Professor Fred Hickling, and his critical perusal of my early written efforts, has been especially influential. Others I must thank include Newlands Greenidge, Woodville Marshall, Rita Pemberton, Bridget Brereton, Tara Inniss, Althea Clarke, and Pat Stafford. Special thanks are due to Shirley Egbunike for endless hospitality and encouragement.

There has been valuable input from several historians of the Caribbean based in Canada, notably Denise Challenger, but also Melanie Newton, David Trotman and Juanita De Barros. Overseas historians of colonial psychiatry, such as Sally Swartz, Angela McCarthy, Julie Parle, and Lee-Ann Monk, have been most helpful. In Britain, a number of people have contributed in various ways, including my Birmingham colleagues Jonathan Reinarz and Rebecca Wynter, as well as Louise Hide, Margaret Jones, Letizia Gramaglia, Peter Barham, Yolande Jessamy, Waltraud Ernst, Hilary Sparkes, Gad Heuman, Keith Fenby, Daniel Smith, and Richard and Wendy Webster.

The Association of Caribbean Historians and the British Society for Caribbean Studies have provided invaluable opportunities for learning and fruitful interaction. I benefitted considerably from presenting papers at their conferences in Curacao and Martinique (ACH), and Liverpool and Warwick (BSCS). I am also grateful to the organisers of conferences or seminars who invited me to present, at the Universities of Exeter, Glasgow Caledonian, Manchester, Leeds, Oxford Brookes, Heidelberg, Toronto, and the West Indies (Barbados). The discussions that ensued at those events gave me important information, as well as the means to develop and consolidate ideas.

Without the contribution of archives and archivists, libraries and librarians, historians lack the means to proceed. The British National Archives at Kew, and particularly its voluminous Colonial Office collections, were invaluable for this study. Important material was located in the Barbados National Archives; my special thanks are due to

David Williams and his staff for their advice and practical help. I also acknowledge the contribution of staff in the Archives of Antigua, Trinidad and Tobago, St Lucia, St Vincent, and Grenada, as well as the libraries of the Barbados Museum and Historical Society, London School of Hygiene and Tropical Medicine, the Religious Society of Friends, University of Birmingham, University of Warwick, University of the West Indies (Barbados), the National Library of Jamaica, the British Library, and the Wellcome Library.

I express my gratitude to the Wellcome Trust for their assistance with a Travel Expenses Grant that facilitated much of the research leading to compilation of this book.

Finally, but not least, my grateful thanks go to my wife June for her endless patience, quiet encouragement, and companionship in journeys around the Caribbean.

Abbreviations

BPP	British Parliamentary Papers
CO	Colonial Office
DG	*Daily Gleaner* (Jamaica)
FCB	Female Case Books (Barbados)
HYR	Half-Yearly Report of the Poor Law Inspector (Barbados)
JG	*Jamaica Gleaner*
LSHTM	London School of Hygiene and Tropical Medicine
MCB	Male Case Books (Barbados)
MSJ	Medical Superintendent's Journal (Barbados)
TAJ	The Asylum Journal (British Guiana)
TNA	The National Archives (Great Britain)

Introduction

The study on which this book is based came about almost by accident. Whilst on a car journey in Barbados in 2006 my attention was drawn toward some attractive old colonial-style institutional buildings in Black Rock, near Bridgetown. My travelling companion informed me that this was 'Jenkins', the colloquial name for the Barbados Psychiatric Hospital, the island's former lunatic asylum. My historical curiosities were aroused. I had researched and written extensively on the history of psychiatric institutions in eighteenth- and nineteenth-century England, including a recently completed book on Georgian 'lunatic hospitals'. The possibility of a new, related study suddenly appeared. On returning to Britain, literature searches and enquiries confirmed that there had not been extensive historical exploration of the early development of mental health services in the former British colonies of the West Indies, in contrast to regions like India, Australasia, and Africa. The challenge seemed too great to pass by. Moreover, it accorded with a long-standing fascination with the Caribbean region and its peoples, nurtured whilst I was a member of the West Indian Society at the London School of Economics during the social and political ferment of the late 1960s.

A study within the history of colonial psychiatry is deeply problematic. It has to engage with several highly contested aspects of historical scholarship – the histories of empire, medicine, psychiatry, and institutions. Each presents its own complex set of problems, challenges and dilemmas, some of which are shared. Inherent within them all is the axis of power and powerlessness. At one end stands the ruling elite, including landowners, politicians, colonial administrators, agents of the law, and doctors. At the other are found indigenous peoples, the labouring poor, sick patients, inmates, and also subordinate staff located slightly apart from those in their charge. It is the diverse and dynamic interactions

1

between these various elements and groups of people that have stimulated a growing literature on the rise and development of colonial psychiatric institutions.[1]

The introduction of an additional dimension, the history of the 'British' West Indies, adds significantly to the complexities of the task. It is a history deeply intertwined with the profound issues surrounding the large-scale transportation and enslavement of African peoples, their legal emancipation in 1838, and the perpetuation thereafter of White colonial dominance in an altered form. Although slavery had existed in other British colonies, notably in southern Africa, it was the sheer scale, permeation, and prolongation of the system in the West Indies that singled out its peoples' experience. The Atlantic slave trade, enslavement itself, abolition, and the post-emancipation Caribbean have all been the subject of voluminous literatures. Apart from the practical difficulty of digesting even a part of this vast body of work, consideration of the region's history poses particular challenges. The overwhelming presence of manifest evil at its core places the historian in a deep quandary. Attempts at historical objectivity can be mistaken for apologism, whilst outright condemnation of key groups of participants and their actions could be construed as poor history. The problems are further exacerbated by the continuing close linkages between the history of the region, its peoples, and their present states of being. The after-shocks and consequences of enslavement, emancipation and colonial domination remain widely evident in contemporary Caribbean societies.

The enormity of the suffering of Black and Brown people in the Caribbean during the era of enslavement, whether transported from Africa or native Creole, was bound to have profound effects on collective and individual mental health, both at the time and subsequently. Direct documented evidence of insanity before the establishment of asylums is sparse, primarily because it was not the sort of thing that the authorities, or even abolitionists, would actively seek and record. Some isolated, though graphic, early recorded testimony from British and White Creole observers is considered in Chapter 2. However, there was widespread adherence to the views of the prominent Scottish physician and lunacy reformer Sir Andrew Halliday, who claimed in 1828 that 'We seldom meet with insanity among the savage tribes of men. Among the slaves in the West Indies it very rarely occurs.... ' Mental disorders, he contended, primarily occurred among people whose 'organs of the mind' had undergone 'greater development' and 'better cultivation'.[2]

In preparing this volume I have been greatly influenced by discussions and exchanges with the eminent Jamaican psychiatrist and historian

Fred Hickling. He has questioned the whole conception of the origins of mental health care in the Caribbean, suggesting that the region's entire history is based upon a 'European Psychosis' that became apparent over three centuries:

> The period 1500 to 1800 represents the initial manifestation of the European Psychosis which has gripped an entire racial group and which indeed, has affected the entire world. This madness is based on a complex collective delusion, which deems all people in the world with white skins superior to all others with non-white skins; thereby justifying systematic eradication and enslavement of millions of people across the globe...

Coupled with the great collective delusion has been a drive to possess the lands and plunder the resources of the non-White peoples.[3] Hickling goes on to posit that European colonisation of the world 'invented psychiatry as a handmaiden for oppression and control of its delusional system'.[4] One extension of this analysis is that insanity among colonised peoples is primarily a European creation. The main vehicle for managing insanity in the British Empire, the lunatic asylum, was consequently perceived by 'colonial Black people' as merely 'a sophisticated extension of slavery'.[5] In this argument, the expressed progressive ideals of reform-minded medical men provided a cloak for the oppressive and exploitative intentions of the imperial power.

The British colonial authorities did not appear to perceive the dissemination of asylums for the insane in that way. Conscious motivations for their gradual establishment throughout the empire comprised both benevolent and controlling intentions. The system emanating from Britain sought to promote both the 'custody' and the 'cure' of mentally disordered people, goals that sometimes co-existed uneasily.[6] The lunatic asylum occupied a liminal space somewhere between the hospital and the penitentiary, retaining elements of each. The balance varied according to time, place and circumstances. In the colonial context the dynamics were sharpened by the complexities and contradictions inherent in governance. The 'civilising mission' was clearly an important element in the development of institutions for the insane. However, in the British Empire it was never articulated to the same extent as in the French colonies.[7] The less demanding concepts of colonial obligation and paternalism appeared to influence the British ruling classes.[8] In regard to the West Indies, that sense of obligation was magnified, at least in the Colonial Office, by a conception of retaining some residual responsibility for the plight of formerly enslaved peoples.[9] This notion of responsibility lay behind the encouragement to develop lunatic asylums in most Caribbean colonies in the 1840s and 1850s. It also helped to determine the Colonial Office's forthright response to the critical scandals emerging at the Public Hospital and Lunatic Asylum of Kingston, Jamaica in the late 1850s onwards, detailed in Chapter 3.

The ostensibly benign approach of the British toward provision for the colonial insane was reflected in the gradual permeation of the organisational framework that became known as 'moral management'.[10] The moral management system encompassed both custodial and curative aspects. It provided structure, order and discipline in the institution, whilst also encompassing interactions and activities geared toward promoting patients' recovery and rehabilitation. In many ways it represented the conflicting rationales of psychiatric care. In the colonial setting, as will be argued, moral management closely resembled a civilising process (even if not overtly part of a 'mission'). At its heart was organised work and occupation. In the British West Indies, with its history of enslaved, unremunerated labour, the system could be expected to have particular significance. This leads inevitably into consideration of the extent to which the lunatic asylum was part of the colonial apparatus of social control, as distinct from a representation of paternalistic benevolence.[11]

A particular challenge for those studying the social history of mental disorder is to ascertain the voice of the patient. The problem is magnified when investigating the experiences of colonised, particularly non-White, peoples. The nature of the historical records is that most information is provided by medical staff and colonial officials, augmented by politicians and journalists. Patients' experiences, perspectives, and concerns can appear periodically in case records and similar material, albeit refracted through the perceptions of medical men. Unfortunately, only a limited amount of West Indian asylum case records appear to have survived. Where possible this material has been utilised to illustrate the circumstances leading people to the asylums, in Chapter 6, and their behaviours after admission, in Chapter 7. It is especially fortuitous that circumstances in mid-nineteenth-century Jamaica produced a good deal of first-hand testimony from people deemed insane. The Kingston asylum scandal of 1858–61 forms a central element of this book. The extensive literature emanating from it included vital contributions from recent patients, as well as relatives and sympathetic staff members. Their evidence is invaluable in gaining an understanding of the institution's complete dysfunction, and the consequences for those incarcerated within its walls.[12]

In preparing this book I have been conscious of the need to cater for different audiences. People with an interest in the history of British or colonial psychiatry and psychiatric institutions will require relevant contextual information on the former Caribbean colonies, their societies and peoples. Similarly, those whose main interest is in the social history of the former British West Indies will want to understand the development of lunatic asylum provision in the wider British and colonial context. Inevitably this may mean that basic information which is quite familiar to some readers will be unknown to others, and similarly material highly relevant for one reader may be of less value to another. These issues are addressed in Chapter 1, which outlines the rise of the asylum system in Britain and its gradual dissemination to the empire, followed by a summary of key economic, social, and political developments in the Caribbean during the decades that followed emancipation in 1838.

Finally, mention should be made of terminology and nomenclature. By and large, terms and descriptions current in the nineteenth century have been employed, as these best convey the contemporary reality. Thus the words 'lunatic' and 'asylum' have been considered more appropriate than their modern variants, as have other culturally loaded terms like 'idiot' and 'imbecile'. In regard to naming countries, 'British Guiana' has been deemed more apt as a historical colonised entity than 'Guyana', a relatively recent creation as an independent nation state, and a similar decision has been adopted for 'British Honduras', now 'Belize'. In relation to race and ethnicity the term 'Black' has been used to describe people of primarily African origin, and 'White' for those of European origin. The terms 'Brown' and 'Coloured' refer to people of mixed ethnicity, according to context. However, unfavourable descriptions like 'Negro', 'Mulatto' and 'Cob' have only been utilised in the form of direct quotes. The word 'Creole' has been taken to refer to people, of whatever ethnicity, born in the West Indies.

1
Caribbean Institutions in Context

A study of the development of provision for people deemed insane in the British West Indies cannot be approached in isolation. It needs to be placed within a wider context encompassing several key determinants, some common to Britain and its empire but others more specific to the region. In approaching the problems associated with insanity, colonial authorities sought to adopt similar remedies to those utilised in Britain where, from the mid-eighteenth century, a growing consensus emerged that mentally disordered people required care, treatment, and containment in a specialist institution. By 1840 the English lunatic asylum system had attained a degree of sophistication.[1] The institutional model was already being disseminated to the empire, based on an assumption that it would meet the needs of diverse colonial societies, as part of the civilising infrastructure. In its transmission to the West Indian colonies, the British approach required adaptation to meet the particular social, political and economic challenges of a region dominated by the complex realities and consequences associated with the prolonged enslavement of African people, and their emancipation in 1838. Social dislocation and dysfunction were endemic throughout the region and persisted into the next century. These conditions produced fertile breeding grounds for the genesis and perpetuation of mental disorders.

The rise of the public asylum in England

Although the great expansion of England's asylum system occurred during the nineteenth century, its roots were older. The origins of Bethlem Hospital in London dated back several centuries. The opening of the magnificent new Bethlem at Moorfields in 1676 proclaimed that the mentally disordered poor constituted a public charitable responsibility. The hospital remained the nation's most prominent institution for the insane for another hundred years, simultaneously a symbol of enlightened, humane care for a deeply unfortunate group of people and also, in its characterisation as 'Bedlam', a place of disorder, neglect, and mismanagement.[2] The emergence of the voluntary hospital movement in the eighteenth century offered an alternative model.[3] The establishment of St Luke's Hospital for Lunaticks in 1751, financed by public subscription, was subsequently followed by other lunatic hospitals or asylums in several provincial cities, including Manchester, York, and Liverpool.[4] These voluntary institutions were portrayed as a benevolent alternative to the growing number of private 'madhouses' catering both for the wealthy insane and, increasingly, for 'pauper lunatics' supported financially by their parishes.[5]

A rationale of these lunatic hospitals, and some better managed private asylums, was the principle that effective treatment required early removal of the patient from the environment in which the disorder had originated to an institution managed by a specialist practitioner assisted by experienced 'keepers'.[6] Michel Foucault has famously construed the eighteenth century as the period of the 'great confinement' of all manner of damaged and disadvantaged people, the mentally disordered among them; all were social misfits incapable of contributing to the labour market and earning a living.[7] He identified the emergence throughout Europe of specialist institutions for the insane, which became increasingly medical in character.[8] Other historians, particularly Roy Porter, have contended that Foucault's perspective over-states what occurred in England, where the numbers confined as lunatics in specialist institutions, though rising, stayed relatively small.[9] Indeed, many mentally disordered people still remained in their families and communities, or in parish workhouses, some well cared for whilst others were subjected to neglect and harsh treatment.[10]

The currents of thinking and activity associated with the era of 'Enlightenment' yielded significant changes in approaches to madness and its treatment. Earlier conceptions of the madman as a wild beast who required taming, or a savage whose excesses had to be forcibly controlled, gave way to an acceptance that insanity constituted a loss of reason and rationality but not of humanity. Treatment became increasingly directed toward the restoration of reason, albeit accompanied by practical measures to minimise risk to the sufferer and those around him.[11] Madness, as a disease of the brain, was seen to require medical interventions, directed by a physician, who might deploy medicines as well as physical treatments like bleeding, blistering, and water-based remedies such as warm baths or cold showers.[12]

The most significant therapeutic development in the latter half of the eighteenth century was the elaboration of psychological techniques. Dr William Battie, the first physician to St Luke's Hospital, and his colleague and rival Dr John Monro of Bethlem Hospital, both highlighted the importance of 'management' in approaching an individual's mental disorder, according it a significance at least equal to that of medicine.[13] 'Management' referred both to the direct interactions between practitioner and patient, and to the regime and environment within which they occurred. Its essence lay in the physician gaining a degree of control over the patient, by the simultaneous or alternating use of stern authority and a humane, kind demeanour. The methods became refined under the heading of 'moral treatment', describing approaches that were primarily psychological or social rather than medical. Although the Quaker-inspired York Retreat, opened in 1796, has been closely identified with the conception of 'moral treatment', the associated practices were already established in several enlightened public and private institutions.[14] Radical critics of the asylum, like Foucault and Andrew Scull, have construed 'moral treatment', as practised at the Retreat, as offering a sophisticated psychological technique to restrain deviant, non-conforming behaviour, different mainly in degree from the physical and mechanical forms of restraint still prevalent in most lunatic asylums.[15]

By the early nineteenth century, an active lunacy reform movement was promoting the extension of public asylum provision. The parliamentary investigation of 1807 led to legislation in 1808 giving county magistrates powers to erect an asylum for pauper lunatics, or to join with voluntary subscribers and provide also for charitable and private patients.[16] Within a decade several counties had opened an asylum, and others were following.[17] Major parliamentary enquiries in 1815–16 and 1827 highlighted abuses in certain public institutions (York Lunatic Asylum and Bethlem Hospital) and some private 'madhouses', particularly in London.[18] Their deficiencies were contrasted with the progressive practices of the York Retreat, the new county asylum at Nottingham, and a few well managed madhouses.[19]

One of the key emerging elements in 'moral treatment' practices was the elevation of work as an instrument of therapeutics. It had been introduced on a small scale at the York Retreat and at St Luke's Hospital, where patients might be engaged in gardening, domestic tasks, or in assisting with building maintenance.[20] The benefits were first propounded by Dr William Saunders Halloran, in 1810, based on his experience at the Cork Lunatic Asylum in Ireland.[21] However, it was William Ellis at the West Riding Asylum, Wakefield, who first implemented an extensive structured programme of work for patients in a public asylum. By 1825 many of his male patients were engaged in agricultural and gardening work, and trades like tailoring, carpentry, shoe-making, and weaving, in addition to domestic work on the wards, in the laundry and the kitchens.[22] Ellis's efforts at Wakefield demonstrated that work could channel patients' thoughts and energies away from distressing and risky symptoms, thereby promoting recovery and preparing them for a return to the community and labour market. The model was gradually disseminated through the public asylum system, Ellis himself replicating it on a grand scale at the Middlesex County Asylum, Hanwell, after becoming its medical superintendent in 1830.[23]

By the late 1830s there was a well-established 'mixed economy of care' in provision for the insane in England. Mentally disordered people might find themselves institutionalised in a voluntary lunatic hospital, a county asylum, a private asylum, or a local workhouse.[24] The nature and quality of care and treatment ranged throughout from the humane and enlightened to the custodial and seriously defective. The lunacy reformers' cause was boosted in 1838 with the proclamation of the 'Non-Restraint System' in two important public institutions. Robert Gardiner Hill published his celebrated lecture, announcing his complete abolition of strait-waistcoats, chains, straps, handcuffs, and all the other paraphernalia of mechanical restraint, in the (voluntary) Lincoln Lunatic Asylum.[25] Within months John Conolly had replicated abolition in the huge Middlesex County Asylum at Hanwell. Their claimed achievements animated medical and asylum discourses, arousing heated controversy between advocates and opponents of 'Non-Restraint'.[26] A new standard had nevertheless been set for the humane management of insanity, to be incorporated into the armoury of 'moral treatment'.

Armed with the philosophies of non-restraint and work-based treatment, Lord Ashley[27] and his fellow reformers secured a nation-wide investigation into all forms of institutional provision, conducted by the Metropolitan Commissioners in Lunacy during 1842–43.[28] Their seminal report of 1844 provided a comprehensive, detailed survey of conditions in public and private asylums, as well as workhouses accommodating insane patients. The commissioners condemned custodial facilities and the excessive use of mechanical restraint, praising places where these practices had ended and patients were kept occupied. The deplorable state of many private asylums was exposed, whilst the advantages of existing county asylums were highlighted, notwithstanding their overcrowded state.[29] The ensuing report and its recommendations proved highly influential. In 1845 Peel's government implemented the legislation that directed provision for years to come. Henceforth, each county or county borough was required to provide a pauper lunatic asylum, either on its own or in conjunction with neighbouring authorities. A national regulatory body, the Commissioners in Lunacy, would systematically inspect public and private institutions, lay down standards, and oversee the licensing of private asylums.[30]

By 1860 most English and Welsh counties had established a lunatic asylum, creating a network of substantial purpose-built institutions, most with large tracts of land to provide both employment opportunities for patients and room for future expansion. Their regimes were intended to be therapeutic and humane, based on three key elements – the employment and occupation of patients; non-restraint; and the classification of people according to gender, nature and degree of their mental condition, and, in some instances, social class.[31] Essentially, the ideas and principles of 'moral treatment' had been incorporated into a comprehensive system for the organisation and operation of large lunatic asylums, increasingly becoming systematised as 'moral management'.[32]

The scale of development of the county asylum network exemplified the great faith placed in these institutions. The years following the 1845 legislation witnessed a period of widespread 'therapeutic optimism'. It was anticipated that the relief and recovery of the evidently growing numbers of lunatics could be accomplished in the new model asylums. Increasingly, however, that optimism came to be misplaced. The numbers of people being admitted continually exceeded the numbers leaving by either discharge or death. Although many recovered to a level permitting return to family and community, a significant number failed to get better and swelled the body of chronic patients whose condition was deemed 'incurable'. By 1870 most asylums were overcrowded to a point where it was both uncomfortable and detrimental to recovery, and the problems continued to magnify. To meet the exigency, additional wings or stories would be added or new buildings erected.[33] In some counties second asylums were built, and in the most populous like Lancashire, Yorkshire, and Middlesex, several very large institutions had opened by the turn of the century.

There has been argument among historians as to the significance of the ever-growing numbers of people incarcerated in British asylums in the later nineteenth century and after, reflecting debates that took place at the time.[34] One incontrovertible factor was the steady rise in national population, particularly in urban areas.[35] On its own, however, this was not a sufficient explanation. There has been a recurrent assumption that societal changes and dislocations, like industrialisation, urbanisation and loosened family ties, increased the prevalence of mental disorder. Despite evidence to support some of these contentions, the case has not been conclusively proved.[36] More significant may have been a greater tendency to identify people as insane, either potentially able to benefit from asylum admission or otherwise to cause harm, distress, or inconvenience to other people. There were two elements to this greater identification. Firstly, as medical and welfare services became more sophisticated, a larger number of sufferers from mental disorder were observed or diagnosed. Secondly, as Andrew Scull has convincingly argued, the range and degree of mental maladies or deficiencies seen to merit removal and detention in an asylum was steadily expanding. As he represented it, the asylum became a place to sequester people whose behaviour was odd or disruptive, and who were socially inadequate and chronically decrepit.[37]

By the early twentieth century the country possessed an extensive system of very large public lunatic asylums, each accommodating anything from several hundred to 2,000 patients. New, separate asylums had been developed, particularly in the London area, for specific groups such as 'idiots' and 'imbeciles' and the 'quiet chronic'.[38] Although the ideals associated with 'moral management' continued to inform the operation of these institutions, the practicalities of maintaining large numbers of dysfunctional people proved daunting. Individualised treatment had become impracticable. Medical superintendents depended increasingly on detailed organisation, structure, order, and routine to enable the institution to function with a degree of efficiency. Certain elements of 'moral management' proved to be effective aids in the process. The classification of patients, according to behavioural presentation, type of disorder, and prospect of recovery, had become increasingly elaborate, with dedicated wards for 'refractory' patients, the 'convalescent', and people suffering from epilepsy. Work and occupation retained their place at the heart of the therapeutic programme; if anything their significance became even greater. As asylums expanded, so too did attached farmland acreages and the range of employment opportunities. Work was still regarded as the most effective means of promoting recovery, both by engendering a sense of self-worth and by the inculcation of normal routines which might prepare the individual for a return to outside society.[39] For asylum managers and staff, the routines and disciplines of work were also an invaluable means of fostering order and conformity in the institution.

Historians of Britain's mental health services have tended to emphasise the negative aspects of the late nineteenth-century lunatic asylum, highlighting the huge, impersonal, bureaucratic, rigidly ordered, isolated and stigmatised institutions, with their populations of chronic, incapacitated, often chaotic, damaged, and blighted individuals. The consensus has been that therapeutic endeavour was minimal, and that the great asylums became primarily warehouses for containment.[40] They have been regarded as increasingly counter-therapeutic, perpetuating all the problems associated with 'institutionalisation'. Whilst there is much truth in that perspective, it has perhaps obscured what was actually achieved. The country had invested considerable resources in the construction, maintenance and staffing of the asylums. Although motivations are always complex, the investment appeared to indicate a benevolent approach to sufferers from mental disorder on the part of the ruling classes, if not the public at large. The asylums, with their fine buildings, provided shelter and relative comfort for damaged people, many emanating from the most socially deprived sections of the population. And, whilst many did swell the ranks of the chronic and 'incurable', significant numbers experienced relatively short periods of incarceration followed by improvement sufficient for a return to family and community. It was this positive aspect of 'asylumdom' that reformers and colonial officials ostensibly sought to transmit to Britain's colonies.

An empire of asylums

The development of British colonial psychiatry and its institutions has attracted considerable interest among historians over recent decades.[41] Scholarly studies have been published on the history of provision for the insane in former colonies on the continents of Australasia, Africa, Asia, and the Americas.[42] The intersections between the histories of empire and of psychiatry have offered fertile ground for consideration of how diverse ideas, motivations, and forces coalesced to form an apparently cohesive system. More specifically, the colonial lunatic asylum emerged from the dynamics of conquest, power, and exploitation, as well as those of benevolence, paternalism, and humanitarianism. The asylum has been construed as a prime representation of the imperial 'civilising mission', providing a site for both beneficial and oppressive aspects of the processes of 'civilisation'.[43] In practice, the 'moral management' programme of the British reformed asylum resembled a civilising process. The key elements of non-restraint, classification and, particularly, organised occupation, were geared to bring order, structure, and a sense of responsibility to disordered people, so preparing them to resume their place in society.

There has been much debate around the historical role of psychiatry, and of the lunatic asylum, as part of the apparatus of 'social control'.[44] The discourse has expanded to encompass the additional dimensions added by imperialism. The asylum clearly formed one element in the instruments by which the colonial state was managed and ordered, along with other institutions like schools, churches, prisons, and hospitals. To that extent it was a recognisable part of the general means of maintaining order and stability. Several writers, among them James Mills, have gone further and interpreted the colonial lunatic asylum as an overt instrument for imposing social control, by the removal and incarceration of the socially deviant and the intended restoration of people to a sufficient level of competency to rejoin the labour force.[45] Other scholars, like Megan Vaughan, have been more circumspect, arguing that in some colonies the numbers admitted to their small asylums were almost insignificant in relation to overall population, rendering the social control argument almost invalid.[46] In reality the lunatic asylum was in a paradoxical position, for it enshrined elements of treatment and care alongside those of custody and public protection. This wider set of roles informs the perspective that a concentration on 'social control' can lead to an over-simplified analysis, though it was certainly relevant in the development of colonial asylums.

The more custodial aspects of managing insanity predominated in the first half of the nineteenth century. The earliest facilities commonly emerged from the penal system. Before an asylum was constructed, lunatics were routinely held in jails whether or not they had committed a criminal offence.[47] In some instances, particularly in parts of Africa, close proximity between asylum and jail was maintained for decades.[48] Physical conditions in early colonial asylums were frequently poor, with limited medical input, sparse dietary provision, and extensive use of mechanical restraint, under the surveillance of staff whose role closely resembled that of prison warders.[49] To an extent these were instinctive responses by colonial authorities, when the main criteria for asylum admissions centred around violence, perceived dangerousness or public disorder, accompanied by signs of apparent derangement.[50]

There is little evidence that concerted pressure from London led the main colonies, one after another, to establish an asylum for their lunatics. As most had some form of local administration exercising a degree of autonomy, any such collective measure was unlikely to succeed.[51] Nevertheless, a pattern emerged with three distinct phases. The first covered the early decades of the century, when lunatic asylums of some description were opened in a few older colonies. The initial developments were on the Indian sub-continent. A small private 'madhouse' opened in Bombay as early as 1670. Lunatic asylums, initially private but later under governmental authorities, were opened in Madras in 1787 and Calcutta in 1793, catering primarily for White British people. By 1820, several specifically 'Native Lunatic Asylums' had also been opened.[52] There was relatively early activity also in the 'settler' colonies. A small asylum for 20 patients was opened in New South Wales, on Castle Hill, in 1811. Within a few years it became overcrowded and was removed in 1825 to Liverpool.[53] In southern Africa the first provision was in 1818, at the Old Somerset Hospital in Cape Colony, when very basic cells were made available for lunatics.[54] Almost simultaneously, there were comparable developments in Jamaica.[55]

The second period of significant development was in the 1840s and 1850s, influenced by events in Britain, particularly the lunacy reformers' endeavours that culminated in mandatory provision of public asylums after 1845. In some key colonies new asylums were opened to replace or supplement existing inadequate institutions. Across India, after major legislation in 1858, there was a period of unprecedented asylum construction.[56] In the Cape Colony, a new asylum was opened on Robben Island in 1846, initially for 70 patients.[57] In Australia, asylums were established in New South Wales in 1838 and 1849, South Australia in 1846, and Victoria in 1848, and later in the territories of Queensland and Western Australia in 1864 and 1865 respectively. In the more populous southern territories, additional asylums had to be built to meet the steadily increasing demand.[58]

In several parts of the empire the first proper lunatic asylums were established during this period. In Canada, asylums opened in the provinces of Quebec in 1845 and Ontario in 1850, replacing previous facilities located within Montreal and Toronto jails.[59] A small, unsatisfactory asylum opened in Newfoundland in 1846, replaced by a new building in 1855.[60] An asylum for 30 people opened in Singapore in 1841, adjoining the jail.[61] In Sierra Leone the first specific facility for lunatics was created in 1844, as part of the colonial hospital.[62] In Ceylon, legislation to establish an asylum was enacted in 1839 and one of Hanwell Asylum's medical officers, an associate of John Conolly, was brought from England in 1844 to superintend its construction and opening. However, faced with apparently insuperable difficulties in adapting an English design and 'moral treatment' approaches to the tropics, he gave up and returned home. The Ceylon Lunatic Asylum finally opened in Colombo in 1847.[63] In New Zealand, the first asylum was established near Wellington in 1854, followed over the next few years by others in different parts of the country.[64] The 1840s and 1850s also witnessed the establishment of asylums in most of the West Indian colonies.[65] A third phase of activity occurred in the late nineteenth and early twentieth centuries. Existing asylum systems were consolidated and expanded in India, Australia, New Zealand, and Canada. In some colonies older unsatisfactory institutions were replaced, as in Singapore in 1887 and Sierra Leone in 1910.[66] Institutions were established in newer colonies, such as Fiji in 1884.[67] Most significantly, asylums appeared across the recently acquired African territories, exemplifying the aim to transport British 'civilisation' to the continent. These apparent benefits were bestowed on the Gold Coast (now Ghana) in 1888,[68] Egypt in 1895,[69] Nigeria in 1906,[70] Rhodesia (now Zimbabwe) in 1908,[71] and Nyasaland (now Malawi) and Kenya in 1910.[72] In what would later become South Africa, several new asylums opened throughout Cape Colony and in Natal.[73] Although the nature of provision in many British African colonies left much to be desired, it generally compared favourably with asylum conditions in French North African territories.[74]

The specific situations and events that propelled people into a colonial lunatic asylum did not differ significantly from those prevalent in Britain. Although the genesis of mental illness remains subject to ongoing debate, there is wide consensus that it stems from some admixture of biological, medical, hereditary, environmental, psychological, and social contributors. Under the premise that an individual's descent into insanity occurs as a consequence of high stress levels reacting with constitutional predisposition,[75] it might be surmised that the colonial experience provided fertile ground for generating serious mental disorders. The case has been powerfully expounded by Frantz Fanon, the radical psychiatrist from Martinique.[76] From a very different perspective, the 'East African School' of psychiatrists, including Gordon and Carothers, advanced a similar contention, acknowledging rapid social change as a precipitant of insanity.[77] Most historians of colonial mental health have drawn strong connections between the complex economic, societal and cultural pressures endured by colonised peoples and the onset of psychiatric symptomatology.[78] Experiences of poverty, displacement, unemployment, social exclusion, oppression, discrimination, and cultural marginalisation all took their toll, even on the psychologically strong. For some people, the breakdown into insanity signalled defeat by overwhelming forces. In other cases, as Jonathan Sadowsky has argued, symptoms of madness constituted a means to express protest and resistance.[79]

The central role of families in the identification of mental disorder, and in subsequent asylum admission, has been highlighted eloquently by Catharine Coleborne.[80] Relatives almost always viewed committal as a last resort, when their material and emotional resources no longer sufficed to maintain and care for the mentally disordered member. In many instances they would only seek admission when the household situation had become almost intolerable, notwithstanding asylum medical officers' assertions that early identification and treatment were essential to increase the likelihood of recovery. This reluctance was due in part to a sense of familial duty, but also to the fear and stigma associated with the asylum. Among indigenous peoples, where the lunatic asylum appeared as an alien British institution, resistance to committal was magnified, partly accounting for the relatively low numbers admitted. The family tolerance threshold for disruptive behaviour was normally much higher than in 'settler' colonies. Recognition of the possible advantage of admission might come later with a realisation that at least their distressed loved one would be fed and clothed in the asylum.[81]

The family's role and influence might not always be favourable, for interactions within it could conceivably produce the stressors leading a vulnerable member toward the asylum. Other factors present within colonial societies were also delineated alternatively as either symptomatic or causal of mental disorder. Long distance migration, for example, represented a significant upheaval in an individual's life. In some instances it was indicative of a restlessness of spirit that might coincide with a mental disorder.[82] Displacement or migration across borders, often in search of work, was undoubtedly a significant risk factor, particularly where associated with separation from family and culture, a phenomenon that became apparent in southern Africa.[83] In 'settler' colonies, where many people originated from Britain and Ireland, the evidence is that immigrants were admitted to lunatic asylums in significant numbers, often out of proportion to their presence in society.[84]

Certain indigenous religious and spiritual manifestations, conveniently perceived by European outsiders as 'witchcraft' or 'possession', were commonly regarded both as symptoms and causes of insanity.[85] These perceptions contributed greatly to the racialisation of diagnosis, management and treatment of mental illness in colonial settings. Several historians have demonstrated how the development of colonial psychiatry in Africa confirmed and perpetuated stereotypes regarding the comparative psychological vulnerabilities and responses of Africans and Europeans. The alleged 'savagery' of Africans, and their lack of 'civilisation', were construed to account both for their proneness to violent forms of insanity and for the relative absence of disorders, like depression, that prevailed among allegedly more intelligent and sensitive Europeans. Whilst paternalistically inclined medical men acknowledged that factors like the clash between native and imposed cultures brought particular stresses to Africans, these were regarded as little more than a mitigation of their evident inferiority. Indeed, as Sloan Mahone has suggested, Africa itself came to be viewed in some quarters as potentially pathological.[86] Although much of the historiography on this theme has concentrated on Africa, similar patterns of perception by White elites were also evident in Asian, Australasian, and West Indian colonies.[87]

Despite a growing emphasis within social hierarchies on distinctions of race, and European expectations of entitlement to superior conditions in public institutions, racial separation was not universal within colonial asylums. In places as diverse as Ceylon, Australia and New Zealand, as in the West Indies, members of different races were admitted to the same asylum, and were likely to share accommodation and facilities, albeit sometimes under protest. In Fiji, separate wards were provided for European men and women.[88] Even in the South African colonies asylums were not initially segregated, though this began to alter before 1900. Segregation then became the norm, either with distinct buildings for Blacks and Whites or entirely separate institutions.[89] In newer African colonies the asylums were primarily for Africans; White patients from southern, central and east Africa were often sent to South Africa. The notable exception was Southern Rhodesia (Zimbabwe), where some White patients were admitted to Ingutsheni asylum from its opening in 1908, with separate, superior accommodation provided for them. Nevertheless, many still went to South Africa until a separate White-only facility opened in 1926.[90] Separation according to race had the longest history in Indian asylums and became entrenched. The only partially integrated facilities in the time of the East India Company, in Bombay, had been superseded by the mid-nineteenth century, by which time several asylums had been established specifically for natives. In effect, two completely separate networks developed, one for White Europeans and the other for Brown Indians.[91]

One corollary of perceptions of White superiority and Black or Brown inferiority was the determination of appropriate treatment and therapeutics in the asylum. European patients were generally considered more likely to benefit from 'moral management' approaches, being expected to possess the intellectual and emotional sophistication to respond to a humanitarian regime founded on non-restraint, kind treatment, and useful occupation.[92] Based on presumptions of lower intellectual and emotional capacity, Black and Brown patients would normally experience a regime orientated more toward custody and containment than active treatment. However, one central element of the moral management ethos was viewed as universally beneficial – productive work. Manual labour was construed as particularly applicable for native peoples, important in promoting recovery and the resumption of assigned roles within colonial society. Consequently, work remained a prominent constituent of asylum regimes throughout the empire.[93]

Moral management became the dominant paradigm in the dissemination of an imperial lunatic asylum system, adapted according to local circumstances. In several instances the standard bearer was a reform-minded English or Scottish doctor who went out to one of the colonies to become medical superintendent of its lunatic asylum. One of the more prominent was Dr Frederic Manning, appointed medical superintendent of the Tarban Creek asylum in New South Wales in 1867, becoming Inspector General of the Insane for the colony in 1879. Edward Paley, after previous English asylum experience, was superintendent of Yarra Bend asylum in Victoria in the 1860s and also progressed into an inspectorial role.[94] In some instances doctors were appointed by the Secretary of State for the Colonies, following recommendation from the Commissioners in Lunacy, like Dr Joseph Plaxton who took over the Ceylon Lunatic Asylum in the early 1880s.[95] Another who carved out a key position was the enterprising Dr W. Gillmore-Ellis, who directed the Singapore asylum from 1888 until 1909.[96] John William Dodds was a particularly influential man. Formerly deputy superintendent of the Montrose asylum in Scotland, Dodds became Inspector of Asylums in Cape Colony in 1889 and also superintended the newly opened Valkenburg asylum from 1891.[97] Dr James Hyslop, another energetic reforming Scotsman, superintended the Natal asylum from 1882 until 1914.[98]

By 1900 most of the larger established colonies possessed at least one substantial lunatic asylum, usually accommodating several hundred patients. These institutions were generally operating on a moral management orientated system, based around the extensive employment of patients. However, the problems endemic in Britain were also appearing elsewhere. As populations grew so too did the numbers and proportions of people admitted to asylums. Inevitably, accommodation and facilities failed to keep pace. Serious overcrowding became a common feature, accompanied by dilapidated buildings, inadequate sanitary arrangements, poor hygiene, and disease. Impoverished or parsimonious colonial governments were often unable or unwilling to take remedial action. Treatment and management of patients became compromised by a lack of trained medical officers and shortages of suitable attendants. In some instances, neglect and ill-treatment became more prevalent, as standards of care deteriorated.[99]

There was always a 'two-tier' system, for in some smaller colonies population size did not justify the development of a substantial asylum operated on moral management lines. They often had only a basic institution offering food, shelter and custodial care, with minimal comforts.[100] This situation had long prevailed in places like Sierra Leone, and also became the norm for most of Britain's newer African colonies. As various historians have shown, in the Gold Coast, Nigeria, Kenya, Rhodesia, and Nyasaland, the initial asylum facilities were rudimentary, usually with extensive use of mechanical restraint, and in some cases conditions hardly improved for decades.[101] By 1914, taking the state of these asylums together with the deterioration occurring in established institutions in South Africa, Australia, Canada and elsewhere, the overall picture in the British Empire had become rather bleak.

Islands of dislocation and despair

It is beyond the scope of this volume to consider in depth the events, movements and forces that shaped the history of the West Indian slave colonies. The agonising story has been recounted, analysed, dissected and appraised many times.[102] The mass theft of human beings from Africa and their transportation into bondage in the Americas by allegedly advanced and civilised Europeans was one of the great catastrophes of the last millennium. The institution of Black enslavement in the British West Indies was perpetuated for almost 300 years.[103] In retrospect, the extent of evil and injustice made manifest constituted a gigantic collective 'insanity' among the perpetrators, as Hickling has powerfully argued,[104] quite distinct from that engendered among victims without number. If the realities of enslavement and slave societies produced conditions conducive to the production of disordered mental states, the upheavals associated with the post-emancipation era applied new stresses and pressures to vulnerable constitutions. The historical experiences of the peoples of the British Caribbean were largely unique within the empire.[105] The societies and economies of the islands, and the associated mainland territories in British Guiana and Honduras, were forged out of enslavement and the plantation system. Abolition and emancipation brought profound changes. For the rest of the nineteenth century and beyond these were societies in a state of transition. However, there were also distinct continuities; certain realities remained relatively unaltered. The basic economic system was maintained, despite coming under severe strain. Political and governmental arrangements continued largely unchanged. Their perpetuation was reflected in the particular configurations of race, class, and culture that continued to hold sway. The following brief contextual summary of the post-emancipation period attempts only to highlight the main elements that determined the colonies' functioning and the lived experiences of their peoples.

The whole rationale for establishing the West Indian colonies, and for the importation of huge numbers of people from Africa, was economic. The intention was to provide cheap, mass produced tropical foodstuffs for the European market, enabling plantation owners and merchants to reap great financial rewards. The ending of enslavement in 1838 seriously threatened these objectives, leading the planters to adopt measures to maintain their position by keeping labour costs as low as possible. This proved particularly problematic in Jamaica, the largest British sugar-growing island, where steps were taken to keep formerly enslaved people on the plantations, either by encouragement or by coercive means, with varying success. In colonies where land was plentiful and labour in short supply, like Trinidad and British Guiana, they later opted to attract workers from elsewhere. Despite the difficulties, the first few years after 1838 proved reasonably prosperous for some sugar colonies and also for people employed in the industry whose labour was in strong demand.[106]

The early period of apparent prosperity, however, proved illusory. The subsequent decades witnessed cyclical fluctuations in the sugar industry, set against a background of steady decline, with all the ensuing ramifications. The first major downturn was precipitated by the British government's 'free trade' inspired abandonment in 1846 of preferential import duties for empire-grown sugar, giving a competitive advantage to exporters of cheap slave-produced sugar from places like Cuba. The measure proved catastrophic for planters in the British colonies, with substantial falls in demand for their produce. Some became insolvent; plantations changed hands, and their numbers were greatly reduced. Ancillary trades, and those employed in them, were badly affected. The consequences were most seriously felt in Jamaica, where about half the sugar plantations were given up between 1844 and 1854, but there were problems too in Trinidad, British Guiana, and elsewhere.[107]

By the mid-1850s the sugar planters' situation was again improving in several colonies, particularly where they had access to adequate supplies of cheap, capable labour. This was certainly the case in Barbados, with its perennial problem of high population density in relation to cultivable land. In Trinidad and British Guiana there were extensive tracts of fertile land but a shortage of people to work on it; their problems were gradually resolved by large-scale immigration.[108] The industry continued in moderate prosperity until the mid-1880s when competition from European beet sugar precipitated a widespread collapse in demand, bringing inevitable repercussions in reduced numbers of estates and forced amalgamations. Wages were lowered, unemployment rose, and great hardship ensued among working people. These depressed conditions persisted into the early twentieth century.[109] In some islands the difficulties were eased by diversification into other more profitable crops, such as coffee, cocoa and bananas. In Trinidad, there were notable commercial and industrial developments, including the beginnings of the asphalt and oil industries.[110]

In the wake of the post-1838 dispensation, the perceived interests of sugar planters and labourers diverged significantly. The planters wanted to keep people attached to the estates, and available for work when required. In contrast freed people aspired to devote more time to work on their provision grounds, or to acquire smallholdings and become independent peasant proprietors. Where possible, men sought alternatives to the hated estate work, whilst many women opted to devote more time and energy to their families as well as to grow and sell provisions. In colonies with a reasonable land supply, planters found themselves unable to retain adequate labour at the price they wished to pay and wage rates were driven upwards. In Jamaica, the tensions were acute in the first years after emancipation. Attempts by some plantation owners to coerce their labourers by imposing rents for provision grounds or threatening evictions created deep resentment among those previously enslaved, leading many to leave the estates despite inevitable hardships. Motivations toward independence prevailed among formerly enslaved people throughout the Caribbean. Those in Trinidad and British Guiana were most favourably placed, even having a choice between acquiring their own land holdings, continuing to work on the estates at favourable wage rates, or perhaps to do both. In crowded islands like Barbados, however, with the lack of surplus land, there were few alternatives to continued employment on the plantations at low wages.[111]

The initial upheavals and displacements within the plantation system brought considerable distress and hardship. Although Black and Coloured people were now nominally free, their economic options were constricted, unless they had a skilled trade or access to land for independent settlement. Their difficulties were exacerbated over the following decades by the deleterious effects of structural changes in agricultural production and prolonged periods of trade depression. Poverty, hunger and sub-standard housing became widespread realities. Some gravitated toward the crowded, insanitary towns to seek alternative employment, perhaps as hucksters or street traders, but often joining the ranks of under-employed casual labourers, living in wretched conditions.[112] Outward migration became a favoured option for many from the overpopulated islands of Barbados and the Leeward Islands, with Jamaica, Trinidad and British Guiana as favoured destinations.[113] In the early twentieth century the construction works for the Panama Canal attracted many thousands of migrants from all over the Caribbean, but particularly from Barbados and Jamaica.[114]

Inward migration became a factor of the greatest significance, sought by planters wanting additional labour at suitably low wages. Immigrants were encouraged initially from within the Caribbean and from Africa, some of the latter having been 'liberated' from slave ships *en route* to countries where slavery persisted. Several thousand White immigrants were attracted to the region from the Portuguese island of Madeira. Asia, however, became the main source. Some indentured labourers were brought in from China, but the bulk came from the Indian subcontinent. Over 36,000 Indian immigrants went to Jamaica by the early twentieth century, compared to 144,000 to Trinidad and nearly 240,000 to British Guiana. Their increasing presence had considerable economic and social effects, particularly in the latter two colonies where the planters achieved their plentiful labour supply and could reduce wage levels. The plantations again began to thrive, but the native Creoles found their living standards reducing and problems of poverty and hardship accentuated. Rural under-employment fostered a drift to the towns. The immigrants themselves faced enormous problems of adjustment, after all the upheavals associated with migration itself. Their experiences contributed directly to the engendering of social and health difficulties.[115]

Attempts were made to address the medical needs of the immigrants, promoted by colonial administrations who reminded employers of their economic interests in maintaining a healthy workforce.[116] However, overall health and welfare provision remained an area of relative neglect. Indeed, in the years after 1838 it even deteriorated, with the abandonment of plantation hospitals and other basic medical services formerly provided by planters on better managed estates.[117] Significant numbers of doctors left Jamaica and other islands after the ending of estate contracts.[118] The provision of health services became the responsibility of a reluctant public, supported by the impecunious authorities. A voluntary public hospital had been established in Kingston as early as 1776, remaining a White-only institution until after emancipation. Although it received financial support from the colony's government, by the 1840s its extremely poor conditions had become notorious.[119] Colonial hospitals were gradually established in most West Indian territories, including Barbados, Trinidad, British Guiana and Antigua, though these were invariably under-resourced with poor facilities.[120] Some specialist hospitals were also established for people with stigmatised conditions, such as yaws, leprosy, and venereal diseases.[121]

For most of the nineteenth century public health conditions in the main towns, like Bridgetown, Kingston and Georgetown, were dire. The slums of Port of Spain, with what Bridget Brereton called their 'notorious barrack ranges', were among the worst examples, but similar circumstances were replicated elsewhere. Appalling sanitary conditions arising from filthy streets, defective drainage, inadequate sewerage, and open cesspools fostered the dissemination of epidemic diseases, like cholera and yellow fever. The ravages of cholera resulted in numerous fatalities; 30,000 died in Jamaica in 1850, and 15,000 in Barbados in 1854. Effective public health and sanitary measures to confront these problems were implemented extremely slowly.[122]

All the health-related problems of poverty and deprivation prevailed, including extremely high infant mortality rates.[123] Wider welfare measures to address the issues, such as systems for poor relief, were minimal. However, by the late nineteenth century, most islands possessed 'almshouses' or poor-houses in some form or other, catering for the destitute infirm and elderly.[124] Cash-strapped colonial authorities usually acted only when matters became so serious that a response could hardly be avoided. Most of the funds available for public services tended to be channelled into education, especially when there was also an element of reformation for delinquent young people.[125] This intersected with the ruling elites' constant emphasis on the maintenance of order.

The consolidation of the mechanisms of criminal justice was always considered the greatest priority, in the face of the constant threat of demoralisation and social disorder.[126] Dilatory, inadequate health and welfare provision reflected wider issues of colonial governance. In the early post-emancipation period the British government acknowledged some responsibility to address deficiencies in the economic and social infrastructures of former slave colonies. In crown colonies, such as Trinidad, British Guiana, and the islands of the Windward and Leeward groups, British governors could directly influence policy, though their scope for activity was limited by scarce material and financial resources. However, in the old colonies of Jamaica and Barbados, with their long-established planter and merchant-dominated assemblies, resource considerations were outweighed by an intent to demonstrate their political independence from, and residual resentment of, the imperial government. They tended to treat interventions by secretaries of state, civil servants, and particularly island governors, as unwelcome manifestations of interference. Assembly members could choose to obstruct any proposed measures designed to alleviate economic and social conditions or address popular grievances.[127]

Those grievances among the Black population emanated largely from disappointed conceptions of the new 'moral economy', under-pinned by considerations of justice and fairness, expected to prevail after emancipation.[128] Resentments built up regarding lack of access to land for independent cultivation, tenancy arrangements, evictions, labour contracts, unemployment, poverty, discriminatory justice administration, and exclusion from political participation. Many even feared attempts by the authorities to re-impose slavery. Issues and concerns such as these underlay periodic outbreaks of violent communal protest throughout the region, as in Jamaica (1839, 1848, 1859), Barbados (1839, 1863, 1872), Trinidad (1849, 1860), Dominica (1844), St Vincent (1862, 1891), St Lucia (1849), Tobago (1876), Grenada (1895), St Kitts (1896), and British Guiana (1848, 1856, 1896, 1905). By and large these manifestations were spontaneous and localised, though several were sufficiently serious – with destruction of property, injuries and even deaths – to provoke military intervention.[129] Occasionally, events escalated into something on a much larger and more dangerous scale, as occurred in Jamaica in 1865 and Barbados in 1876.

The Morant Bay rebellion in Jamaica in October 1865 was a defining event, for the island itself and for the British Empire as a whole. The events of the uprising, and particularly its brutal suppression instigated by Governor Edward Eyre, aroused both widespread alarm and bitter controversy at the time, and have since produced a considerable literature.[130] The disturbances were sparked by a relatively insignificant court dispute in the coastal town of Morant Bay. Its rapid escalation into a quasi-revolutionary situation indicated the depth of seething disaffection among Jamaica's Black labouring population, barely a quarter of a century after emancipation. Inter-racial suspicion and ill feeling, to the point of hatred, was evinced in the spontaneous violence and the level of brutality occurring among all sides in the conflict. Twenty White people were killed by angry mobs, and many Black people were shot dead by the militia and 'volunteers'. Two hundred were subsequently executed in Morant Bay itself, as part of Eyre's ruthless repression. A further 600 were publicly flogged, and 1,000 houses were burnt down by the authorities as punishment. The identified ringleader Paul Bogle and the prominent Coloured politician George William Gordon were executed for their alleged roles. In the aftermath, Jamaica's constitutional arrangements were suspended and its troublesome assembly abolished, to be replaced by direct rule in the form of crown colony government.[131]

Although the Barbados 'Confederation Riots' of 1876 did not match the scale of Morant Bay, they were momentous enough for the island's people. As in Jamaica, the widespread poverty, discontent, and pervasive sense of injustice among the Black labouring population was easily ignited, as already shown in disturbances in Bridgetown in 1872.[132] The catalyst was a move by the new liberal-minded Governor John Pope-Hennessy, backed by the British government, to promote formation of a Windward Islands Federation that included Barbados. There would be shared arrangements for policing, prisons, and justice administration, as well as for access to the lazaretto and lunatic asylum. The members of the Barbados assembly and executive council, not without reason, saw Pope-Hennessy's 'Six Points' as a means to severely curtail their powers. They marshalled large public protests against the governor and his proposed reforms. Much of the Black labouring population rallied in his support, perceiving him as sympathetic to their many grievances and opposed to the power and privileges of the hated planters. Disturbances broke out on 18th April and lasted for a week, with attacks on property and some destruction of crops and livestock. At least seven rioters were killed and 30 wounded. In the end, amidst fears of a Morant Bay situation, Pope-Hennessy called in military assistance to restore order. Over 400 people were arrested. Pope-Hennessy was soon transferred out to Hong Kong, and the prospects of restricting the powers of the Barbados plantocracy were put back several decades.[133]

The dynamics of racial and class conflict were clearly evident in the various outbreaks of public disturbance in the post-emancipation era, and in the responses to them.[134] Entrenched perspectives and stereotypes were perpetuated, notwithstanding the steady emergence of a moderately prosperous, well-educated Black and Coloured middle class. Indeed, conflicts related to ethnicity were much more complex than merely between Black and White. Shades of colour came to be highly significant in determining social standing, particularly within the ranks of the intermediate classes.[135] Inevitably, a disparate middle class found its interests, loyalties and sympathies divided on key issues. Some sections tended to align with the White ruling class, pursuing goals of 'respectability' with a view to advancement. Other members, as represented by people like the unfortunate George William Gordon in Jamaica and Samuel Jackman Prescod in Barbados, recognised their uncomfortably anomalous position in a highly discriminatory society and sought to uphold the rights and interests of the underprivileged, impoverished Black masses.[136] The whole picture of racial and ethnic conflict was further complicated in Trinidad and British Guiana, and to a lesser extent in Jamaica, by the mass immigration of people from the Indian sub-continent, who not only competed for work and for land but also brought their entirely different manners and customs.[137]

Ongoing distinctions between races and social classes in the West Indies were manifested in the diversity of cultural forms.[138] European- or British-based customs and cultural patterns remained most influential, both among the White elites and the non-White middle classes who aspired toward social betterment, respectability and acceptance. These groups worked hard to disseminate 'civilised' standards downwards, particularly through the vehicles of the educational system and the Anglican churches. By 1900 they had achieved a measure of success in these endeavours, apart from among people of Indian origin relatively untouched by Europeanisation.[139] For the masses of Creoles of African origin, however, the situation was much more complex, particularly in islands like Jamaica and Barbados subjected to longer spans of British dominance. African-based forms and patterns continued to be a strong element, mediated by all the dislocations and disruptions associated with enslavement and its aftermath. Thus, formal Christian worship persisted alongside the numerous fundamentalist and revivalist churches, chapels and meetings. The influence of quasi-religious spiritual activities, like 'obeah' and 'myalism', remained considerable among large sections of the Black and Brown population, despite repeated endeavours by the authorities to suppress them. Patterns of sexual relationship and family life also continued to reflect the cultural dualities. Growing numbers of people opted for formal marriage and settled family units, whilst others continued with patterns of cohabitation, children from different partners, and unstable family arrangements, engendered during enslavement.[140]

By 1914 an uneasy cultural consensus had emerged between the races and classes. The 'Creole society', with a common adherence to a specifically West Indian identity, had attained a degree of maturity.[141] Within it, British-based values, traditions and customs had retained predominance, even enhanced by the embedding of a widespread loyalty to Empire, Crown and Christianity, as well as the dissemination of common sporting pursuits like cricket. At the same time aspects of Black popular culture had become incorporated into the mainstream, like the annual carnivals in Trinidad and comparable events elsewhere.[142] Creole culture advanced social cohesion, helping to link some of the disparate elements of fractured societies. However, many harsh realities persisted. Plantation-based economies remained vulnerable to the vagaries of the trade cycle and to competition from other parts of the world. Emerging new industries in Trinidad and Jamaica offered some promise but had thus far brought little benefit to the bulk of the people. Problems of under-employment, poverty, and hunger were still pervasive. Riotous responses were periodically evident, along with some emergent trade union activity. Political structures, however, remained weighted toward the White elites, despite limited concessions to elements of the Black and Coloured middle classes.[143] Three quarters of a century after enslavement, the British Caribbean colonies could still be construed, politically, economically and socially, as most unhealthy.

Conclusions

Despite some areas of progress after 1838, the British West Indies remained a materially deprived region for the majority of its populations. Although the greatest evils associated with enslavement had been removed, Black and Brown people still faced patterns of living riddled with conflicts and dilemmas, quite apart from the mere struggle for survival. The White elites were not insulated from painful adjustments to a post-slavery society. Given these complex dynamics, and their power to create severe stress, there was ample scope for the emergence of many psychological and emotional casualties among all sections of society.

The legislation of 1845 confirmed a public responsibility to provide regulated institutions for the insane poor of England and Wales.

These were largely based on humane treatment principles, as exemplified by the practices of 'moral management', with productive work at the heart of the system. It followed that a similar acceptance of state responsibility was required for British subjects in the colonies. Versions of the British system were gradually disseminated throughout the empire, nominally adapted to meet local needs and conditions. There were divergent motivations behind provision in Britain, notably considerations of custody and containment set against expectations of benevolent, curative treatment. Such concerns were magnified in the colonies, for the administration of empire itself required a balance between control and benevolence. The colonial lunatic asylum, in the West Indies as elsewhere, represented the imperial project in action.

2
The Early Lunatic Asylums

The economic exploitation, racial oppression, and endemic violence that characterised the formation of the British West Indies from the seventeenth century onwards created inherently pathological societies. The consequential effects on the physical health of individuals received a good deal of attention during the eighteenth and early nineteenth centuries, from local medical men and abolitionists, as well as from plantation owners concerned to maximise the productivity of enslaved people working on their estates.[1] Physical ailments, whether emanating from disease or from bodily injury, were often plain to see and might be amenable to diagnosis and treatment. Mental or psychological disorders, however, were another matter. Their likely prevalence among the enslaved population did not constitute a priority for the planter classes, who had more directly pressing matters to contend with. In any case, actual insanity must have been difficult to detect among the various manifestations of distress or unpredictable behaviour, especially when there was reassurance from presumed experts like Sir Andrew Halliday that insanity 'very rarely occurs' among 'the slaves in the West Indies'.[2]

Emancipation and the ensuing upheavals brought a heightened awareness of the massive social problems confronting colonial governments with minimal resources to address them. Mental disorder was merely one of a range of pressing health-related issues. It was, however, accorded a degree of priority because of the capacity of sufferers to pose risk to others and create disorder or nuisance. Essentially, the management of insanity was considered a public order issue. As a consequence, measures to provide remedies focused on containment and protection of other people rather than serious attempts to offer curative treatment. In the absence of a viable alternative, the means to deal with mental disorder were initially located mainly within the penal system.

Insanity and criminality inevitably became associated within public consciousness, reinforcing a custodial approach. This orientation would persist when designated asylums for the insane came to be established in the West Indies.

As elsewhere in the empire, subsequent developments of facilities to deal with victims of insanity were influenced by what had earlier happened in Britain. A lunatic asylum became the accepted institutional solution, and its provision one of the recognised early calls on public expenditure. However, the philosophical journey toward viewing the asylum as potentially an enlightened, curative institution had not yet been undertaken in the West Indian colonies. Consequently, the asylums opened throughout the region during the 1840s and 1850s were hardly comparable to those recently established in England. Colonial governments, short of funds as they claimed to be, were not able or prepared to contemplate an expensive modern facility to address one particular social problem, however awkward. The sense of duty toward formerly enslaved people, advocated by liberal elements in Britain, was of only marginal influence. The authorities would only consider the creation of small asylums at minimal cost, to meet basic requirements for custody of the inmates and protection for people outside.

Antecedents

The earliest written references to mental disorder and its treatment derive from the Spanish occupation of Jamaica in the sixteenth century. A settler in 1531 described the use of herbs and incantations by the indigenous Arawak people to treat the 'mind-riven' and those who wandered.[3] A facility for European clergy and laypeople experiencing mental disorders was opened in the 1570s, under the guidance of the Jewish *converso* physician Beniamo de Caceres, consisting of a four-room building adjoining the monastery in Santiago de la Vega (later Spanish Town). This early lunatic hospital may well have continued to operate at least until Jamaica changed hands in 1655.[4] The British-sponsored epoch of sugar and slavery that followed, with all its associated conflicts, pressures and dilemmas, provided fertile breeding grounds for manifestations of psychological distress and imbalance, among both enslaved people and their 'masters'.

The genesis of insanity among enslaved people could be firmly located in the deeply painful processes of forcible removal from home, family and Africa itself, followed by the equally traumatic long voyage to an unknown destination in squalidly inhuman, crowded, insanitary conditions, shackled in the lower decks of a slave ship.[5] The consequences for the mental health of those experiencing the 'middle passage' were apparent to any benevolently minded, or conscience-stricken, surgeon practising on British slave vessels. The abolitionist campaigner Dr Thomas Trotter had been a surgeon on one of the ships.[6] When asked by a parliamentary select committee in 1790 whether the slaves appeared 'greatly dejected when they first come on board', he testified that:

> Most of them, at coming on board, shew signs of extreme distress, and some of them even looks of despair; this I attributed to a feeling for their situation, and regret at being torn from their friends and neighbours; many of them, I believe, are capable of retaining those impressions for a very long time.

Trotter related how he had often heard slaves below deck making 'a howling melancholy kind of noise, something expressive of extreme anguish'. The women appeared to be worst affected, and he observed many in 'violent hysteric fits'.[7]

Mental distress was identified as a key contributor to physical sickness and disease on the ships. The surgeon Alexander Falconbridge,[8] a veteran of several voyages to Africa and the West Indies, contended that, along with appalling sanitary conditions, 'a diseased mind' was one of the main causes of Africans succumbing to dysentery.[9] This argument was extended by the naval surgeon Isaac Wilson, based on experiences from his slaving voyage on the 'Elizabeth' out of London during 1788–89.[10] Six hundred Africans were taken on board, and Wilson observed that their countenances were pervaded from the outset by a 'gloomy pensiveness', and the 'appearance of melancholy' persisted. There were 155 deaths during the voyage, mostly from dysentery; Wilson attributed two thirds of these directly to the effects of 'melancholy'. He heard people say in their native language that they 'wished to die'. Their state of 'despondency' led to refusal of 'sustenance' and consequent debilitation, reducing resistance to 'the flux'.[11] He detailed a direct linkage between mental and physical symptomatology:

> The symptoms of melancholy are lowness of spirits and despondency; refusing their proper nourishment still increases these symptoms; at length the stomach gets weak, and incapable of digesting their food: Fluxes and dysenteries ensue: and, from the weak and debilitated state of the patient, it soon carries him off.

Despite strong hostile questioning regarding medical evidence for his conclusions, Wilson refused to budge.[12] Isaac Wilson also gave instances where severe mental distress led directly to death. In one case, an individual's 'despondency' had produced 'madness'. He initially appeared 'well' when coming on board, but soon began to look 'pensive and melancholy', and 'a certain degree of wildness appeared in his countenance'. He became 'noisy', demanding to be removed from his irons. His behaviour continually disturbed both crew and other slaves, until he eventually died 'insane'.[13] Wilson described several apparent suicides. In two instances African men jumped overboard, intending to drown themselves, despite high netting around the decks to prevent such attempts. One of them was seen 'making every exertion to drown himself, by putting his head under water, and lifting his hands up, as if exulting that he got away'.[14] Two women separately managed to kill themselves below deck by hanging or self-strangulation utilising a rope.[15] One young man starved himself to death, after refusing 'any sustenance whatever'. Despite attempts at mild persuasion and then use of the 'cat', he 'always kept his teeth so full shut that it was impossible to get anything down'. His death followed nine days of food refusal.[16] A common factor in these cases was the strong element of conscious resistance, conceivably outweighing any indications of mental disorder.

The implications of transplantation from Africa for the mental wellbeing of enslaved people were highlighted in 1801 by the unusual abolitionist St Kitts planter Clement Caines.[17] He observed that 'Though few die in their passage from Africa, multitudes die in consequence of it.' During the journey, the 'seeds of debility and despondence, of sickness and of death' were sown among them, never to be eradicated. Care and medicine were 'of no avail' for the 'brood of latent disorders' fostered. Any attempts by the 'master' to encourage people or 'reconcile them to their lot' was bound to be in vain, because their 'melancholy and despondence cannot be rooted out'. They were likely to 'pine and droop, linger rather than live, and shortly sink into the grave'. Caines suggested that over a quarter of imported Africans died during the initial 'seasoning' period, as a direct consequence of removal from their homeland. For dramatic effect, Caines brought four recently acquired African slaves to the Leeward Islands Assembly, to demonstrate their debilitated state. One had 'never raised his head or smiled, since I purchased him. There he is. Melancholy has marked him for her own.'[18]

Manifestations of melancholy, despondency, and depression were highlighted by some medical men who wrote about health and disease in the West Indies. Several drew links with the common habit of dirt-eating. According to the medically trained planter David Collins in 1811,[19] the symptoms of eating chalk, clay, or other substance could be spotted when 'a negro is languid and listless, and so much indisposed to motion, as to require to be impelled to it by threats; when he is short-breathed, and unable to ascend a hill without stopping'. Among possible causes, he identified the 'power of the passions' as being well known, for 'we find that negroes, labouring under any great depression of mind, from the rigorous treatment of their masters, or from any other cause, addict themselves singularly to the eating of dirt'. Collins pointed out to his fellow planters that the remedy might lay in looking to their own self-interests:

> If the disorder arises from a depression of spirits, or from the vexatious treatment of the master, that must be corrected; for negroes are sentient and sensible beings; with cheerful minds, they are capable of doing a great deal; but, if broken-hearted, you are to expect only impotent efforts and mortal disease.[20]

Dr James Thomson also drew attention to the deleterious effects of dirt-eating among enslaved people, suggesting that 'melancholy' was a chief cause and that the outcome was often fatal. He drew close linkage with adherence to 'obeah', arguing that 'With adult negroes the most frequent cause of their being addicted to dirt-eating is the temporary relief given to the melancholy, attendant on the idea of their being under witchcraft.'[21]

Thomson interestingly commented that 'madness' was one of several 'hereditary derangements' that were 'almost strangers to the inhabitants of tropical climates'.[22] Dr Thomas Dancer appeared to have come to similar conclusions in 1810, suggesting that both melancholia and mania were 'rare Diseases in this climate', as were other mental disorders like hysteria and hypochondriasis.[23] Their views were presumably based on a narrow conception of what constituted insanity, especially bearing in mind Thomson's notice of the effects of 'melancholy'. More specific evidence of mental distress, often to the point of insanity, among enslaved people can be inferred from incidental sources like the writings of the infamous eighteenth-century planter Thomas Thistlewood.[24] As Trevor Burnard shows, the atmosphere of oppression, cruelty, and sexual violence that pervaded his plantation regime was often reflected in traumatised behaviours. The unfortunate case of 'Sally', bought by Thistlewood in 1762 and sold on in 1784, exemplifies a woman whose painful experiences engendered a listlessness, dejection, and complete absence of spirits characteristic of severe depression.[25]

Thistlewood's diaries show numerous instances of people whose distress led them to try to harm themselves or commit suicide, some succeeding in their intentions.[26] Suicide was, in some circumstances, a rational response to unsupportable circumstances. Nevertheless, it was a measure of extreme mental anguish at the time of commission, and a source of concern to others within and beyond the victim's circle. As Vincent Brown has argued, the combination of oppressive working conditions, loss of status, social isolation, and hunger, in addition to 'the longing to return to ancestral lands', led many enslaved Africans to 'destroy themselves', by hanging, throat cutting, or other more covert means like food refusal. Recently arrived people were always the most prone to suicide.[27] Contemporaries like Dr David Collins noted the vulnerability of new arrivals to suicide during the 'seasoning' period, contending in 1811 that it was a key cause of mortality. Some resorted to it in response to 'severity in the form of harsh rebukes, threats, and chastisements', which 'created disgust and terror'.[28]

Suicides were not restricted to the seasoning period, or indeed to the enslaved, as is well illustrated by some Jamaican press cameos. In December 1817 'a negro named Joe, belonging to Mr Pacinco' hanged himself. He evidently prepared well, having 'deliberately tied a handkerchief over his eyes, got upon a barrel, and suspended himself to a beam, then jumped off the barrel'.[29] Earlier in the year William Smith, a plantation overseer for ten years, 'in a fit of mental derangement, put a period to his existence by shooting himself'. This was also well planned, for he removed his shoe and used his toe to pull the trigger of the musket placed in his mouth.[30] A particularly tragic and intriguing story was that of an unnamed 'free woman of colour' from Spanish Town, who hanged herself in her bedroom, leaving three orphaned children including one only two months old. The reporter pointed out that 'the poor woman bore a good character, was in comfortable circumstances, and no reason can be discovered for her having committed so desperate an act'. The coroner's jury could only conclude that 'she hanged herself in a fit of phrenzy or delirium'.[31] In at least two of these cases there had been an assumption of insanity, even if only temporary in nature. Their reporting may have been significant, occurring at a time when specific hospital provision for the insane in Kingston was being developed.

Madness in jail

In the years prior to establishment of a designated lunatic asylum, Black and Coloured enslaved people displaying signs of mental disorder would mainly have been contained within their families or plantation communities. Where behaviours became too threatening or disruptive, the options would be a period in the dungeon-like 'hothouse' of an estate hospital,[32] or incarceration in a local jail. The latter solution came increasingly to the fore following the demise of most estate hospitals after 1838, and in response to the growing problem of mentally disordered people creating a public nuisance in the streets. It was also the method commonly employed in other British colonial territories.[33] It reflected the authorities' perceived linkage between insanity and criminality. For the sufferers it signified a regime based on custody and punishment rather than care or treatment.

The use of the jail as a place to incarcerate lunatics had been established well before emancipation, and its shortcomings were well recognised. An enquiry commissioner noted that, in St Vincent in 1826, 'persons of every description' were confined in the jail – 'debtors, criminals, runaways, and lunatics', all held in 'dark and damp' cells.[34] He was dismayed by the 'miserable condition' of the jails in all the islands, and was particularly concerned that convicts, debtors, prisoners-of-war, slaves, and lunatics were all 'very promiscuously' confined in the dungeons.[35] A similar situation prevailed in the Trinidad jail, with three lunatics confined alongside 40 other prisoners who included six runaway slaves and nine 'capital offenders'.[36] In some instances incarceration continued for many years, without prospect of an end. In 1830 'the negro Addison', aged 52, was reported to have been in the jail in Berbice (British Guiana) since September 1819. He was described as 'perfectly harmless, under no restraint in the gaol-yard, and confined in the gaol merely because there is no establishment in the colony for the reception of lunatics'.[37]

In the old colonies of Barbados and Jamaica, some lunatics were being housed in the jails even though designated asylums had been established for some time.[38] In 1835 three insane people were reported to be in Bridgetown jail; a few months later the Barbados governor was advised that the jail was 'at present burthened with two maniacs'.[39] In 1839 some 'maniacs' were reportedly confined in Middlesex County Gaol at Spanish Town, Jamaica, in close proximity to prisoners under sentence of death. An observer was greatly concerned about the noise created by the 'maniacs', which 'must distract the minds of unfortunate creatures condemned to death'. Three lunatics, two female, were also stated to be in the St James's House of Correction near Montego Bay.[40] Similarly in British Guiana, the establishment of an asylum in the 1840s did not prevent the continued confinement of some 'private Lunatics' in Georgetown jail several years later.[41]

In other islands use of the jails was standard practice by the 1830s and persisted after emancipation. In Antigua the prison authorities were particularly concerned about the numbers of lunatics in the jail. It was observed in 1837 that some were convicted prisoners, whilst others were sent there 'merely as a place of safety and security, to prevent their doing mischief without'. The unusual step was taken of creating an 'asylum for lunatics' within the jail itself, situated at one end of the basement floor. The rooms had only apertures for ventilation and small gratings in the doors to admit light. The inmates were allowed access to the prison yard for exercise. They were under the care of the jailer, assisted by a nurse 'to cook and attend them', their diet regulated by a visiting surgeon. Prisoner returns for the 'Common Gaol', completed in March 1838, showed 11 lunatics, three male and eight female, nine being 'black' and two 'coloured'. The sharing of facilities was considered highly undesirable for other prisoners, it being lamented that 'even the lunatics are allowed to exercise in the same yard, some of whom were covered with loathsome disease'.[42] Nothing more than protective custody was offered to them for, as Dr Adam Nicholson later recalled, 'care was only taken to prevent them from injuring themselves or others and no means used for curing their mental malady'.[43] When Antigua's first asylum opened in February 1842, 14 people were transferred from the jail.[44]

In Grenada the problem was on a smaller scale but the approaches were similar. By 1847 the financially strapped authorities were considering their options. The St George's common jail was still the only 'place of security for dangerous lunatics', three people being reportedly confined there, one apparently fit for discharge.[45] By March 1850, the number of 'dangerous lunatics' in the jail had risen to five. After £100 was set aside, the lunatics were removed in May 1851 to a rented house in St George's.[46] In some Windward and Leeward Islands the use of jails persisted for rather longer. In St Vincent, despite 'repeated solicitations' by local officials, several lunatics were confined in the jail in 1851, even though 'the place is by no means proper for them as regards either themselves or the prisoners'.[47] In fact, they were not finally removed to an asylum until 1873.[48] In Dominica two lunatics were in the jail in 1852, adding to difficulties in conducting a punishment regime. The island's lieutenant governor unsuccessfully sought their transfer to the asylum in Antigua, and the jail continued as Dominica's only option until 1858.[49]

In St Lucia a different approach was adopted, probably influenced by earlier occurrences in Antigua. It was reported in 1847 that there were 19 lunatics on the island, 17 of them females. Although most were 'harmless and unoffending', four were 'under restraint', presumably in the jail.[50] In 1852, a 'house for the reception of lunatics' was erected within the Royal Gaol precincts, comprising 'two large cells' for confinement of those deemed dangerous. In July 1854 there were two female patients. One was a 'monomaniac' who believed herself to be a nun, committed following complaints from the convent's Mother Superior about her constant intrusions; the other had attempted to set fire to her father's house.[51] The lunatics were removed from the jail during 1855.[52] Nevertheless, as late as 1888 there were again 15 insane patients in the jail, after the island's asylum closed down.[53]

In Trinidad use of the jail persisted longer than in many places. The governor, Lord Harris, complained in 1848 that the channelling of government expenditure toward immigration had serious consequences for public services. He despaired about the consequences for the insane:

> The lunatics and idiots wander at large about the streets, to the annoyance and disgust of all, except when they at times become violent then, if by chance room may be found either in the gaol or the hospital, or the police station, they are confined. Daily, during the last year, have I desired to commence building only a few strong rooms, in which they might be housed, but the want of funds has stopped me.[54]

Harris managed to arrange for the creation of a temporary lunatic asylum within the Royal Gaol, in Port of Spain.[55] Its custodial nature was clear from the equipment ordered through the Colonial Office in London, which included 'One dozen strait Jackets, for men and women' and 'One dozen setts of irons for desperate and violent maniacs'.[56]

John Candler, an English Quaker who had until lately spent several years as superintendent of the York Retreat, visited the jail in early 1850 and was actually quite impressed with its lunatic asylum. He described the rooms occupied by the 14 lunatics as 'good', and saw none under mechanical restraint. He considered the asylum 'very creditable to the Colony', as an initial attempt 'to improve the condition of this afflicted class of society'.[57] Another observer, however, condemned the facility in 1852 as 'unfit for purpose'.[58] Despite its inadequacies, numbers increased steadily. In January 1857 there were 36 lunatics (16 female), accommodated in 24 cells in 'the lower passage of the old Prison'. The jail's keeper observed that 'in consequence of the noise that is kept up by them at night, it is almost impossible for the Night Watch to hear anything else that is going on'.[59] When the Trinidad Lunatic Asylum finally opened in late 1858, 40 people were transferred from the Royal Gaol.[60] This, however, was not the end of the story, for the new asylum was soon filled. By December 1862 there were again 12 male lunatics in the jail, a situation only remedied in 1864.[61]

Mentally disordered people in neighbouring Tobago were still less fortunate. There was no specific provision at all on the island, and as late as 1875 'violent cases' were being held in 'a portion of the gaol'.[62] Two years later, officials including the Chief Justice protested that the 'common gaol is not a suitable place for these unfortunate people', and that it was a 'barbarity'. Apart from three or four dangerous lunatics 'in durance in the gaol', there were four or five more who 'ought to be locked up' but were not sufficiently dangerous. Concerns for the general prison regime were greater: 'The confinement in a jail of these madmen is most objectionable and highly prejudicial to discipline, as from the nature of their case they set all rules at defiance'.[63] The situation was considered completely unacceptable, requiring early remedy, though it was not until 1879 that arrangements were made for the most dangerous to be transferred to the Grenada asylum.[64] There were still two female lunatics 'lodged in the Common gaol' in Tobago in 1882.[65]

The prolonged use of jails to house the insane undoubtedly cemented the perceived connections between madness and criminality. Since the time of enslavement prisons formed an integral part of the colonial repressive structures. Sufferers from mental disorders found themselves confined within an overtly punitive institution, mixed with all manner of stigmatised people, in crowded, insanitary conditions. Even where there were attempts to designate a separate part of the jail to accommodate them, there was often little difference in the conditions they experienced. The best they could expect would be some preferential treatment in regard to dietary needs and basic health care. Protests about the confinement of mentally ill people in a common jail tended to focus more on their undesirable presence among other prisoners than any inappropriateness for dealing with mental disorder. The prison regime, and the attitudes under-pinning it, remained influential in most of the lunatic asylums subsequently developed.

Troubled beginnings

Whilst enslavement persisted, the problems of insanity only received real notice when they presented within the ruling White population. In the old colonies of Jamaica and Barbados, early attempts to establish specific provision were directed accordingly. In the aftermath of abolition, colonial governments reluctantly accepted responsibility for the most pressing casualties among formerly enslaved people. The jail solution was no more than a temporary expedient, for it conceivably created more problems than it had solved. It was left to each individual colony to determine its approach. The consequence, from the early 1840s onwards, was a piecemeal *ad hoc* development across the region of rudimentary lunatic asylum facilities. In most instances, these were set up at minimum cost and with little reference to principles and standards becoming established in Britain.

As the largest, most populous of Britain's sugar islands, Jamaica was the first to establish formal medical facilities beyond what were provided on the plantations. A public hospital was founded in Kingston in 1776, offering in-patient treatment primarily for the White community. It was financed partly by voluntary subscription, supported by additional funding from the Jamaica Assembly. Staffing in 1791 included a part-time physician, two surgeons and two apothecaries, in addition to a White matron assisted by five enslaved women.[66] Returns for 1793 show nine insane people admitted to the hospital during that year.[67] Separate provision had been made for them by about 1815.[68] In late 1818 it was agreed that 'the accommodation for lunatics should be sufficiently extended to admit those unhappy persons from all parts of the island'. In addition, there were to be 12 'wards for those, for whom coercion is necessary', located 'out of the hearing of the rest of the patients'. This required acquisition of land adjoining the hospital for people 'at that stage of their disorder which must distress or incommode other patients'. It needed to be surrounded by a high wall, 'to admit air and exercise without the risk of escape'.[69] Agreement was reached in December 1819 to purchase a site to the east of the hospital, after 'the necessity' for a 'range of twelve additional cells for maniacal patients' had again been 'strongly urged'.[70]

The re-named Public Hospital and Lunatic Asylum shared both management and medical officers. The asylum, located in West Street opposite the hospital, consisted of two parallel structures, one for males and one for females, separated by a wooden fence. Each was 120 feet in

length and 16 feet wide, and contained 12 single rooms. The buildings were surrounded by an open piazza and enclosed within high
walls.[71] Patient numbers grew fairly rapidly. Figures for the year 1827/28
show 78 admissions and 33 discharges; by 1835/36 this had risen to
112 admissions and 30 discharges.[72] There were now 49 patients in
the asylum, in addition to a further 30 quiescent lunatics accommodated in the main hospital.[73] Numbers continued to climb steadily,
reaching 71 in 1845 and 127 in 1849.[74] The original patient population had been mainly White, augmented probably by some 'free
Coloured' people. The racial composition started to change before
emancipation and had altered significantly by 1840. It is highly conceivable that the shift toward a predominantly Black and Brown patient
group underlay the authorities' apparent indifference to deteriorating
conditions.

The Kingston Lunatic Asylum had become a problematic institution
by the mid-1830s. In 1836 the hospital's house surgeon lamented the
'want of accommodation' for the numbers of patients, claiming that
the asylum was not in a 'fit state to receive or benefit maniacs' and
that it was 'totally difficult to pursue any mode of treatment but that
of restraint'. Other medical officers concurred, pointing out serious deficiencies in the buildings and a shortage of 'keepers'. Overcrowding was
exacerbated by the need to place some patients in single rooms, due
either to their higher 'station in society' or because they were particularly 'violent and mischievous'. As a consequence, three or four people
were crowded into the remaining rooms.[75] The effects of its crowded
condition were also noted by the British Quaker abolitionist Joseph
Sturge in 1837.[76]

The situation in the asylum continued to deteriorate. Dr James Scott,
who visited in 1839, later recalled that he had never 'witnessed such
a horrid spectacle'. Nearly all the inmates were confined in their cells,
with most 'in stocks, chained'. Many were fed through window bars, and
the 'odour from the cells was of the most pestiferous kind imaginable'.[77]
Another Quaker abolitionist, John Candler, condemned the asylum in
early 1840 as a 'very unfit receptacle for the insane' and a 'wretched
prison' that was 'dreadfully mismanaged'. Having visited several 'madhouses', he 'never saw or formed a conception of one in which misery
and neglect so cruelly predominate'.[78] Matters had become so critical
that in 1840 the hospital's physician, Dr Edward Bancroft, published a
pamphlet exposing the asylum's defects in detail.[79] He likened it to a
'prison of the worst description'. Patients were 'shut up in their rooms,
both day and night', partly to prevent 'illicit intercourse between the

sexes'. A lack of supervision at night resulted in fights which, Bancroft alleged, had directly caused two deaths over the previous five years.[80] Bancroft's protestations led to some improvements. The assembly legislated in 1840 to clarify criteria and arrangements for admission to the asylum, and in early 1842 for appointment of commissioners to superintend and inspect it.[81] Activities toward developing a new asylum were accelerated. A grant of £10,000 made in 1838 was supplemented by a further £10,000 in 1843, and tenders were invited for a suitable site. An area of land near the penitentiary at Rae's Town outside Kingston was selected, and an experienced asylum architect from England appointed. The foundation stone was laid on 13 April 1847.[82] Within the existing asylum, there were moves to reduce mechanical restraint and introduce employment for patients.[83] However, fundamental problems remained and would contribute greatly to the scandals that erupted in the late 1850s.[84]

Table 2.1 Chronological establishment of lunatic asylums

Colony	Date(s) of opening
Jamaica	c1815; 1860
Barbados	c1820; 1846; 1893
Antigua	1842; 1907
British Guiana	1842; 1866
Montserrat	1846
St Kitts	1846
British Honduras	1846; 1881; 1913
Grenada	1851; 1855
St Lucia	1855; 1874; 1912
Trinidad	1858; 1900
Dominica	1858
St Vincent	1873

Source: Various, see individual endnotes.

In Barbados there was some direct provision for insane patients by about 1820, under the auspices of St Michael's Parish Vestry (Table 2.1).[85] A small 'Madhouse' was situated near its almshouse on Constitution Road, Bridgetown. William Price, the parish clerk, was overseer to both institutions, among a range of other responsibilities. He received £75 per annum specifically for 'his care of Maniacs' and for 'finding White and Black Nurses' for the two 'Establishments'.[86] The madhouse could accommodate six patients in individual cells. Admission was initially restricted to White parishioners, though it began to accept Coloured patients in 1829.[87] In 1831 the 'Lunatic Asylum', as now designated, sustained considerable damage in a hurricane which almost destroyed the almshouse. Repairs were carried out and, despite serious deficiencies, it remained in operation until 1846.[88]

The British military, whose main West Indian headquarters was in Barbados, also reluctantly made provision for insane personnel during the 1820s. A building behind the garrison was set aside for 'Lunatic and Invalid Black Soldiers & Military Labourers'. In 1829 it was required for additional barracks accommodation. Complaints had also arisen about the annoyance caused to soldiers in the adjoining barracks from the 'noise and continual disturbance made by the unfortunate madmen'. The commanding officer sought the removal to England of nine men 'considered as Incurables'.[89] He was rebuffed, with a suggestion that they be removed 'into some of the Colonial Establishments for People thus afflicted'. An enquiry was consequently made to St Michael's Vestry as to whether the 'Lunatic Asylum in Bridgetown' could receive 'about eighteen' Black lunatics, for payment. The response was that the asylum catered solely for parishioners and only had one vacant room, but the request could not be acceded to 'under any circumstances'.[90] The records do not indicate how these men were subsequently dealt with.

In 1836, in response to concerns about disruptions caused by disorderly insane people in Bridgetown jail, Governor McGregor reported that serious thought was being given to providing a new 'Institution' for lunatics in Barbados.[91] Legislation to establish an asylum was passed in June 1840.[92] Consideration was given to placing it on the existing St Michael's site, but this was rejected as unsuitable.[93] The assembly and legislative council, supported by Governor Grey, gained agreement to utilise surplus monies from the 'Slave Compensation Fund' toward the cost of providing an asylum, a hospital and a lazaretto.[94] Building work on the asylum commenced in June 1844 and it opened in 1846, having cost approximately £4,000.[95] The new Barbados Lunatic Asylum was located within a four acre site on a hill about a mile outside Bridgetown, adjoining District 'A' police station. It consisted of two parallel ranges that were 130 feet long and 15 feet wide, each divided into ten rooms, linked by the superintendent's dwelling house. There was also a separate small building 'intended for the more refractory patients'. The buildings formed a quadrangle around an enclosed garden.[96]

The first six people were admitted on 30th March 1846, transferred from the St Michael's parish asylum. They were followed two days later by three from Bridgetown jail. Numbers reached 20 by the end of May.[97]

The new asylum's inadequacies soon became apparent. John Candler, recently superintendent of the York Retreat, noted in December 1849 that there were 32 patients, with some required to share rooms. A further 18 people were awaiting admission, having been refused due to 'want of room'. There was no proper day area for female patients, insufficient exercise grounds, and an absence of separate accommodation for the 'noisy and dirty'. He was shocked to observe that 'Two of the Lunatics were naked!' However, staffing levels were relatively high, there being a superintendent, a matron, and four male and three female assistants. A 'Medical visitor' attended at least twice weekly.[98]

By 1852 the Barbados asylum was in a 'crowded state', with 43 patients. Some additional accommodation had been provided by adaptation of a nearby building intended as a lazaretto. Other identified problems included the apparently 'incurable' state of the patients and a lack of facilities for occupation.[99] Detailed returns submitted to the Colonial Office in 1863 indicated a process of steady deterioration in the institution. Numbers had risen to 60 and the overcrowding was worse. Ventilation was poor and there was no proper sewerage. Several cess pits for collection of 'night soil' were scattered around the grounds. These were cleared 'annually'; the 'effluvia' was meanwhile kept down with lime, soil, and dry leaves.[100] The asylum's difficulties continued to mount steadily over the following years.[101]

Antigua had been a British colony almost as long as Barbados.[102] As previously noted, an asylum of sorts had been created within the jail by 1837. The legislative council agreed in 1840 to establish an asylum for 18 lunatics in a former military fortress, 200 feet up on Rat Island, a barren rock promontory linked by a causeway to the capital St John's.[103] Empowering legislation was enacted in 1841.[104] Rules and regulations for the asylum's management were drafted, and amended following advice from the Metropolitan Commissioners in Lunacy in London.[105] It opened on 1 February 1842, with 13 people transferred from the jail, supplemented by some new admissions. A year later it was badly damaged during the severe earthquake that devastated the island, and the patients were transferred temporarily to an unsuitable, unhealthy house in the town.[106]

The isolated location and custodial nature of the Rat Island asylum rendered it inappropriate from all perspectives. Governor Higginson in 1847 recognised the 'insuperable obstacles' to a 'comprehensive system of treatment' being carried out there, noting particularly the absence of means to occupy the patients.[107] Dysentery and diarrhoea were prevalent, attributed by the asylum's medical officer, Dr Adam Nicholson, to its crowded state. He lamented his inability to pursue 'any systematic plan of moral treatment', due to the 'want of Attendants' or even of 'Common Nurses'. The asylum could offer nothing more than a 'place of safety'.[108] In March 1850 John Candler described the buildings as 'very insufficient in point of room', with 31 patients crowded in to accommodation intended for 23. The sexes were not separated during the daytime. The drainage was bad, and there were numerous other 'faults and inconveniences'.[109] The colony's straitened finances were blamed for the inability to recruit sufficient suitable staff or build additional accommodation.[110]

Complaints continued about the Antigua asylum building's inadequacy and the need for essential repairs.[111] An inspection in 1857 concluded that repairs, alterations and additions were 'indispensable and immediately necessary', with 44 patients now confined within rooms intended to accommodate only 36. Under pressure, the Antigua government found a modest £250 for some immediate work.[112] In 1864 Dr Nicholson blamed the 'defective structural arrangements' and lack of properly qualified attendants for a high usage of 'mechanical coercion'.[113] Returns for 1869 showed little change. Governor Pine conceded that the asylum was 'simply ineffective' as a place for 'curing lunatics', offering nothing more than confinement and basic care.[114]

Pine's criticisms also encompassed the other Leeward Islands, where arrangements were yet inferior. A small asylum had been established in St Kitts by 1846, adjoining the hospital and the poor-house.[115] John Candler noted in early 1850 that it contained ten patients, 'under the care of a keeper and his wife'.[116] The three facilities were managed as one single institution. By 1865 it was seriously overcrowded, and the medical superintendent expressed concern about the difficulties of providing treatment for 'cases of mental derangement', who were liable to escape or assault other patients unless subjected to 'close and constant restraint'.[117] Similar arrangements prevailed in Montserrat; an 'asylum for lunatics' had been created in a former fort by 1846, when it accommodated five patients.[118] In 1849 it was taken over as an isolation hospital for smallpox, and the lunatics were relocated to a place providing also for lepers and paupers.[119] By 1867, 'The Chateau' comprised a poor-house, hospital, and lunatic asylum. The buildings presented 'a most lamentable spectacle of inadequate accommodation and general disorganization', with few beds and no latrines; inmates wandered into the adjoining fields to relieve themselves. Without means of classification or separation, the lunatics mixed freely with other patients.

Escapes were frequent, and there were instances of female lunatics giving birth to 'illegitimate children'. The island's administrator blamed the situation on crippling debts and a lack of co-operation from its assembly.[120] In Dominica, following criticism for lack of provision, a small asylum was established in 1858, adjoining the poor-house in a former military prison on Morne Bruce, 400 feet above the main town Roseau.[121] The Colonial Office in 1864 judged it among the most deficient in the whole region.[122]

Circumstances differed little in the Windward Islands. In Grenada legislation to establish an asylum was passed in 1847, but the authorities could not find a suitable building. In May 1851 a rented house at Bain's Lot, on the waterfront in the capital St George's, was licensed as a temporary asylum. A set of rules was produced, covering arrangements for management, staffing, and treatment of the patients.[123] During 1855 a more permanent asylum was opened in a former military barracks on Richmond Hill, high above the town.[124] The accommodation was basic, without proper drainage or sewerage, and with inadequate facilities for occupying the patients.[125] Some improvements were made during the 1860s, but restraint and seclusion continued in extensive use.[126] In St Lucia a small building was adapted as an asylum in 1855. Six people were confined there, described mostly as 'quiet and inoffensive creatures'.[127] The Colonial Office concluded in 1864 that it 'would be difficult to find any institution more defective in almost every requisite than this'. There were no sewage or sanitary facilities, and nowhere for exercise. Chains and strait-waistcoats were in general use.[128] Sir William Des Voeux, appointed St Lucia's administrator in 1869, later recalled the asylum building as completely unsuitable, being 'entirely without any enclosure, so that the wretched inmates were necessarily confined to the house and its small verandah'. They were subjected to forms of restraint 'long abandoned in more civilised communities'. He managed to arrange for the asylum's replacement.[129]

Early developments in the two mainland colonies were equally inauspicious. In British Guiana, a lunatic asylum was established in June 1842, contiguous to the colonial hospital in Georgetown. At the end of 1846 it contained 25 patients, mostly of African origin.[130] John Candler was scathing about the asylum's defects in late 1849, contrasting it with the hospital which was 'a fine building, admirably suited to its purpose'. The 24 lunatic patients were placed in a 'wretched prison yard and cabins surrounding it'. Their situation was 'miserable' and there was 'nothing whatever to contribute to the physical comfort' of either males or females.[131] By 1859, a new building with 55 'cells' had been provided.[132] However, the improvement was only marginal, for the Colonial Office's report of 1864 placed it among the worst in the region: 'No condemnation could be too strong for the present structure; a collection of confined cells wholly unsuited for a tropical climate, almost without means of ventilation'. The sewerage and latrines were deemed 'faulty in the extreme', and it had no baths. There was no provision for employment or occupation, and only 'some small covered yards' for exercise.[133] The lunatic asylum in Georgetown was evidently beyond saving. In 1866 the patients were transferred to a new institution, in adapted buildings on the site of a former fortress near New Amsterdam in remote Berbice province.[134]

The Central American colony of British Honduras produced a flagrant example of neglect and mismanagement. As elsewhere, the common jail in Belize housed lunatics alongside ordinary prisoners, before a small asylum was established in 1846.[135] Unusually, this building served also as a dispensary. Within a few years the experiment had 'entirely failed in both objects', and in 1854 the asylum again became part of the jail.[136] During 1855 the wooden building housing the lunatics was removed from the jail site to the public hospital, though the patients were still offered no more than 'the careful custody of watchful servants'. The colony's governor conceded it to be only 'a mere place of personal restraint', without attempt to offer a proper treatment system.[137] In 1865 the asylum was described as 'unfit for the purposes to which it is devoted', reflecting no credit on the colony.[138] A damning report by the Public Medical Officer in 1867 condemned the building as 'in every particular ... so objectionable, and so deficient in every comfort, and almost every convenience'. There was almost no ventilation, and no way to keep the cells clean. The woodwork in the roof and windows was dilapidated and rain entered everywhere. The wooden floors were sodden, most patients having lost 'all sense of decency and cleanliness'; an 'offensive odour' pervaded throughout. There was nowhere to seclude dangerous patients, and little scope for exercise or occupation.[139] These circumstances persisted for another decade at least.[140]

Despite Trinidad's relative importance, it was late in establishing a lunatic asylum, due allegedly to the poor state of its finances. As early as 1843 the Colonial Office agreed to make £2,800 available from the surplus Slave Compensation Fund, but this offer was not taken up.[141] In 1847 Governor Harris expressed exasperation about the problems created by wandering lunatics and the shortage of funds that prevented him from making proper provision. He proceeded anyway, and in 1852 was criticised by Sir John Pakington, the Secretary of State, for starting the work without sufficient money.[142] Harris's attempts to divert money from funds to promote Indian immigration were thwarted by the Treasury in London.[143] The Trinidad Lunatic Asylum finally opened in September 1858, located at Belmont in the hills outside Port of Spain. It provided for 40 patients (28 males, 12 females), all transferred from the Royal Gaol. Dr Thomas Murray, the medical superintendent, noted the 'marked exhilaration of spirits and interest' they displayed and the rapid improvement in their mental states. Murray had ambitious plans for developing employment and occupation. However, he acknowledged the great limitations facing the institution, such as insufficient staff, lack of furniture, the bad state of the surrounding grounds and related problems of inadequate water supply, poor drainage, and absence of proper sewerage.[144]

The Trinidad asylum had clearly been built as hastily and cheaply as possible. Within a few months there was insufficient accommodation to meet demand, particularly for violent patients. Overcrowding, already evident by early 1860, became a perennial problem. Murray warned that, without construction of an additional wing, the jail might have to be utilised again,[145] and within a year this came to pass. New buildings were completed in early 1864, increasing capacity to more than 70. Bathing and laundry facilities were provided, and some of the water-supply problems addressed.[146] Over the next 15 years, a pattern of piecemeal additions and improvements continued, instigated by Murray.[147] The asylum nevertheless retained considerable deficiencies.

Conclusions

Two essential elements characterised the response to the insanity problem in the post-emancipation British Caribbean. The first was parsimony. Colonial governments were notoriously short of money, whilst being faced with daunting problems of economic and social readjustment after the ending of enslavement. The metropolitan government offered little assistance, having concentrated its own resources on giving 'compensation' to former slave owners.[148] In these circumstances health and welfare services were not prioritised, despite pressure from London. The particular inconveniences caused by mentally disordered people made their situation unusual. In each colony the authorities felt compelled to take action and to allocate limited funds. However, provision was rarely more than basic. Consequently, as Governor Pine of the

Leeward Islands explained in 1870, the small local asylums could do no more than 'confine the lunatics with more or less comfort, and to feed them and attend to their bodily health'.[149] In some instances even those conditions were not met, for the dominant picture that emerges is one of squalor, neglect and mismanagement.

The second key element might be described as security. The main priority for ruling elites was preservation of order, and this accounted for the preparedness to invest some resources in asylums to which disorderly people could be removed. It also helped to explain the nature of the regime in the early asylums. The custodial system of the jail, though not completely replicated, remained influential. The lunatic was seen to require separation into secure surroundings, for the sake of other people and maybe also himself. Any therapeutic considerations were merely secondary. Apart from ideological aspects, however, additional staff, comfortable surroundings, better food, and active occupation required rather more expenditure than a basic custodial facility. In the aftermath of enslavement and emancipation, the colonial authorities considered people who were poor, Black, and insane hardly worthy of such indulgence.

3
Scandal in Jamaica – The Kingston Lunatic Asylum

> The next day I was told to go and wash and the nurses laid hold of me...they called the men from the other place, and one came out...They stopped my nose and shoved me under the water, and I said they send me here to kill me rather than get me better...I did not mind the ducking so much but the water was not clean, some of the people filth in the water...and then you have to drink all that water when you are being ducked...They were all treated as I was, and some worse, those that could not go to the tank and laid down were dragged along like dead dogs...[1]

Between 1858 and 1861 Jamaica was shaken by a scandal surrounding the operation of the Public Hospital and Lunatic Asylum in Kingston.[2] A series of revelations about conditions and the treatment of patients received extensive coverage in England as well as Jamaica. Much attention focused on the lunatic asylum, from whence the more serious exposures of abuses emanated. The principal actors in the drama included doctors, administrators, present and former members of staff, and recent asylum patients. The organs of colonial government in Jamaica and imperial administration in London were drawn in to the ensuing crisis. In Jamaica, resolution of the problems involved dismissals of key medical and other staff, the closure of the disgraced asylum, and its replacement by a new institution. In the wider imperial context, the scandal became a defining event. The stark and graphic evidence from Kingston was viewed as an indicator of the likely state of other colonial hospitals and asylums. It had considerable ramifications for future health provision, particularly for the insane, throughout the British Empire.[3]

The critical situation exposed in the Kingston Public Hospital and Lunatic Asylum was neither new nor unique. Its antecedents preceded emancipation, following which the institutions were subjected to unprecedented pressures of demand for their services. The buildings and infrastructure, particularly drainage and sanitary arrangements, were not adequate to meet changed circumstances. The site's cess pools and other health hazards, however, differed little from what prevailed in the surrounding district. A description of Kingston in 1853 referred to ill drained, unpaved streets that became like muddy streams after heavy rain. Due to the absence of sewerage, 'the poorer classes avail themselves of the current caused by the periodic rains to cast the accumulated filth of their dwellings and yards into the streets'. Streets were 'littered over with rubbish'; hogs wandered around 'grubbing up the rubbish for food, bestrewing the surface with their ordure', and dead dogs were often to be found. The air was 'tainted with the most baneful effluvia'.[4] Given this sort of environment, the poor sanitary state of the hospital and asylum was hardly unexpected, and it nurtured a disastrous cholera outbreak in 1850.

The crisis of the hospital and lunatic asylum played out against Jamaica's particular political background, whereby the House of Assembly frequently worked in opposition to the views and wishes of the imperial government, represented by the governor. Politics were highly factionalised, characterised by shifting, unstable alliances between interest groups. Party allegiances, to the 'Country Party', the 'Town Party', or the Liberals were unstable, but provoked high levels of rancour and personal animosity.[5] Race constituted a significant underlying factor; several of the leading politicians were Coloured, exhibiting an inconsistent tendency toward support of liberal reforms.[6] Certain governors, notably Charles Darling, the incumbent throughout much of the hospital and asylum affair, embroiled themselves in the political hurly burly, not always in a manner meeting with approval from the Colonial Office.[7] Political perspectives, and their incident personal quarrels, were significant both in management of the institutions and in the manner of tackling the issues.

Looming problems

The Kingston hospital's physician, Dr Edward Bancroft, had brought the state of the lunatic asylum into public view in 1840.[8] He highlighted its crowded state, necessitating confinement of four or five people in a room barely fit for two. The inevitable consequence for 'people of this description, when thus over-crowded', was that 'bickering and quarrels very often arise, as well among the women as the men, which lead to blows, that sometimes inflict severe bodily injury, and always disturb their peace of mind'.[9] There were no means for classification according to mental condition or behaviour, it was difficult to keep the sexes separate, and facilities for exercise and recreation were lacking. The prison-like arrangements were compounded by people being kept locked in their cells for long periods due to low numbers of 'keepers'.[10] Bancroft contended that the asylum was totally inadequate for the island's needs and that it was 'incumbent' on the assembly to 'do away entirely with the present Establishment' and erect a new one.[11] The assembly, however, concluded that the hospital and asylum no longer required their own physician and Bancroft's services were dispensed with.[12]

The lunatic asylum's difficulties, and Bancroft's critique, were in 1843 brought to the attention of Lord Stanley, the Secretary of State, in the context of allegations of ill-treatment of a female patient. George Ossett, an assembly member, visited the asylum and declared it 'a Disgrace to the Country'. Stanley called on the reluctant governor, Lord Elgin, to ascertain the facts and take appropriate action.[13] Elgin duly inspected both institutions and confirmed some of the problems. He pointed out that attempts had been made to regulate the asylum in early 1842, by appointing commissioners to oversee its management, but the legislature was unable to implement Bancroft's proposed reforms due to the expense. Nevertheless, he assured Stanley that 'no endeavour on my part shall be wanting to remedy the existing defects in these Institutions'.[14]

By the time of Elgin's undertaking in late 1843, conditions in the asylum had conceivably worsened. Restraint implements, consisting of 'chains, stocks, hand and feet locks', were extensively used. In some instances 'men were chained on their backs and never allowed to rise'; they were allegedly 'fed like wild beasts', by means of food placed at the end of a stick.[15] Some men would 'eat their own ordure'. Males and females 'had nearly free access with each other'.[16] Inmates were bathed in the open yard, with females washed by a male attendant.[17] Some reforms were implemented, first by the hospital's house surgeon Dr Alexander Campbell, and then under his successor Dr James Scott. Mechanical restraint was discontinued; proper baths were installed. Following David Ryan's appointment as warden of the asylum in late 1845, arrangements were initiated for 'putting the lunatics to labor', on supposed English asylum lines. A piece of land was acquired to provide work for male patients, whilst some women were employed washing, making and repairing clothes. Ryan later claimed considerable therapeutic success from these measures.[18]

Although there had been modest improvements in the manner of treating the asylum patients there was little evidence of significant change in the physical conditions. The medical officers, Drs Joseph Magrath and James Scott, made regular representations about the state of both the hospital and the asylum. In 1846 they pointed out that the cesspools were 'in the immediate vicinity of some of the wards, and the drains are badly constructed'. It was impossible to prevent a 'noisome malaria', leading to 'very deleterious effects'. The problems were compounded in the asylum, with its 'small, ill-ventilated and always over-crowded' cells. A 'considerable number' of the lunatics had died of bowel complaints.[19] In January 1847 a group of medical practitioners inspected the hospital and condemned the inadequate, unsuitable buildings, the accumulation of 'decaying animal and vegetable matter', the 'uncovered gutters', the large cesspool, and so on.[20] Later in the year, Magrath and Scott reported the asylum more crowded than it had ever yet been.[21] In 1849 they once more drew attention to the 'malaria arising from the privies and cesspools' surrounding the hospital and the diseases prevalent both there and in the asylum. Again they noted that the asylum was 'becoming more and more crowded'.[22] Much worse was still to come, however.

John Candler, the former superintendent of the York Retreat, re-visited the hospital and asylum in May 1850. He observed that, although the lunatic asylum was 'much more commodious' than in 1840 and the patients no longer subjected to mechanical restraint, it was still 'little better than a prison'.[23] In March 1850 the medical officers had openly warned the management board of 'the necessity that exists for the removal of the cesspools, and for establishing a proper system of drainage, to carry off the filth and surplus water of the institution'. With its overcrowded state, they dreaded 'the most serious consequences' if drastic measures were not taken.[24] Their worst fears materialised in October 1850, after cholera hit Jamaica.[25] In the hospital a quarter of the patients were attacked with the disease and 39 died. These, however, were modest figures compared to the disaster striking the lunatic asylum, where most inmates contracted the disease; out of 145 patients, 82 died.[26] It had been more than usually overcrowded just before the outbreak, and the first case occurred in a ground-floor dormitory in which 14 people slept. The room was badly ventilated and the 'nuisance' from its open tub for excretions was 'often horrible'. Dr Gavin Milroy, the respected Scottish physician and epidemiologist who reported to the

British government, asserted that conditions in the asylum were directly responsible for the deaths. He condemned it as 'most unsuitable as an abode for the insane' and called for its immediate evacuation.[27]

Even the cholera catastrophe brought little real improvement in either institution, to judge from medical officers' reports. In 1852 Magrath and Scott complained that the conditions fostered diseases like scurvy and, particularly, diarrhoea and dysentery, which 'carry off a large number of those that die, being always most prevalent during and after heavy rains'. They pointed out that the drains would burst and the cesspool filled so rapidly as to become 'utterly useless', causing unpaved areas to become 'a perfect swamp'.[28] During the year 1855/6 some repairs were carried out in both hospital and asylum, though these hardly addressed the hazards of rotting roofs or the drains and cesspools.[29]

Other concerns specifically relating to the lunatic asylum surfaced in the early 1850s. Allegations of ill-treatment led to a House of Assembly enquiry in 1852. The asylum's warden, David Ryan, was accused of having recalcitrant patients ducked or repeatedly plunged into the bathing tank, causing them to become 'exhausted and speechless', or even 'half drowned'. Female patients were ducked (or 'tanked', as it became known) by male staff. Charges of cruelty were upheld against Ryan and he was dismissed in 1853.[30] Nevertheless, his wife Judith Ryan retained her role as matron of the female part of the asylum. It would later emerge that she perpetuated and extended practices associated with her husband.

By the late 1850s, despite attempts to address abuses and ameliorate conditions, the Kingston Public Hospital and Lunatic Asylum was in a parlous state. The Jamaican government was certainly aware of the situation and reluctantly took a more direct role in management of the institution. Legislation in 1855 empowered the governor to appoint commissioners, comprising three members of the island's executive and four others, to oversee the operation of the hospital and asylum, and the appointment (and dismissal) of officers.[31] These new arrangements, however, were not adequate to meet the exigencies arising in 1858.

Four years of turmoil

The circumstances of the hospital and asylum in early 1858 hardly differed from preceding years, when Dr Lewis Bowerbank launched his outspoken attack on the institutions, their officers and systems of management. Bowerbank was an established Edinburgh-trained, Jamaican-born physician, with a particular interest in public health issues. His critics claimed that his initial interventions were motivated by disappointed ambition.[32] In February 1858 he was one of three candidates applying to become the public hospital's consulting surgeon, at a salary of £200 per annum. The hospital commissioners, however, opted not to fill the post, advertising to that effect in the Kingston newspapers in March. Bowerbank, presumably angered by the rejection, published a trenchant pamphlet addressed to the commissioners, detailing extremely poor conditions in both institutions and alleging that serious abuses had occurred. He demanded a full enquiry.[33]

Bowerbank's pamphlet caused a stir, drawing a swift response from Dr James Scott, the house surgeon for the hospital and asylum.[34] Scott considered his own work and motives had been impugned, and his endeavours to expose the institutions' deficiencies and effect improvements ignored. He claimed the lunatics' condition had been 'greatly ameliorated' whilst he was in post and that Bowerbank's criticisms were based on old evidence, as he had not visited the hospital since 1854.[35] Scott's attempts to rebut the allegations appeared extremely defensive, leading Bowerbank to counter with another combative pamphlet.[36] He dismissed Scott's *Reply* as 'from beginning to end a tissue of personal abuse and misrepresentation', full of 'unfounded denials, irrelevant allusions, arrogant assumptions'.[37] Scott was accused of denying or concealing evidence of serious occurrences in the asylum, including extreme violence and 'improprieties' among patients, escapes, and exploitation of their labour by staff.[38]

Bowerbank's exposures and the resulting exchanges received wide press coverage in Jamaica, arousing great public interest. Governor Charles Darling was drawn in, as was the Colonial Office in London. Darling visited the hospital in February 1858 and declared himself 'highly pleased with the clean and comfortable aspect of the place', and 'gratified' by the appearance of the lunatic patients, which 'presented the most marked and happy contrast' to a previous visit some years before. He overlooked the advance preparations made for his arrival.[39] His complacency contributed to an acrimonious meeting with Bowerbank in early April. Darling wanted firmer evidence in support of the allegations before even considering a commission of enquiry.[40] Their relationship steadily degenerated, with Bowerbank insisting he would only submit his evidence to an independent enquiry.[41] Darling complained to Bulwer Lytton, the Secretary of State, about Bowerbank's attitude, claiming that his actions were 'not altogether free from political ends or Party prompting'.[42] As the quarrel became increasingly personal, Henry Taylor, the Colonial Office's leading official on West

Indian affairs, privately expressed exasperation at Darling's avoidance of the key issues.[43] Meanwhile, Bowerbank issued another pamphlet, addressed to members of the legislature. It summarised hospital and asylum affairs over the last 20 years, highlighting their appalling sanitary state, overcrowding, high death rates, and the cholera disaster.[44] Despite having received some specific evidence by December 1858, Darling remained unmoved and Bowerbank threatened to carry his case directly to the British government and public.[45] Their well-publicised dispute rumbled on for several more weeks.[46]

Bowerbank had already attracted attention in England. His initial pamphlet was noticed in *The Lancet*, which summarised the hospital's sanitary evils and urged perseverance in his reforming endeavours.[47] His subsequent 'Letter' to members of the legislature was transmitted to the journal, which in December proclaimed that it had rarely met with 'more convincing evidence of the terrible evils of continued sanitary neglect in a public institution'. It called upon Bulwer Lytton to use his influence with Jamaica's governor to establish a formal enquiry.[48] On 17 March 1859, a question in the House of Commons brought an assurance that Lytton had instructed Governor Darling to make a 'searching inquiry' into the 'state and management' of the hospital and lunatic asylum.[49]

Having meanwhile determined on more forthright action, Bowerbank sailed to London, bringing the campaign to the heart of imperial government. After an apparent rebuff by the Colonial Office, he gained the attention of the Commissioners in Lunacy, the regulatory body for the country's lunatic asylums.[50] He attended one of their weekly meetings in April 1859, making a statement and presenting copies of his publications.[51] The commissioners, under the Earl of Shaftesbury's leadership, showed great interest in Bowerbank's revelations about the Kingston asylum, and he was summoned to several meetings.[52] They adopted his call for an independent enquiry led by someone from England and urged the Duke of Newcastle, the new Secretary of State, to take action. Newcastle communicated his wishes to Governor Darling, pointing out that he expected the Jamaican government to meet the costs.[53]

Bowerbank's presence brought further British press coverage of the asylum's troubles. Condemnatory articles appeared in *The Times* and *The Anti-Slavery Reporter*.[54] In late August *The Times* published a lengthy letter from 'B' (probably Bowerbank himself), which characterised asylum conditions as 'discreditable to civilization', with 'gross mismanagement and cruel treatment'. He castigated the Jamaican authorities' incompetence, insisting that any enquiry be headed by someone from Britain, preferably a Lunacy Commissioner.[55] A correspondence ensued, with several people expressing outrage at the revelations.[56] Coverage continued in *The Lancet*,[57] and reports appeared also in the *Medical Times and Gazette*[58] and *The Journal of Mental Science*.[59]

The Duke of Newcastle was more interventionist than Bulwer Lytton. Following further complaints from Governor Darling about Bowerbank, he expressed 'feelings of little respect for any of the disputants'. He regarded Darling as 'intolerant of differences of opinion', unscrupulous, and lacking in dignity or sense of impartiality.[60] However, despite Newcastle's firmer approach, Darling still resisted an enquiry, arguing that Jamaica's legislature would not be prepared to pay without 'other and stronger grounds', and suggesting that most of Bowerbank's charges were groundless. Although conceding that 'the present asylum is wholly unadapted to the curative treatment of Lunatics', and was 'nothing more than a place of confinement', he insisted that its well known defects required no investigation. He also condemned Bowerbank's renewed 'attempts at agitation' since his return to Jamaica in late September 1859.[61]

Darling's perspective had been strengthened by a report from Daniel Trench, appointed in January 1859 as Inspector and Director of the Public Hospital and Lunatic Asylum, a post created in response to the controversies.[62] Trench reviewed a critique produced by the English Commissioners in Lunacy, based on Bowerbank's claims. Addressing each criticism, he concluded that most alleged abuses were historical and unrelated to present circumstances.[63] He suggested that Bowerbank's allegations were largely based on testimony from employees, whose 'class' fitted them only to provide 'very imperfect data'.[64] Trench acknowledged the defects of the asylum buildings, the privies and the cess pools, but claimed these had been partially remedied.[65] He denied that female patients were left naked in their cells, or exposed to public view whilst partially clothed.[66] He minimised claims of frequent fighting and serious injuries among patients, and categorically denied beatings and ill-treatment by staff.[67] He rejected claims about medical staff shortcomings, and asserted that the alleged inadequacies of the matron, nurses, and other staff had been addressed.[68] Of particular significance was Trench's dismissal of suggestions that patients were forcibly ducked or 'tanked' in a stone bath. He claimed that properly supervised bathing took place each morning, and occasionally at other times when part of 'treatment'.[69]

It was probably the direct intervention of the Earl of Shaftesbury, chairman of the Commissioners in Lunacy, which weakened Darling's resistance.[70] In October 1859, a supporting letter from Shaftesbury to Bowerbank appeared in the *Jamaica Tribune*. Evidently anxious to defend himself against strong criticism from the commissioners, Darling advised Newcastle that, if the legislature agreed to an enquiry, he would recommend a grant of the necessary monies.[71] However, the assembly refused to agree to an enquiry headed by someone from England.[72] As the *Journal of Mental Science* earlier concluded, it had become a 'party question' in Jamaica.[73] Arguments between protagonists had become so heated and personal that Newcastle openly reprimanded Darling for allowing public officers to publish letters in newspapers. He demanded cessation of the practice, as the public were concerned not with personalities but whether abuses had occurred.[74]

Under mounting pressure, Governor Darling became increasingly active. Following advice from the Commissioners in Lunacy about the asylum's 'extraordinary' mortality rates during 1859, Newcastle required scrutiny of the cause of each death, and sought specific explanations regarding the deceased baby of Elizabeth Green, an asylum patient. Darling transmitted the demands to Daniel Trench, who then passed them to Dr James Scott. Scott's equivocal responses downplayed the problems, portraying them as part of a personal attack on him. By now, Darling himself had become sceptical about the information he was receiving.[75]

Events came to a head in the summer of 1860. Ongoing suspicions regarding ill-treatment of asylum patients were confirmed on 26th June when Caleb Hall, the hospital's 'Purveyor', reported an incident to Trench. Having heard a loud noise, he went to investigate and saw a nurse, Nancy Lloyd, 'dragging a Lunatic Patient along towards the Bath in a most unceremonious manner'. When the naked patient, Deborah Lloyd, tried to escape she was forcibly thrown down onto the stone pavement. Blows were exchanged between several people. Trench mounted an immediate investigation, which drew contradictory evidence between Caleb Hall and Judith Ryan, the female asylum matron, who sought to explain things away.[76] Trench suggested to Governor Darling that Mrs Ryan had displayed 'much want of proper surveillance and...neglect in the performance of her duties'.[77] Darling's response was unusually decisive. Nancy Lloyd was immediately dismissed, but more significantly he concluded that Mrs Ryan's 'veracity is not to be relied on' and instructed that she be replaced by a 'more fit person' as asylum matron.[78]

Worse was still to come. Dr Lewis Bowerbank had renewed his role as crusader against abuses, and on 14th July he advised the governor of a 'most extraordinary and frightful case', referring to an unnamed 'poor female' recently discharged from the asylum.[79] Within a few days she was identified as one Ann Pratt, and was invited to present her evidence to Trench on 27th July[80] Meanwhile, the whole saga was placed firmly in the public domain with the publication of Ann Pratt's explosive pamphlet *Seven Months in the Kingston Lunatic Asylum, and What I Saw There*.[81] Although much was in her own words, Bowerbank clearly participated in its production.[82] It detailed harsh and oppressive conditions in the asylum, violence and ill-treatment by staff, and particularly the practice of 'tanking'. Ann portrayed graphically how she and other patients were often forcibly submerged in a tank of water filthy from multiple use, with nurses summoning male 'labourers' and other patients to assist when unwilling or recalcitrant victims resisted.[83] Bowerbank astutely despatched copies of the pamphlet to England, compounding the pressure on the Jamaican authorities.[84]

The hospital and governmental authorities were wrong-footed by the pamphlet. The initial reaction was to treat it as a fabrication, engineered by Bowerbank. Daniel Trench advised Governor Darling that Ann Pratt was a 'poor ignorant woman' and the pamphlet a 'disingenuous contrivance', designed to corroborate Bowerbank's charges.[85] Information regarding Ann's background was gathered from her parish of Hanover, intended to discredit her as a witness. She was described as a fair-skinned 'mulatto' woman, of loose morals. It was claimed variously that she had two illegitimate children from a Jew named Levi, she was cohabiting with another man, she was a prostitute, and she had attempted to stab her children. She was alleged to have falsely accused several (White) men of rape, but could not appear in court to give evidence against them as she was 'raving mad'. Whilst in prison awaiting removal to the asylum, she reportedly smeared and ate her own excrement.[86] Trench maintained that Ann Pratt's narrative was 'from first to last a disgusting overdrawn picture of falsehoods and exaggerations', whilst acknowledging that some staff might have used 'greater roughness and severity' than proper toward violent lunatics.[87] Governor Darling was inclined to accept the denigration of Ann Pratt and dismissal of her evidence, especially as she refused to meet with him. He sought legal guidance on bringing an action for libel against the pamphlet's publishers, and instructed Trench to publish all the relevant papers.[88]

However, Jamaica's Attorney General opted for a different course. After receiving new information regarding the death of a Black patient

named Matilda Carey, in addition to Ann Pratt's charges, he initiated prosecutions against Judith Ryan and other asylum staff for
manslaughter.[89] The case came to the magistrates' court on 17th August
and lasted four days. Evidence was taken from several people, including Mrs Carey's husband and parents, former patients Ann Pratt and
Henrietta Dawson, and Bowerbank. Mrs Ryan denied guilt and claimed
the prosecution had been instituted by 'designing parties'. The magistrate sent the case for full trial.[90] Darling now had little room for
manoeuvre. On 1st September he initiated his own enquiry into 'tanking', and whether its main purpose was bathing or punishment.[91]

Following publication of Ann Pratt's narrative, others entered the fray.
A pamphlet issued in late August 1860 contained extracts from the
diaries of the late Richard Rouse, warden of the hospital from 1854 until
late 1858 when he was summarily dismissed.[92] Rouse had been one of
Bowerbank's chief informants regarding conditions and abuses in the
institutions.[93] His papers, posthumously edited by his son and 'a friend',
provided clear corroboration for Ann Pratt's allegations.[94] The pamphlet
presented graphic detail about overcrowding, sanitary deficiencies, and
fighting among patients.[95] It contained more allegations of violence and
cruelty perpetrated by staff, particularly Mrs Ryan, as well as claims
regarding violent deaths, sexual misdemeanours, and exploitation of
patient labour.[96]

In the increasingly febrile atmosphere patients reacted to events, especially news of staff prosecutions. The new matron reported an upsurge
of violent, obstructive and threatening behaviour among the women,
which she and her staff could barely contain. According to Dr Scott,
previously manageable people had started to 'disregard all rule, to set
at defiance the orders of the Matron, and to resist the attendants in
the discharge of their ordinary duties'. Both patients and staff were,
he claimed, at risk of serious injury. He conceded that 'moral management' approaches were ineffective in these circumstances. Much blame
was placed on Dr Bowerbank, with patients 'constantly ejaculating' his
name. Trench ordered the matron and staff to be 'firm and determined
in the performance of their duties'.[97] Governor Darling demanded the
'utmost exertions' from Trench to preserve order in the asylum.[98]

The enquiries initiated by Darling into tanking, undertaken by Daniel
Trench, brought conflicting responses. Some former staff upheld allegations of cruelty against Mrs Ryan, whilst others testified differently.[99] Dr
Scott denied there was any deliberate process called 'tanking', though
conceded that 'a certain amount of force' was sometimes necessary to
deal with violent patients who resisted 'ordinary bathing'.[100] The most

telling evidence came from Mary Clarke, recently a nurse in the asylum, who resigned after injury caused by a severe blow from a patient. She knew no distinction between bathing and tanking, describing in simple but graphic terms how she held patients' heads between her knees whilst water was poured over them. When they resisted she was 'compelled to fight with them to force them into the bath', and would call for assistance from the 'washerwoman', 'quiet sensible patients', or even one of the male 'labourers'.[101] Trench sought to play down Mary Clarke's evidence, but Darling was taken aback and demanded to know whether Scott and the other medical officers had sanctioned these methods. Scott equivocated, professed ignorance, and then agreed to end the forcible pouring of water over patients' heads.[102]

By late October 1860 many of the papers and pamphlets had arrived in London and were passed to the Commissioners in Lunacy. The outcome of legal proceedings in Jamaica was keenly awaited.[103] However, the trial of Mrs Ryan and several nurses in April 1861 resulted in acquittals.[104] On receiving the news the Duke of Newcastle was outraged. He was 'not satisfied' with Darling's statements and instructed him to forward the judge's evidence notes, accompanied by a full report on the trial and related facts. He demanded information on whether 'tanking' was resorted to 'for purposes of punishment, discipline or intimidation', and whether Drs Scott and Keech had sanctioned or known about it.[105] A chastened Governor Darling carried out his orders. He believed that tanking had been used both for punishment and intimidation; although the medical officers had not sanctioned it, they ought to have been more vigilant. Illustrating his resolve, Darling advised Newcastle that the acquitted nurses had not been reinstated. He also confirmed that the legislature had now established a formal Commission of Enquiry into the 'past and present management' of the Public Hospital and Lunatic Asylum, 'armed with the fullest powers', and that its deliberations had commenced.[106]

The Public Hospital and Lunatic Asylum Commission convened at the Kingston Court House on 14th May 1861, chaired by Judge Alan Ker, along with two members of the Legislative Council, one being Dr Robert Hamilton, a physician not associated with the hospital.[107] For three months the commission considered voluminous evidence regarding both institutions, from doctors, administrators, ministers of religion, trades-people, neighbouring inhabitants, present and previous staff members, former patients, and their relatives.[108] The first two days comprised a detailed presentation by Dr Bowerbank.[109]

The commission was later unapologetic about the initial concentration on his testimony and emphasised the value of his contribution, despite inherent shortcomings.[110] Other prominent witnesses included Dr James Scott,[111] Daniel Trench,[112] Mrs Judith Ryan,[113] the 'labourers' John Williams[114] and Mary Bell,[115] and former patients Elizabeth Scott,[116] Ann Pratt,[117] and Henrietta Dawson.[118] Notwithstanding Dr Scott's defensiveness and denials, Trench's self-justifications, and Mrs Ryan's outright defiance, the evidence was overwhelming. The final report, published in November 1861, constituted a devastating indictment on the hospital and the asylum, their management and oversight, the medical officers, and various members of staff.[119]

A degraded institution

The mountains of evidence emerging after Bowerbank's initial exposé in 1858, culminating in the 1861 enquiry report, revealed that Kingston's Public Hospital and Lunatic Asylum exemplified all that could be defective in a colonial medical institution. Some of the main problems were common to both hospital and asylum. Their location in an unhealthy part of town magnified the difficulties associated with inadequate drainage, sewerage, and sanitation, exemplified by open cess pools, all especially hazardous for sick patients in close proximity.[120] Medical staffing was insufficient to meet the needs of both institutions, exacerbated by the distractions of external private practice. Individual failings were characterised by financial and sexual misconduct as well as poor practice.[121] There were too few subordinate staff, particularly nurses, and most were badly paid, lacking experience, and of poor calibre.[122] Financial malpractice and corruption had prevailed, involving staff members and external contractors, some associated with members of the board of management.[123] The general lack of oversight, regulation, and inspection of all aspects of the institution's operation was only partially alleviated by Daniel Trench's appointment in 1859.[124]

Notwithstanding the hospital's many shortcomings, the lunatic asylum attracted most of the critical attention. The commission fully accepted the graphic evidence regarding physical conditions, particularly on the female side.[125] The testimony of former patient Henrietta Dawson was particularly telling. She described cells with 'only one window each without glass or jalousies, there are iron railings and outside wooden shutters'; they had stone pavements, and 'in each cell there is one or more wooden tubs without covers, these smell very offensive'.

Some cells had iron bed frames, with canvas bottoms, a sheet and pillow, but:

> There were other cells however into which from ten to twelve women were put at night, old and young, sick and well, in these were no bed frames, but two or three wooden inclined platforms for some of the people to lie down on, into these at locking up time the patients were turned with only the chemise on, which they had worn during the day, they were allowed no sheet, no rug or mat, no pillow, in some of these cells there was no tub put, the calls of nature being performed on the stone pavement, these cells are in a sad filthy state in the mornings, the platform pavement and walls being smeared with blood and filth, at particular periods more especially the place presented the appearance of a slaughter house...

These cells had to be washed out each morning, and were constantly damp in the rainy season; 'the very platforms on which the people lie down are soused with water'. To make matters worse, 'the cells are loaded with bugs which devour the people and drive away sleep'.[126]

The asylum's seriously overcrowded state was considered a key factor in precipitating violent outbreaks among patients.[127] Henrietta Dawson described 'fearful battles' in the packed cells. In the morning 'when the people come out of such a cell, their naked, daubed, bloody, disfigured and wounded bodies tell too plainly what they must have suffered'. She claimed that the dead body of Elizabeth Martin, a woman in poor physical health, had been removed from a cell one morning in late 1859.[128] Elizabeth Scott, another former patient placed in a crowded room, described how at night 'the people raised and commenced to fight each other'; she became afraid and sat in a corner.[129] Similar evidence regarding frequent fights, sometimes fatal, came from John Williams, a 'labourer' in the asylum. He confirmed also that particularly violent women would be locked in a cell on the male side.[130]

Fights occurred also during the daytime. The commission concluded that the 'keepers' condoned the violence, even encouraging it, finding 'brutal pleasure in these shocking encounters'.[131] From the publication of Ann Pratt's pamphlet onwards there were numerous allegations of direct staff involvement, contributing to the prosecutions for manslaughter. The commission fully accepted evidence that:

> Not only by the subordinates but by the matron with fists, feet, sticks, broomsticks, straps, umbrellas, or any other weapon which happened

to be in the way they were habitually beaten, sometimes to the effusion of blood. Water was constantly dashed in their faces; they were often violently dragged along the pavement till their bodies were bruised and torn, becoming afterwards a mass of sores...[132]

Henrietta Dawson both witnessed and experienced this sort of treatment by Mrs Ryan and others acting under her orders.[133] According to the 'labourer' Mary Bell:

> I have seen Mrs Ryan beat the people with broom stick and with her Umbrella. I see her jam her umbrella in their skin all about. She kick the people with her foot. She do it to all. If I name one I must name all. She take the turn stick (the stick they prepare the salt-fish with) and beat the people with it. One day she take it and lick a woman on the forehead with it, the place raise up, and the blood spill out into her eye and over her face and dress.[134]

Despite Mrs Ryan's vehement denials, corroborated by Dr Scott,[135] the commission was inclined to believe the testimonies of Mary Bell, Henrietta Dawson, and others.

It was the element of violent coercion that aroused general outrage about 'tanking', regarded by the commission as the chief abuse perpetrated in the asylum.[136] There were lucid descriptions of the process, by people who had experienced it, observed it, and carried it out. One of the most powerful came from Henrietta Dawson, who clearly distinguished between bathing, with the patient's consent and co-operation, and tanking. In the latter:

> the person is seized by nurses or others, generally four or more in number, one holds one leg, another the other leg, another takes one or both arms, and another the head, the extremities are then extended and the person plunged into the bath, and kept under the water, frequently one or more assistants getting into the tank also, and one sitting on the chest or shoulder of the person, while the one who has hold of the head seizes the person by the throat, and sometimes especially when there is hair on the head to be grasped, swings the head from side to side of the tank...

The 'unfortunate' would be kept submerged 'till all resistance ceases'. She had repeatedly seen Mrs Ryan 'standing by crying out with apparent delight' to ' "give it her well" '.[137]

Elizabeth Scott recounted being forcibly put into the tank by nurses, assisted by a man. Afterwards, unable to stand or sit, she had to be laid down on the ground. Her strongest objection was to the filthy state of the water, which she involuntarily drank.[138] John Williams, a 'labourer' regularly ordered to assist, confirmed that the water was changed daily and numerous people were bathed in it. He acknowledged its use for punishment; 'it was customary to tank the Lunatics when they were troublesome or noisy', for they 'stood in great fear of tanking'. He described his own role:

> I would get into the Tank, and hold the two hands of the Lunatic, a person outside would hold her feet, and she would be shoved under the water. I held the nose and pressed down the chest, and when the people fought I had to press them down in the water – there used to be great struggling – the women wore the chemise only, and in the struggle would expose themselves shamefully.[139]

Ann Pratt's much disputed claims had been fully verified.[140] The commission, without reservation, condemned outright all aspects of the 'tanking' system. Although it accepted that the realities might have been concealed from Drs Scott and Keech, it denounced their grave dereliction of duty in not exercising proper oversight.[141]

Other lunatics were enlisted to assist with tanking resistant patients. Ann Brown told how, in punishment for a misdemeanour, she had been taken to the tank 'by a mad woman like myself'.[142] Ann Pratt earlier described how, on Mrs Ryan's orders, she was tanked by three female staff members, assisted by a patient named Mulgrave. She was plunged face down into the water, and held down in the tank. According to Ann, 'I struggled hard; I fought for my life. Mulgrave sat upon my chest, and when I threw her off she sat upon my shoulders'.[143] On another occasion three lunatics had assisted in tanking her.[144] The involvement of patients in restraining or coercing others went beyond tanking. Ann Brown reported that she had assisted the night nurse to separate fighting patients.[145] One particularly controversial figure was Nicholas Steele, an epileptic lunatic patient allocated menial tasks like emptying cesspools and cleaning out cells in the mornings.[146] The former hospital warden, Richard Rouse, described Steele as 'ferocious' and prone to alcohol. He was often allowed to supervise other patients, male and female. Rosa Henry alleged that she was beaten by Steele across the forehead with a stick, leaving a bad scar. There were numerous instances when Steele had beaten female patients, sometimes with a cat o' nine tails.[147] He was summoned to assist with the tanking of females.[148] The commissioners concluded that Steele became an object of 'implacable terror and hatred' to other patients.[149]

Numerous assertions were made regarding the misuse of patient labour. As earlier noted, measures to occupy them were initiated in the 1840s, with ostensibly therapeutic intentions. James Scott still maintained in 1861 that employment was 'a means of the speedy recovery of patients' and that he had 'never seen any harm result from this'.[150] However, the original aspirations had evidently become distorted. According to Reverend George Trueman, who lived opposite the hospital, the 'Lunatics have been made to work very hard in the institution, in the hot boiling sun'.[151] He had seen instances when patients refused to wash clothes and 'the nurses go after them and hold them by the back of their necks and pitch them back to go and work'.[152] There was corroboration from Elizabeth Scott, who recalled that she and others were kept in the heat of an open piazza: 'we were compelled to sit all day and the lady gave us clothes to mend'.[153] Ann Canning claimed that Mrs Ryan forced them to sweep the yard, carry water, and wash clothes. If they refused, she would order them to be tanked.[154]

Allegations multiplied that Mrs Ryan, the matron, benefitted personally from patients' work. Henrietta Dawson did needlework for her and dressed her hair. She stated that those working for Mrs Ryan or the institution received an additional half loaf of bread with their rice water. Henrietta described comprehensive arrangements, whereby patients undertook household and other tasks at Mrs Ryan's home, some distance from the asylum, including making clothes for sale.[155] According to Charlotte Campbell, Sarah Carter did much work for Mrs Ryan, including making her grandchildren's clothes. She and others were kept back in the asylum for their work, long after their recovery.[156] Henrietta Dawson confirmed that people 'who are in their senses' were prevented from leaving, and that 'good seamstresses and persons who made themselves useful to Mrs Ryan have frequently been kept thus'. Relatives seeking discharge had been advised that they were still insane.[157] The commission concluded that patients' labour had been exploited and condemned the detention of people 'after cure' for the value of their services. In Sarah Carter's case, the circumstances had 'so wrought on her mind, that, being otherwise subjected to tanking, and general ill-usage, she died'.[158]

The commission's scathing report provided extensive detail about all the abuses prevalent in the asylum over several years. Its catalogue of environmental deficiencies, building defects, lamentable sanitary arrangements, appalling living conditions, management incompetence, medical inadequacies, staff malpractices, and individual shortcomings added up to an incontrovertible denunciation. Although 'tanking' and the associated practices and behaviours were identified as the greatest of the abuses, they were clearly symptomatic of a much deeper malaise. The commissioners could find virtually no redeeming feature in the Kingston Lunatic Asylum, other than that its days were numbered.

A racial dimension?

The complex social delineations associated with race and shades of colour that emerged in Jamaica during the centuries of enslavement were perpetuated in the decades following emancipation. In some ways, the distinctions, stratifications, and discriminations became accentuated.[159] These phenomena were reflected throughout the hospital and asylum saga. The records rarely refer directly to people's race, colour, and related social class, but incidental evidence indicates that considerations associated with these factors permeated all aspects of the asylum's operation. It was clear that the main medical participants, such as James Scott and Lewis Bowerbank, were White, as were other prominent figures like Governor Charles Darling, Daniel Trench, and the three members of the Commission of Enquiry. At the other end of the societal scale, most asylum patients were Black or Coloured. The racial backgrounds of people in between were more indeterminate. Middle-ranking figures like David and Judith Ryan were probably White, though they may conceivably have been Coloured; Mrs Ryan was presumably a Jamaican Creole, to judge from the syntax of her reported conversation.[160] The former hospital warden Richard Rouse, however, was identified as a 'Negro'.[161] Subordinate asylum staff members were almost certainly Black or Coloured. Inevitably, the common usage of English surnames makes definite identification of race and colour difficult. In some instances there is only sufficient data to draw inferences regarding individuals' ethnicity.

Governor Darling's attitudes regarding race and class illustrated prejudices emanating from his English upper class background and others learned during service in Jamaica.[162] In carrying out official duties he was in periodic conflict with prominent Coloured politicians and newspaper publishers like Edward Jordon and Robert Osborn, and this had degenerated into outright hostility by 1860.[163] Darling's reluctance to take a firm stand in response to the hospital and asylum's problems was probably influenced by disdain for many of those involved. This became apparent in 1859 when he denigrated some ministers of religion, who submitted allegations of malpractices in response to a letter requesting information.[164] He disparaged the Reverend Mr Hyams as a 'Native of Colour and Minister of the United Methodist Free Churches', after he declined to give evidence about abuses other than on oath 'before what he considers a competent Tribunal'.[165] George Trueman was characterised as a 'self-constituted "Preacher of the Gospel"', who was both 'ignorant and illiterate', and an 'intimate associate' of the dismissed Richard Rouse.[166] A police inspector informant of Darling's described Trueman as a 'Mulatto Man', never properly ordained but 'one of those self constituted preachers of which there is such a number in this Country', and an itinerant 'Native Baptist Preacher' without church or congregation.[167]

Darling's preconceptions surfaced directly in 1861 when accounting to the Duke of Newcastle for his refusal to reinstate nurses acquitted in court from charges of ill-treatment. He acknowledged having been guided by a spirit of 'precaution and expediency' rather than the 'dictates of strict justice'. There were, Darling considered, 'few people for whom so much allowance should be made in respect to occasional loss of temper and exhibition of force as those whose natural passions have not been moderated by civilising influences'. They were, after all, 'of what may be called an African temperament', and they had the onerous duty to handle and restrain 'violent lunatics' who were mostly from the 'same Excitable Race'.[168] Darling's views were informed by advice from Daniel Trench, who in September 1859 dismissed Bowerbank's sources of evidence as principally 'persons who have been about the Institutions as Employees', drawn from 'a class proverbially inapt at discriminating or fixing dates'. Consequently, 'under any Circumstances' they could only provide 'very imperfect data, either for comparative purposes, or for the consideration of by gone events'.[169] Although Trench had not directly referred to race, the implications were clear.

Dr Lewis Bowerbank's perspectives were rather more complex. Early in the controversy he had to deny accusations of being motivated by 'dislike to colour' and 'hatred to race'.[170] Mrs Ryan, for example, alleged that Bowerbank had refused to board a private female lunatic patient of 'highly respectable connexions' in the house of a Black man.[171] Bowerbank, however, was careful to portray his liberal, humanitarian credentials in dealings with the British press and Commissioners in Lunacy.[172] He pointedly enlisted people of colour, such as Richard Rouse, Ann Pratt, and Henrietta Dawson, in his quest to expose the affairs of the hospital and asylum. Ann Pratt even stayed as a guest in his house after her discharge from the asylum.[173] Nevertheless, Bowerbank's dealings with them might be interpreted as patronising or even exploitative.

In the female asylum Mrs Ryan oversaw a strict stratification by colour and class, most apparent in the bathing arrangements. Henrietta Dawson described a 'distinction of color', whereby 'the whites bathe first and then the blacks'.[174] Another former patient, Catherine Fare, recalled Mrs Ryan saying that 'the brown people must bathe first', and consequently Ann Pratt was bathed before 'the black people'.[175] Coloured patients enjoyed other privileges. Henrietta Dawson reported that 'The hair of all the negroes is cut', but others only had theirs cut when they were violent. Both she and Ann Pratt were permitted to cut their own hair and wear their own clothes.[176] Black patients had to wear standard asylum dress with 'PUBLIC HOSPITAL' and 'F.L.A.' marked on in large letters.[177] A lighter skin, however, was little defence against Mrs Ryan's punitive inclinations, as Ann Pratt's and Henrietta Dawson's testimonies confirmed. It may even have increased their perceived role as troublemakers, for Mrs Ryan derided Ann as 'that impudent mulatto thing'.[178] Margaret Ravenshill, a former night nurse, was distressed by the violent treatment of another Coloured patient, Sarah Adolphus, described as a 'fair lady I do not know if she was white',[179] and Ann Pratt claimed that Louise Cochran, a Coloured woman of French extraction, had been treated with particular brutality.[180]

Patients were well aware of Mrs Ryan's attitudes. One Black woman who incurred her particular wrath was Harriet Jarrett (or Gordon), born in Sierra Leone and known as 'Port Royal Woman'. She was considered troublesome and subjected to several beatings. On one occasion she was tanked until 'half dead'. In another instance, having been beaten by Mrs Ryan herself with fists, an umbrella and a stave, she retorted that 'you beat me as if you were beating your old nigger'. Harriet subsequently became ill in the asylum, and died after removal to the hospital sick room.[181] Mrs Ryan's hostility was not only directed toward patients. A deep enmity developed between her and Richard Rouse, the hospital warden, whom she sought to undermine by every means possible. He considered himself a 'marked-man', and claimed to have overheard Mrs Ryan saying that she would do anything she could to 'get that negro out'.[182]

A keen awareness of racial and colour differences permeated relationships between staff and patients, and even among patients themselves. Ann Pratt alleged that she had been repeatedly abused, threatened, and manhandled by Frances Bogle, a Black nurse, who called her variously a 'damned negro man's wife', a 'damn'd mad mulatto brute', and a 'damn'd mulatto bitch'.[183] Richard Rouse recalled that, in response to racial abuse by a patient named Miss B., nurse Mary Clarke told other patients that 'if that white bitch...box any of you, to return it, as she curse negro too much'.[184] Mary Clarke later sustained a violent blow beneath the breast from Sarah Adolphus, 'one of the Brown Patients', which caused her to cough up blood and led to her resignation on grounds of ill health.[185] Rouse also highlighted the case of James Logie, a powerfully built 'colored gentleman', who died from his injuries in May 1854 after a severe beating. Whilst in a dying state Logie claimed that 'these Estate negroes raise upon me and murder me', though it was not entirely clear whether he was referring to other patients or their 'keepers'.[186] There were also suspicions of a race and class element surrounding the violent beating and subsequent death in 1860 of Benjamin Rodrigues Da Costa, an insane Jewish man from a 'respectable' family. Because of overcrowding, Da Costa was placed in a cell with his alleged assailant, a 'convict Lunatic' awaiting trial for larceny.[187]

Apart from instances of strife related to distinctions of colour and ethnicity, little other evidence emerges of clear manifestations of racial consciousness. However, one tantalising cameo has survived among the voluminous records created by the asylum crisis. Ann Brown, from Montego Bay, was admitted as a patient in October 1859 and remained for a year. By her own admission she was 'very saucy', and was tanked several times. On one occasion she had fought with a 'Coolie' patient and 'boxed her'. She proudly informed the Commission of Enquiry in 1861 that, 'at Christmas time I called upon the other black sisters to sing and dance, and we did so from eleven to seven o'clock'. As a consequence, 'the night nurses had no rest'. The following day Ann was taken to the tank and ducked three times.[188]

Aftermath

A crisis as serious as the Jamaica hospital and asylum scandal was bound to have profound consequences in both short and longer terms. The clear, unequivocal conclusions produced by the Commission of Enquiry, published in November 1861, left little room for argument about responsibility or the need for drastic change. Although highly critical of the ineptitude of the hospital's board of commissioners and the supervisory shortcomings of Daniel Trench, the strongest criticisms were directed toward the medical officers, Dr James Scott in particular. Whilst it was accepted that he was not deliberately complicit in the abuses, he had demonstrated 'dereliction of the gravest character'. Mrs Judith

Ryan was singled out as responsible for commission of the worst abuses and their concealment. Scott was further implicated because of financial indebtedness to Mrs Ryan and her husband, and a consequent lack of enquiry into her management, leaving control of the asylum 'almost entirely in her hands'.[189]

The task of dealing with Scott fell to Governor Darling. In January 1862 he was charged with culpable neglect for lack of supervision of Mrs Ryan and her staff, in relation to their 'violent treatment and ill usage' of the female lunatics.[190] Scott, however, continued to vigorously defend his role and actions.[191] He contended that the commission's proceedings were neither legal nor just and disputed much of the evidence, claiming it consisted mostly of hearsay statements by illiterate people. He condemned Ann Pratt as a 'complete maniac', incapable of either being a witness or giving reliable evidence.[192] Darling was 'dissatisfied' with Scott's defence statements and placed the case before Jamaica's Privy Council.[193] It upheld the charges and recommended Scott's suspension, which Darling implemented prior to taking leave from the colony.[194] Scott submitted a protest letter to the Duke of Newcastle, seeking review of the whole proceedings.[195] Newcastle was unimpressed and he was subsequently dismissed.[196]

Daniel Trench was strongly criticised by the commission for his lack of awareness of misdemeanours in the asylum.[197] In January 1862, Darling called upon him either to resign or show reason why he should not be suspended.[198] Trench, however, survived despite arousing the ire of Henry Taylor at the Colonial Office, who could not understand 'how it was that that for a year & a half his services were so truly ineffective to detect the daily & habitual commission of the most cruel & revolting crimes'.[199] He was eventually given a severe reprimand and reprieved, subject to strict conditions.[200] Trench continued as Inspector and Director of the hospital (and the new lunatic asylum) for several more years. More junior staff members were less fortunate; all the attendants 'implicated in the disgraceful cruelties' were dismissed.[201]

Dr Bowerbank gained considerable celebrity in Jamaica for having exposed wrongdoing. He was rewarded with a public address and presentation of a valuable piece of plate, and later by erection of a statue. He continued to play a prominent role in Kingston medical circles, becoming chairman of the board of visitors of both the public hospital and the new lunatic asylum. Bowerbank was regularly embroiled in heated intra-professional quarrels between medical men. In January 1865, he reluctantly became the public hospital's medical officer because all the other possible candidates had either resigned or refused to stand.[202] His reputation as a reformer of abuses had been rewarded by election to the House of Assembly in 1860. In 1862, Bowerbank became custos (chief magistrate) of Kingston,[203] in which capacity he played a part in the ruthless suppression of the Morant Bay rebellion in 1865. He personally arrested the Coloured politician George William Gordon and oversaw steps toward his summary trial and execution. An unapologetic supporter of controversial Governor Edward Eyre, Bowerbank sent the Secretary of State a full report on the events of Morant Bay.[204] In the early 1870s he was heavily involved in confronting government failings in dealing with smallpox.[205] He subsequently retired to England and died there in 1880.[206]

Within Jamaica, one major consequence of the scandal was accelerated activity toward completion and opening of a new lunatic asylum. Progress since the original vote of money in 1838 had been painfully slow, the foundation stone being laid in 1847. Thereafter construction was dilatory and then stalled completely due to factors like design faults, political disputes, financial mismanagement, and probable corruption.[207] Governor Darling sought to move matters forward after his arrival in 1857. By early 1858, though, he was inevitably blaming delays on objections raised by Bowerbank, who had criticised ventilation arrangements and proposed appointing an experienced medical superintendent to oversee the project.[208] In October 1859, Darling cited the colony's inability to afford the original plans, when confessing that work was proceeding with 'tropical languor'.[209] By now, however, the Kingston asylum exposures were prompting demands for urgent action. In January 1860 Darling secured assembly agreement for £2,500 to be spent on preparing part of the new building.[210] The male patients from the old asylum were removed there in November 1860, despite its lack of readiness.[211] The females followed eventually, in August 1862.[212] The disgraced Kingston asylum could now finally be closed down.

The scandal had wider ramifications, following the active interest of the British press, the Commissioners in Lunacy, and the Colonial Office. Concern grew to ascertain whether the revelations about appalling conditions and abuses were restricted to Jamaican medical institutions or exemplified a more general problem through the empire. In July 1862, Henry Taylor obtained Newcastle's agreement to gather information from all British colonies regarding their hospitals and asylums.[213] Advice was taken from the College of Physicians and the Commissioners in Lunacy on specific issues to be addressed.[214] The stated origins of the circular despatch to colonial governors of 1st January 1863 were the 'evils and defects' disclosed in the public hospital of Kingston, Jamaica, and the 'flagrant abuses and cruelties of long standing' detected in the lunatic asylum. The questionnaire sought detailed information on the operation of their hospitals and asylums, including location, buildings, sewage and drainage, patient numbers, treatment methods, staffing, medical attendance, and governance arrangements.[215] The returns to Newcastle's circular formed the basis of a comprehensive report published in 1864. It showed that standards and conditions varied considerably between colonies. Although the Jamaica example was the most blatant, there were several colonies across the empire where the state of the hospital or asylum was equally deplorable. In the West Indies those of British Guiana, St Lucia, and Dominica were condemned as being 'conspicuously the worst',[216] whilst significant defects were noted in the asylums of Antigua, Grenada and Barbados.[217] The report provided a series of recommendations regarding sanitary standards, management, medical oversight, and the need for regular inspections.[218] As Sally Swartz has argued, the report marked a new Colonial Office approach, whereby it sought to implement a framework of regulation, surveillance and promotion of good practice from the centre.[219] If nothing else, the Jamaican asylum scandal had shifted the British government's stance toward more direct intervention in matters hitherto regarded as the responsibility of colonial administrations. Here was an opportunity to uphold the ideal of the enlightened, benevolent imperial power.

Observations

The Kingston Lunatic Asylum scandal was a pivotal episode in the history of imperial psychiatric provision in the nineteenth century. There had been other public asylum scandals, both in Britain and the colonies. Revelations regarding the York Lunatic Asylum and Bethlem Hospital, highlighted in the parliamentary select committees of 1815–16, were of great importance in advancing asylum reform in England, as were further exposures of abuses in private lunatic asylums in 1827 and 1844.[220] The dissemination of reforming ideas to the colonies led some forthright medical officers to make representations about the unacceptably bad state of their local institutions. A scandal arose in Cape Colony in 1852, after concerns about poor conditions and lack of classification of patients were compounded by allegations of cruelty in the asylum on Robben Island.[221] Also in 1852, in the Victoria Lunatic Asylum at Yarra Bend, revelations including neglect and brutality toward patients precipitated a major scandal and an enquiry by the colonial parliament.[222]

Although important enough in their regional contexts, these episodes reverberated little beyond them. A number of factors combined to account for the unique notoriety achieved by the Kingston asylum. One feature was the vast amount of published and other written material produced, much entering the public domain, providing detail about every aspect of the institution's operation. The extent of documentation reflected the scale of the crisis. With well over 100 patients, although crowded into a confined area, the Kingston asylum was a relatively large institution by regional standards. Its problems were of long standing, many having been exposed originally in 1836.[223] Dr Edward Bancroft's well-publicised revelations in 1840, the regular representations made by other medical officers (including Dr James Scott) from the late 1840s onwards, and the circumstances of David Ryan's dismissal in 1852, suggest a wide awareness of conditions and practices in the asylum. Yet, there was no serious attempt at remedial action until the situation erupted into outright scandal. Even then, two years elapsed between Bowerbank's initial foray in 1858 and the dismissals of Judith Ryan and other staff, and more than another year before Scott followed.

The lengthy gestation period of the asylum crisis calls into question why the situation was allowed to continue for so long. Henry Taylor at the Colonial Office laid the blame squarely on the Jamaica Assembly for its 'apathy and niggardliness', as well as the corruption of some members.[224] Certainly, the colony's continuing economic difficulties strained government finances and partly accounted for delays in constructing the new asylum.[225] The administration's inherent parsimony was compounded by the obstructions and delays resulting from interminable wrangling between the shifting political factions. Behind the financial and political impediments to action, however, it is difficult not to conclude that the 'apathy' noted by Henry Taylor bore some relation to general perceptions of the asylum's patients. Most were, after all, doubly stigmatised. They not only presented the most extreme and noticeable manifestations of insanity, but the great majority were also Black or Coloured. Less than 25 years since the ending of enslavement, the Jamaican ruling elites had not yet concluded that all shared a common humanity.

Despite the apparent low regard for asylum inmates' welfare, however, one singular element became increasingly apparent during the saga – the active participation of former patients and lowly staff members in the gathering and revelation of damning information. Whether or not abetted by Lewis Bowerbank, Ann Pratt's published and verbal testimonies represented a critical tipping point, leading the Duke of Newcastle, prompted by his Colonial Office officials, to demand effective action from Governor Darling and the Jamaica Assembly. Subsequently, formal evidence given by people like Henrietta Dawson, Ann Brown, and Mary Bell, as well as Ann Pratt, clearly influenced the Commission of Enquiry's proceedings, notwithstanding Dr Scott's strident protestations about their unsuitability to give credible evidence. The degree of attention given to what today might be construed as the 'patient voice' was remarkable and almost unprecedented.[226] At that historical juncture it would even have been highly unusual in England itself, where psychiatric enlightenment was arguably several decades in advance. The articulate, graphic testimonies of ordinary Jamaicans have provided an unusually valuable insight into the perspectives of people receiving 'treatment' for their mental disorders in the mid-nineteenth century.

4
Reform – The Jamaica Lunatic Asylum

Any hopes that measures adopted in the wake of the Kingston asylum scandal would resolve the problems of managing Jamaica's insane were initially dispelled. The new lunatic asylum was surrounded with difficulties from the outset. It was quickly evident that a new building, even if properly finished and fully equipped, was not sufficient. A complete change in approach would be required to all aspects of the institution's operation. The skills and expertise to accomplish this, however, were not available on the island, and the colonial authorities accepted reluctantly that they would have to be imported. Once that happened the Jamaica Lunatic Asylum began a remarkable transformation. By the late 1860s it had emerged as a model institution, providing an example for emulation throughout the British Caribbean. The achievements, though, proved difficult to sustain. Toward the end of the century there were critical challenges, familiar enough in Britain, related to ever-growing patient numbers and increasingly inadequate facilities.

Birth pains

The new asylum had a prolonged period of incubation, for it was 1838 when the Jamaica Assembly first voted £10,000 to purchase land and erect a building.[1] Legislation in 1843 facilitated its construction, and a further £10,000 was made available. Following a prize competition, the experienced British asylum architect J. Harris was commissioned to design a modern institution for 250 people. His plan largely replicated an English public asylum. An imposing central structure containing medical superintendent's accommodation, administrative offices, and rooms for private 'upper class' patients, was to adjoin three parallel buildings on either side, for males and females respectively, with airing

Figure 4.1   Plan of Jamaica Lunatic Asylum
Source: John Conolly, *The Construction and Government of Lunatic Asylums and Hospitals for the Insane* (London: John Churchill, 1847), adjoining pp. 182–3.

courts located between the buildings. The asylum was to 'possess all the conveniences of an European establishment'. The main concession to tropical conditions was a network of covered verandas, to provide shelter and access to sea breezes (Figure 4.1).[2]

The slowly materialising asylum was located by the sea at Rae's Town near Kingston, on land belonging to the nearby penitentiary.[3] There was criticism from the outset. One prominent local medical man considered the site a serious mistake resulting from 'ill-judged parsimony'. Apart from the negative connotations of proximity to a large prison, the location was 'decidedly malarious', being on low ground in a particularly hot area. The risk of contracting diseases like dysentery was high.[4] Other critics subsequently highlighted ventilation arrangements unsuited to the local climate, and inadequate drainage.[5] Adverse climatic conditions contributed to long delays in construction and the steady deterioration of unfinished buildings. The 'folly' of a badly planned, expensive, unfinished structure was widely condemned. It remained as little more than a shell for over ten years, other than when utilised briefly during cholera outbreaks in 1851 and 1854.[6]

Eventually, Governor Darling and the assembly were impelled by the old asylum's crisis to finance a partial opening, the men being transferred in November 1860 and the women in August 1862.[7] Neither buildings nor grounds were ready; less than half the originally planned structure had been completed, giving accommodation for only 120 patients. However, by September 1863 it housed 159 and matters were becoming serious. The acting medical superintendent, Dr Charles Lake, complained that 'the Asylum is incapable of comfortably accommodating the present number'. Saloons on the male and female sides were converted into sleeping rooms, and the infirmaries appropriated as dormitories. There was no scope for classification, so the 'quiet and inoffensive Lunatic, as well as the refractory and dangerous are unavoidably associated together'. Attendant numbers were 'utterly insufficient for the requirements of the Institution'. They struggled to maintain control over patients 'many of whom evince a refractory and destructive disposition, and frequently attempt to effect their escape'. Indeed, 11 people had escaped during the preceding year, mostly by climbing over insecure fences. Despite these difficulties, Lake claimed that the principle of treating patients with 'care and gentleness' was being upheld, even though the attendants were frequently being injured by them. He warned, though, that expensive building additions or alterations were essential to enable proper classification, which he recognised as fundamental to the 'modern treatment of insanity'.[8]

With no previous asylum experience, Lake was having difficulty dealing with the challenges. Governor Darling had mooted as early as July 1859 that an experienced doctor from England be sought to superintend the new asylum.[9] However, prolonged wrangling ensued between Darling, the Jamaica Assembly, the Colonial Office, and the Commissioners in Lunacy, before agreement in early 1862 on an annual salary of £600. Darling's request that the Duke of Newcastle select 'some fit and proper person' was referred on to the Lunacy Commissioners.[10] They recommended Dr Thomas Allen, Assistant Medical Officer at Lincolnshire County Lunatic Asylum since October 1854.[11] Opened in 1851, that asylum contained 500 patients by 1863. It operated on moral management and 'non-restraint' principles, based around an extensive programme of indoor and outdoor employment.[12] Allen's diligent work there had attracted praise.[13] In July 1863 Newcastle appointed him medical superintendent of the Jamaica Lunatic Asylum, his new salary representing a great advance on his previous meagre earnings.[14] Allen secured agreement for Lincolnshire asylum's head male and female attendants, John Freshney and Ann Brown, to accompany him to Jamaica as warden and matron.[15]

Thomas Allen quickly demonstrated his single-minded approach. When the ship docked in Kingston on 5th October 1863, he proceeded directly to the asylum. He was taken aback; no preparations had been made for their arrival and suitable accommodation was not provided. He graphically described his first impressions to Governor Edward Eyre:

> I was much pained and shocked at the utter state of neglect which I saw all around me. The approach to the Asylum was wretched – at one Corner of the Estate are some dilapidated hovels, there were two brick pillars for the Entrance gate way, but no gate, there was free Ingress or Egress to any person or for animals to stray about the Asylum Grounds; the fences which consisted of Cacti were broken down in places – there was no proper road, and the whole of the Land surrounding the Asylum was uncultivated and covered with high weeds.

The building had a 'neglected and unfinished appearance'. The airing courts, surrounded by iron bars, had 'a most prison or menagerie like appearance'. Rubbish was strewn around, and offensive smells emanated from patients urinating through the bars. He noted defects and sanitary deficiencies throughout the asylum's interior. The patients appeared 'exceedingly noisy, most disorderly, & evidently beyond Control'. He told Eyre that 'Deep disappointment was felt by me with everything I saw, either as regards patients, buildings, or system pursued', and he foresaw 'very many serious difficulties in developing a sound modern system, such as is pursued in English asylums'.[16]

Even before formally commencing duties on 21st October, Allen set to work energetically, meticulously inspecting every corner of the asylum and recording his findings in great detail.[17] His report betrayed signs of the stubbornly obsessive approach that critics later castigated. Graphic, detailed descriptions of conditions covered many manuscript pages, providing a comprehensive picture of a squalid, failing institution.[18] He entered every room, examining beds and bedding, particularly noting if they were soiled or infested with insects. For example, for male No.9 dormitory he recorded: '1st. Hay mattress, stained, and bloody, a filthy central stain. 2nd. Hay mattress, swarming with bugs, bedstead filthy.' '4th. Mattress filthy and pillow dirty, filthy stains on under side of mattress; swarming with bugs, and thousands at head of bestead.'[19] For female dormitory No. 10 Allen noted: '3rd Bed. Sheet bloody and stained. Mattress very stained, and foul looking, under side mouldy, and in a filthy state – bugs about the mattress.' There was worse to come in dormitory 6, first bed: 'Mattress rotten torn, and swarming with bugs, upon the bedstead are masses of bugs. I never saw such a sickening sight.'[20] Dirt and foul smells pervaded the institution. Describing the area around the male 'saloon', he wrote: 'Passage and steps, stained, and dirty, offensive smell of urine from patients making water on the landing, or from the Water Closet, the Exhalations from which are carried directly into the Saloon.' Access to the airing court was through a 'dangerously narrow' entrance, and then down 'dark, confined, bad smelling, and dirty' stairs.[21] Sanitary arrangements were crude and frequently defective, with blocked or leaking urinals and overflowing tubs.[22]

The dreadful state of the rooms, bedding, and sanitation was reflected in standards of personal care. Allen found many patients wearing dirty clothes, including women whose shifts were 'filthy and stained black with menstrual fluid'.[23] Bathing arrangements were haphazard and towels rarely used. Men were bathed on Sundays, some in the sea and some in the asylum's baths. Those refusing might be washed under a tap, forced into a bath, or have 'a few pails of water thrown over them'.[24] Some women utilised the airing court gutters:

> After a shower of rain when the gutters contain a good quantity of water, they wade about and wash themselves, and upon two occasions which I witnessed myself, a patient stript off her clothes in a moment, and lay in the gutter quite naked.[25]

An inspection of male patients predictably revealed infestations of head lice. These included 'A Coolie, hair long, very disordered, filthy and full of ... lice', and another man whose hair was 'in a filthy state. I am informed that his head is full of lice and that he will not allow any one to touch his head.' Allen was shocked when informed that 'one attendant looks after the whole of the patients hair. For this purpose he has one comb, which he keeps in a bag in the pantry. No hair brushes are used.'[26]

Apart from the grim physical environment and dirty, unhygienic conditions, Allen was greatly exercised by the challenging, violent behaviours of many patients. When combined with insufficient and ill-prepared staff, the situation verged on anarchy.[27] He vividly described a male inmates' dinner scene:

> They are very noisy. A patient is now fighting, and others are chattering and quarrelling.... The patients sit or not at the table as they choose. They come up and walk off with their food, and take it when, and where they please: they are eating distributed about the Airing Court, in all places, and corners, and some on the steps of the single

rooms, some on the East portion of the Airing Court, others on the top of the Watercloset wall. Others near the pump, and the fence of the Airing Court. They all eat with their fingers. No Grace is said – No Tablecloths were used, nor the slightest social decency observed. It is a most savage banquet and a dreadful sight.[28]

Allen was equally shocked by dinner on the female side, noting that: 'Patients take their meals in all directions, all eat with their fingers, and like savages. There is terrible disorder, some are standing on benches, others are carrying food on their heads.'[29] In general, he thought the women more disorderly than the men. After inspecting their single rooms on 17th October he commented that 'All smell offensively. The patients are noisy and there is general turmoil. Several patients are eating their dinners..., many patients are most indecent, and some are nearly naked, having on only a loose shift.'[30]

Allen observed that fighting among both males and females was common, and staff frequently had to intervene.[31] This was partly a consequence of inadequate facilities to separate patients. The use of saloons as dormitories posed particular problems. In the female saloon:

There are some Extremely violent, and dangerous patients, sleeping here, who are utterly unfit to be there, but in consequence of the want of adequate accommodation, and a greater number of single rooms, there is no other room in which they can be placed. Two night nurses are obliged to be constantly present to prevent patients from fighting or quarrelling, but at times, notwithstanding, that these persons have courage, Experience, & forbearance, they are attacked and exposed to danger.[32]

Apart from the asylum's many practical limitations, the particular nature of the inmates presented great difficulties for those attempting to manage them:

A large proportion of the patients, especially on the female side, are ungovernable, troublesome, dirty, and half savage in their habits, and others are but little removed from a state of barbarism. Whether this be due to the absence of the restraining influence of Education, and civilising associations, I am from my short Experience in the Colony, unable to say, but it must be obvious that, the reactions of such Elements of disorder, and violence, one upon the other, must be opposed to cure, comfort, tranquillity, and freedom from danger unless separated.[33]

Allen's portrayal of many patients as essentially savage and barbarous did not specifically refer to race and colour, but the implications were apparent. This became a recurring theme in his depictions of the institution's inmates over the following years.[34]

In summarising the asylum's state and the challenges posed, Allen emphasised the inadequacies of buildings and accommodation and their unfitness for curative treatment. He complained that the bedrooms, particularly for males, were 'gloomy, cheerless, badly lighted, badly ventilated, unfurnished', and the sanitary conditions produced 'a low standard of health', leading to 'uneasiness, irritability, suffering and mortality'. There was 'dangerous overcrowding' and lack of single rooms meant people being placed together in dormitories, even if 'utterly unfit' to be safely mixed. Because of 'indiscriminate association', patients with particular needs could not receive 'individualized treatment'. Proper classification according to behavioural presentation was unachievable, removing a main element in promoting recovery. According to Allen, with only one male and one female ward, the asylum 'cannot possibly accomplish that which is absolutely required'. It was 'nearly impossible' to achieve 'Rest and the preservation of order, and control'.[35] In several major respects, lessons from the old Kingston asylum had not been learned and many of its evils had been replicated.

The model institution

Having thoroughly analysed the asylum's problems, Thomas Allen made comprehensive recommendations for their rectification. The main proposals were for additional buildings to accommodate patients and an immediate increase in single room provision. He wanted a new kitchen, a surgery or dispensary, a bath-house in the sea, and farm buildings. He called for effective external boundary fences, disguised with a 'ha ha'. He sought major adaptations and improvements in ventilation, water supply, drainage, sanitary arrangements, and security measures.[36] In considering enlargements, Allen detailed the arrangement and furnishing of the accommodation.[37] For example, he was concerned about the floors; in the single rooms means were needed to prevent transmission or reverberation of sound, 'particularly as regards the flooring, upon which I find that so many of the African Lunatics are prone to stamp or beat with their bedding'.[38]

By the time Allen submitted his report to Governor Eyre in April 1864, without waiting for approval, he had already initiated reforms and improvements. Overcrowding was partially addressed by rearranging or changing the use of rooms.[39] Steps were taken to increase security and minimise escapes, by boarding up airing court boundary fences.[40] Various measures were geared toward improving patients' health and comfort. They were given night wear and prevented from sleeping in their ordinary clothes. Sheets and rugs were provided for bedding, and the manufacture of mattresses and pillows was commenced.[41] The practice of a male 'labourer' assisting with female patients, transported from the old asylum, was stopped and the duties allocated to a female nurse. A training programme was initiated for attendants, 'according to the practice of English Asylums'.[42] One highly unpopular reform, however, was an attempt to reduce the patients' dietary allowance. Allen considered they were receiving too much food, without distinction between males and females or whether working or not. The quantity of bread was reduced, but an attempt to limit the meat allowance had to be abandoned 'on account of the violence of some of the patients, and a spirit of insubordination manifested'.[43]

Many of Allen's early endeavours were directed toward the lack of occupation, indoor and outdoor. Apart from some patients assisting with cleaning or washing, or doing needlework, no work was being done. He noted that most 'when not engaged in quarrelling, or in loud soliloquy' were 'listless, indolent and difficult of control'. An employment programme was crucial to his reforming project:

> There is nothing more important in the moral treatment of the insane, than the proper use of means, which contribute to their employment, both mentally, and bodily, and which tend to withdraw their attention from thoughts, & feelings, connected with their disordered state. The value of outdoor occupations for affording opportunities of extended exercise, is too obvious to need enforcement.[44]

By March 1864, 28 men were employed daily in the asylum's grounds, 'trenching, weeding and levelling and removing the large quantities of brick bats and rubbish'. Another five were working as tailors or mattress makers, as well as nine helping in the wards and three in the kitchen. In all, 45 out of 76 male patients were regularly employed. Of the women, 30–40 were occupied with needlework and house cleaning.[45] This, however, was merely the beginning of the employment scheme.

Structured religious observance was another important element of the British system that Allen wanted to introduce. Although the asylum's chaplain attempted to read prayers each Sunday, the arrangements had met with little success. The services, located in the male and female airing courts, lasted for a few minutes. The 'shocking' scene was characterised by the chaplain as a 'solemn mockery'. He recorded his experiences: 'Howling on both sides such as to render my strong voice utterly inaudible. Fowls kept in the next yard to the place of service crowing continually – Patients attempting to climb the wall to look at the Chaplain, and making loud remarks.' Allen arranged for a detached house at the rear to be used for Sunday services. Soon more than 40 patients of both sexes were regularly assembling, all demonstrating 'perfectly quiet & orderly conduct'.[46]

These changes and upheavals provoked resentment in some quarters. Dr Charles Lake, Allen's immediate predecessor, was especially put out. He took criticisms of the asylum's state as a personal slur, claiming the problems were greatly exaggerated. He detailed his own attempted improvements, despite the hindrance of insufficient suitable staff. He insisted that 'order and discipline' had since been replaced by 'riot and disorder', claiming that Allen was 'excessively timid and nervous' and unable to exercise sufficient 'moral control and influence' to ensure 'quiet and discipline'.[47] Perhaps more serious than Lake's discomfiture was the active resistance offered by staff members. According to Daniel Trench, many of Allen's troubles were attributable to the 'wilful negligence of dissatisfied Attendants'. He advised Governor Eyre that both 'old servants' and recently recruited staff had demonstrated a 'manifest spirit of combination to annoy, harass, and if possible injure the reputation of Dr Allen and his English Warden and Matron'.[48] Eyre observed to the new Secretary of State Edward Cardwell that Allen, like all 'reformers of abuses' in the West Indies, had much to contend with and was inevitably subjected to 'much undeserved obloquy and misrepresentation'.[49]

Eyre was immensely impressed by Allen's early achievements, as well as the quality and depth of his reports. He advised Cardwell that Allen was 'a most conscientious, upright, zealous and able Superintendent', and anticipated that under his 'judicious management' the asylum would become a 'credit to the Colony'.[50] Eyre endorsed all the proposals for building adaptations and additions, seeking a substantial contribution from the Colonial Office through its 'Perpetual Annuity Fund'. He carefully pointed out that the additional accommodation would only meet current requirements, with further provision needed for future increases in patient numbers.[51] Extensive building works were under way within months.[52]

Allen's reforms proceeded throughout 1864.[53] He sought to transform every aspect of the asylum's operation, toward the goal of an institution run on modern English lines, enshrining moral management principles. Pressing practical issues were addressed. Sanitary improvements included replacement of ineffective urinals with earthenware utensils and 'night chairs'. Accumulations of lime on dormitory walls, emanating partly from 'absorption of human excretion' around the urinals, were scraped off. Ventilation was improved by removing the glass from some windows. Boarding up of airing court fences prevented 'noisy, troublesome and refractory' patients throwing things through the dormitory windows.[54] However, Allen reiterated the limitations on what could be accomplished without major structural work. The asylum was now grossly overcrowded with 198 patients in a building intended to accommodate 120. Classification remained impossible and the difficulties of containing violent, dangerous patients had increased. He lamented that the existing single rooms were not sufficiently strong to prevent a violent patient breaking down the wall. The inability to classify them was 'opposed to recovery, order, and good habits'.[55]

Much of Allen's effort was directed toward changing staff and patient behaviours, which he identified as closely inter-related. Within a strict non-restraint policy, he sought to impose 'order, and discipline'.[56] Recalcitrant attendants were summarily dismissed for striking patients, purchasing food from them, and drunkenness. Others received suspensions for 'neglect of duty and disobedience of orders'.[57] Following training and instruction received from Allen, Freshney (the warden), and Miss Brown (the matron), attendants were expected to relate to patients in accordance with humane, therapeutic principles:

> Every endeavour is made to induce the Patients to look, think, act, and speak, like persons of sound reason – they are individualized and surrounded by such kind, and civilizing influences, as will break up their morbid train of thought, tend to exercise their self control, as well as to excite their feelings of self-respect.

It was anticipated that staff would become role models:

> The Attendants are instructed that, the Patients placed under their charge, are all more or less deprived of correct feelings, and powers of understanding; and that their peculiar habits, perverseness, outbursts

of anger, illfeeling, or violent conduct, are only the results of disease or infirmity, and that they must on no occasion resent either intemperate language, or unruly behaviour, but must exhibit towards the patients uniform kindness and forbearance.

Allen considered patients more likely to imitate the attendants than follow their instructions, and stressed the importance of examples of 'order, quietness, punctuality, personal neatness, and general propriety of behaviour'.[58]

The new approach was most noticeable at meal times. The chaotic scenes Allen witnessed after his arrival were gradually superseded by patients dining 'decently and with the strictest propriety'. Tables were covered with cloths and grace said, 'to which all reverently attend'. After the evening meal, a Bible portion was read and hymn sung before the patients dispersed quietly to bed. Allen expressed pleasure at the overall 'marked improvement in the conduct, and habits' of the patients. The general 'distressing turmoil, and insubordination' had largely given way to 'order and tranquillity'.[59]

The evolving programme of 'useful occupation' was central to creating this more settled and ordered environment. By late 1864, 25 acres had been weeded, cleared, and planted with vegetables, including sweet potatoes, cotton, pumpkins, and peas. Supervised by three or four attendants, male and female patients were 'cheerful and willing agents'. Men also cleared rubbish, dug trenches, built embankments, and made a 'carriage drive', whilst others worked at trades like carpentry, bricklaying, and painting. All the asylum's clothing and bedding was being made or repaired by patients. Many cleaned the wards and did other domestic work, whilst numerous women engaged in needlework. Allen had great plans for other enterprises. He pointed out that, as well as directly benefitting patients, their work brought considerable financial savings to the asylum.[60]

Recreational activities were initiated. Newspapers were placed on the wards. Smokers were supplied with tobacco 'as an indulgence and incentive to employment'. Walks outside the asylum grounds took place on Sundays, with 30 or 40 men and 20 women allowed the privilege; the claimed benefits included a reduction of perceived 'coercive confinement' and promotion of a 'habit of self control'. Allen's hope was that, once 'general order' became fully established, a wider range of social activities and recreations could be provided 'as in English Asylums'.[61]

The rapid transformation in every aspect of the asylum's operation received strong approbation from Governor Edward Eyre. He prophesied that 'under Dr Allen's able and zealous supervision', it would become 'the most complete and best conducted institution of the kind in the West Indies'.[62] Eyre was evidently undaunted by Allen's reiterated demands for building improvements, or his calls for fundamental reforms in Jamaica's lunacy laws and changes in the asylum's management structure to consolidate power and responsibility in the medical superintendent's hands.[63]

Word spread about the improvements taking place in the asylum and visitors came to see for themselves. A group of Presbyterian ministers visiting one evening in March 1865, after the patients had gone to bed, were struck by 'the decency, the order, the cleanliness, the quiet that reigned through every department'. Reverend James Watson noted the absence of smells in contrast to when the sleeping rooms had contained 'filthy urinals'. The 'decorous order and quietude' reminded him of 'a well conducted dormitory in some private institution'. Above all, he was impressed with the 'law of kindness, gentleness, and forbearance' that characterised the conduct of those in charge of the patients.[64]

A visiting journalist provided a vivid account in 1867. Having participated in exposing the old Kingston asylum's evils, he viewed its successor as now '*par excellence*, the redeeming feature of the colony'. He could hardly believe the prevailing 'spirit of cleanliness, order, decorum and humanity', being struck most by the patients' demeanour at dinner, where 'over two hundred of God's afflicted creatures' showed 'a propriety of manner that would have shamed many a society of people outside the walls'. More than 80 'madmen' behaved with 'as great propriety and order as any well regulated family'. When Dr Allen removed his hat, all rose and a patient recited grace 'in a voice of deep solemnity'. Allen appeared to exercise a 'moral lever', whereby attendants treated patients with kindness and respect. The patients apparently regarded him with reverence:

> It is grand – it is glorious – it speaks of Christian influence – to see the Doctor's reception by his patients. They come round him as he enters among them, some to welcome him with shaking hands, others kneel to him, all look up to him with a trustful devotion beyond my descriptive power to portray. One asks a flower and he gives it, another begs him some slight indulgence, a third places her hand (she is a revival monomaniac) upon his head and blesses him. And thus he travels on, the guiding spirit of the whole, the guardian angel of those entrusted to his care. So among the men. Their listless eyes

dwell upon him as he comes near them, for all appear to recognize in him a common father.[65]

It seemed that a paternalistic micro-community was being created within the asylum. Two years after the Morant Bay disaster,[66] this provided Jamaica's ruling elite with a welcome example of benevolent treatment of disadvantaged poor people. By early 1865 over £4,300 had been spent on building work. Two new refractory wards, each comprising 31 single rooms, were provided. The original single rooms were converted into dormitories, creating more capacity. The amount of airing court space was increased significantly. Major improvements were implemented in ventilation, lighting, water supply and drainage. New farm buildings and workshops were projected, to be constructed by patient labour.[67] However, as Eyre predicted, these measures soon proved insufficient due to continually growing patient numbers. By late 1868 overcrowding had again become critical, and Allen sought to restrict admissions. Proposals to discharge 'harmless cases' to make room in dormitories for the 'violent and dangerous', were strenuously resisted by Allen, who protested to Governor Sir John Grant. The issue went via the Colonial Office to the Commissioners in Lunacy, who endorsed Allen's principled stance. The Secretary of State urged Grant to expedite building additional wards, with single rooms.[68] Money was found to erect two blocks of 32 single rooms 'on the separate system', in 1870, one for males and another for females. Something resembling the asylum's original plan had finally been completed.[69]

The 1870s largely saw consolidation. Patient numbers continued to increase steadily, from an average of 260 in 1872/73 to 381 in 1879/80.[70] Problems associated with overcrowding recurred, despite building extensions and adaptations.[71] At times the solution accentuated the problem, as in early 1876 when patients were crowded together in some wards to permit work elsewhere. Their health was adversely affected, compounded by deficiencies in water supply, with 21 deaths from dysentery as a consequence.[72] Tropical weather conditions brought further hazards. A severe storm in October 1879 caused flooding on the female side, the airing court described as 'like a large lake'; patients were carried away on attendants' backs. A hurricane in 1880 proved yet more serious, with extensive damage inflicted on parts of the asylum.[73]

Occupation continued to be the central plank of the institution's operation. As well as agricultural enterprises like pork production, patients were employed in a commercial sea fishery established in 1869, utilising a large net manufactured in the asylum. The venture proved highly successful, with substantial quantities being caught. At its height in 1875 the annual catch was over 75,000 lbs. Curing rooms were provided within the asylum. Most of the fish was used to feed the patients or farm animals, but a surplus was sold off cheaply to Kingston's poor.[74] Allen reiterated in 1879 that employment 'continues to be regarded as of the highest importance in their treatment'. Every patient, apart from those incapable from 'old age, mental defect, or bodily infirmity', was expected to participate:

> This feature in the management of the Asylum continues to be carefully observed and comprises needlework, coirmaking, tailoring, cocoanut oil manufacture, netting, hat-making, mat-making, farm and garden occupation, and a variety of other branches of industry. Patients act as clerks, porters, messengers, cooks, house cleaners, besides being painters, bricklayers, shoemakers, plumbers, carpenters and fishermen.[75]

In addition to reduced operating expenses, profits paid for recreational activities. Educational classes were provided. A weekly dance was held in the dining hall, with music played by the asylum band, which also performed on some wards, including those for 'refractory' patients. Twice weekly the band led some patients to the sea shore, where they danced and sang. Other diversions included 'magic lantern' shows, and special Christmas celebrations.[76]

By the late 1870s Thomas Allen's goal of a well-ordered institution, functioning on moral management lines, had been largely achieved. His success in Jamaica attracted the notice of the Colonial Office. Anticipating that the example could be disseminated more widely, it was proposed in 1874 that Allen become Inspector for Lunatic Asylums in the British West Indies, similar to roles established elsewhere in the empire.[77] However, an expectation that each colony should share the costs made several reluctant to agree, despite the Secretary of State's repeated attempts at persuasion. Consequently, the formal role of inspector never materialised, though in 1875 Allen conducted inspections in British Honduras and, most notably, Barbados.[78] His excessively detailed and tardy report on the Barbados Lunatic Asylum underlined a reputation for being fussy and dogmatic.[79] It also confirmed the views of other colonial governments that they did not require his services as inspector. [80]

Decline and stagnation

The rebuff to Allen's expectations of professional advancement probably affected his motivation levels. By 1880 the Jamaica asylum's functioning had passed its high point. Over the next few years there were distinct signs of decline and deterioration. Even in the crucial area of patient occupation, the unbounded optimism of the 1860s had given way to a more sanguine outlook. The lucrative fishery enterprise fell away quite dramatically, with only 33,000 lbs of fish caught in 1879/80.[81] Four years later it was abandoned completely, no longer deemed sufficiently productive to justify the outlay on nets and 'fishermen attendants'. The expansion of commercial cattle farming and beef production provided a partial substitute.[82] Nevertheless, reports on work within the asylum became increasingly matter-of-fact, concentrating on financial savings achieved rather than therapeutic benefits.

Allen increasingly made excuses and complaints to explain poor results or other difficulties. In 1880 he criticised Jamaica's lunacy legislation for leaving people in the community until they became dangerous or unmanageable. The asylum consequently became filled with 'a concentrated class of dangerous and incurable patients'. At the same time, he resented being compelled to accept many unsuitable 'aged, imbecile, idiotic' patients, or those with serious physical diseases, because of the lack of parish poor-houses.[83] He made similar complaints the following year, adorning them with his entrenched social and racial perspectives:

> the Lunatic Asylum continues to act as a receptacle for criminals, aged and disgustingly infirm Patients, who, if in England, would be retained in workhouses; as well as for a class of brutal negroes whose natural mental condition is so low, as to justify their being considered irresponsible and of unsound mind.[84]

He lamented that 'respectable' middle or upper class patients were 'compelled to associate with criminals and debased pauper negroes', due to the absence of separate facilities.[85] Allen reiterated this theme in 1882, claiming that some private patients 'related to persons of the highest respectability in the island' had to mix with 'the general body of patients, consisting of criminals, coolies, and a low type of the insane poor'.[86]

After having emphasised the patients' shortcomings, Allen observed in 1883 that the asylum's 'servants' were largely recruited from 'the same ranks'. There were consequently 'immense difficulties' in main-taining 'proper efficiency and discipline', particularly because of their lack of training and undisciplined, often defiant, nature. The result, he protested, was a 'slavish burden' of responsibility falling on him as head of the institution.[87] Staffing difficulties were exacerbated by an exodus of attendants to work on constructing the Panama Canal. They could only be replaced by untrained people, precipitating a significant increase in dismissals for neglect or ill treatment of patients.[88]

Allen's discontents multiplied, reflecting increasing disillusion. He complained about being overworked, under-paid, and not having suf-ficient medical assistance. He contrasted his position, in 1885, with Dr Robert Grieve's at the British Guiana asylum, with similar numbers of patients though many were 'Coolies' and 'therefore more manageable than the class of patients we admit'. He resented pressures to reduce expenditure, particularly on the patients' food, claiming that 'I strive to ensure the greatest efficiency with the greatest economy.' He consid-ered that he was being deliberately prevented by local regulations from exercising 'freedom of speech', and sought intervention from the Sec-retary of State.[89] Allen's grievances grew with a series of conflicts with officials and politicians, often about petty matters.[90] He also aroused the hostility of the Jamaican press, particularly the *Gleaner* whose attacks became quite merciless.[91] By 1886, after 22 years in post, Thomas Allen was exhausted by his labours, frustrations, and increasing criticism. He retired from the asylum in October 1886, returned to England, and was dead within a few months.[92]

The appointment of a successor was conducted through the Colonial Office, supported by the Commissioners in Lunacy. The leading con-tender, Dr George Seccombe of the Trinidad Lunatic Asylum, declined the post once it became clear he would receive little extra pay for manag-ing a larger institution. The successful candidate was Dr Joseph Plaxton of the Ceylon Lunatic Asylum.[93] Plaxton, having previous asylum expe-rience in England, had been in Colombo since 1878. He achieved some success there, demonstrating a critical approach on issues like overcrowding and the problems created for native staff by European patients.[94] His subsequent career in Jamaica confirmed his reforming inclinations. However, Plaxton lacked Thomas Allen's doggedness in demanding resources from a reluctant colonial government. He was pre-pared to state his opinions firmly, but was essentially a pragmatist who accepted having to work within certain parameters. In regard to the asylum's patients, he did not demonstrate the racial prejudices of his predecessor.

Plaxton took up post in April 1887. Although paying tribute to Allen's achievements he quickly drew attention to the asylum's shortcomings. He was realistic about the constraints restricting attempts to occupy the patients. Suggesting that employment was 'the main holdfast in the treatment of the insane', he observed that it was a 'heart breaking task' to 'get the women to employ themselves'; their 'shifts to escape employment' were 'laughable, but intensely wearisome'. This was hardly a ringing endorsement of established practices. He was clearly disappointed at the limited recreational activities and amusements, pointing out the absence of suitable spaces for them. A covered area utilised as chapel, dining room, and amusement room was, during bad weather, an unusable 'miserably wet and dirty place'.[95]

Plaxton was soon confronted with the familiar problem of overcrowding, despite completion of a large new ward in 1886.[96] Patient numbers continued climbing steadily, from an average of 381 in 1879/80 to 465 in 1889/90.[97] Sanitary problems, related to 'soakage of the precincts of the Institution with foul matters', led to an upsurge in dysentery and diarrhoea, accounting for 17 deaths in 1887/88. Extensive building work began during that year, aiming to increase the accommodation to contain 600 people and provide new airing courts. The works brought problems for, though they employed some patients, the female wards became more disorderly from the presence of workmen, rendering 'peace, quietness and good order' impossible.[98] Disruption persisted into the following year, Plaxton complaining that, due to compressing large numbers into insufficient, unsuitable wards, the female side remained 'in a simmer of excitement'.[99]

It became apparent to Plaxton that radical measures were needed to deal with rising patient numbers. Although admission rates remained well below those in England, they had doubled in 20 years whilst Jamaica's overall population had only gone up by one third. This was, he concluded, due to more 'enlightened' use of the asylum and the island's improved internal communications, rather than any significant increase in the incidence of insanity. He expected the trend to continue and anticipated accommodation being required for 800 people within ten years, by when a second asylum would be necessary.[100] The prediction turned out entirely accurate, for there were 819 patients in March 1900.[101]

The 1890s proved critical in the Jamaica asylum's history, when competing dynamics required reconciliation. Considerations of therapeutic effectiveness and patient comfort increasingly conflicted with those of economy and retrenchment. The issue of a second asylum became central to the debate. All parties accepted that additional accommodation was required, but no consensus existed on how to achieve it. Two main options emerged. The first, advocated by Plaxton and his supporters, was for construction of a new, separately administered asylum on the west of the island, meeting the needs of people living far from Kingston and providing ample land for employment and other therapeutic purposes. The second option, favoured by those concerned about expenditure, was to build a second asylum adjoining the present one, enabling savings from joint administration. The differences of approach ensured considerable delays both in decision-making and execution. Meanwhile, the institution once again descended toward crisis.

A newspaper correspondent in 1890, though noting the asylum's clean and satisfactory state, observed that numbers greatly exceeded capacity.[102] Complaints about overcrowding continued to mount. The board of visitors made several representations to the colonial government during 1891, calling for urgent action.[103] The government responded by instructing parochial authorities to refer only the most urgent and dangerous cases.[104] Plaxton became more outspoken in his protests, claiming in 1892 that there were 100 patients more than could be properly accommodated, with airing courts too crowded and effective classification increasingly difficult.[105] Critics also highlighted other problems, such as the unsuitable old asylum buildings and insufficient suitable land for employment and recreation, partly the consequence of its original location on a virtual 'desert'.[106]

In March 1893, the Jamaican government succumbed to pressure and sanctioned a loan of £45,000 for a new asylum, without specifying its location. Plaxton's strong preference for another part of the island was clear, based primarily on therapeutic considerations. However, the colony's Superintending Medical Officer, Dr Pringle, supported the erection of new buildings on the present site to minimise medical staffing expenses.[107] The debate became increasingly animated, with Plaxton arguing that mortality from consumption, diarrhoea and dysentery had risen sharply due to the effects of overcrowding. In 1894 the government established a select committee to consider the issue. Pringle was, in the meantime, replaced by Dr Leonard Crane, who differed even more sharply from Plaxton. He contended that 150 of the asylum patients were 'imbeciles', who could be looked after elsewhere, while Plaxton maintained there were only 35 within this category.[108] Inevitably, the decision was taken to utilise the site at Rae's Town and construct a new set of buildings next to the existing institution, despite Plaxton's active disapproval.[109]

During 1895 two substantial adjoining properties, Chelsea and Belle Vue Penns, were acquired to provide land for expansion.[110] Piecemeal additions were meanwhile made. A new refractory ward was built in the grounds, near a public road, precipitating complaints from local residents about 'shrieks and wails' and 'hideous yells' heard over a wide area.[111] The problems continued to worsen. During 1897/98 the number of admissions exceeded 200 for the first time, and total numbers reached 785. Deteriorating conditions brought another marked increase in mortality rate, attributed to the ravages of diarrhoea, dysentery, pulmonary disease, influenza and yellow fever. Work was at least seen to be proceeding on new buildings. These were designed to accommodate 420 female patients, with scope for further addition, whilst 550 males would be housed in the existing premises.[112]

Any optimism created by the progress of construction works was to prove illusory, in a situation reminiscent of the 1850s. An American asylum superintendent visiting during 1899 noted that the 'new, plain, but substantial and well planned building' was almost ready to receive patients.[113] However, because of constraints on the public finances, there was no money to furnish and equip the asylum and work came to a complete standstill. Patient numbers continued rising sharply to 886 in March 1901, creating 'extreme over crowding'. Nothing had changed a year later, apart from increased numbers. An upsurge of deaths from pulmonary diseases led the *Gleaner* to warn that the congestion had literally developed into a matter of 'life or death'.[114] It evidently all became too much for Dr Plaxton, who spent several months in England recovering from illness. He returned to Jamaica in January 1903, but his health again worsened and he died in January 1904.[115] He was replaced by his senior assistant, Dr D.J. Williams, who received a lower salary.[116]

The asylum's agony persisted some time longer, with the new buildings remaining unoccupied. By March 1904 patient numbers reached the 1,000 landmark. Several coroners' inquests drew attention to the contribution of overcrowding to inmate deaths from disease.[117] Representations to the Jamaican government gradually produced results. During late 1904 part of the new female block was opened and 100 male patients moved in temporarily, providing some immediate relief. The situation on the female side, however, remained desperate. Dysentery claimed 24 women (and one man) during 1904/5, in addition to numerous deaths from tuberculosis.[118] The rest of the new accommodation eventually opened in March 1906, more than ten years after the decision to build it. Following transfer of the female patients, the vacated wards in the old building were quickly filled with males, finally relieving the congestion.[119]

The relative comfort was only short-lived, unfortunately. In January 1907, the asylum succumbed to the great earthquake that virtually destroyed Kingston and killed hundreds of people. Eleven patients died and considerable damage was done. Several original buildings were 'razed to the ground', with almost all the male accommodation 'totally destroyed'. Even some new buildings were flattened or badly damaged. Many people were forced to sleep in the open air for an extended period. At least 16 patients escaped and a larger number of 'harmless' people were sent home to districts with more plentiful food supplies.[120] As might be expected, repairs and restoration took a long time. The work was still not completed by April 1910, although the authorities managed to establish distinct accommodation for private patients.[121] The buildings were all finally restored in 1912.[122]

The growth of patient numbers continued to accelerate. In April 1911 there were 1,169 people in the asylum, with no fewer than 324 admitted during the past year, the highest number in its history.[123] The whole scale of operations was altering. A new ward for 100 females opened during 1912, relieving the overcrowding again prevalent.[124] By 1914, some wards contained even greater numbers; an inquest revealed 120 patients on 'O' ward, housing men regarded as 'harmless'. Wards for 200 men and 200 women were being planned. It was acknowledged that drastic measures were required to cope with the ever growing demand, there being 1,314 patients by March 1914.[125] Dr Williams observed that numbers had doubled in 20 years, whilst Jamaica's overall population had again increased by only one third.[126] The trend predicted by Plaxton in 1890 had continued. By 1914 the Jamaica asylum had become a very large institution, with all the hazards entailed.

Conclusion: Fluctuating fortunes

The circumstances surrounding the origins and early development of the Jamaica Lunatic Asylum at Rae's Town reflected the country's economic and political state, as well as complex, ambivalent attitudes toward insanity and people suffering from it. Humanitarian revulsion at conditions within the Kingston asylum created the impetus to finance and build a replacement. However, the parlous state of Jamaica's plantation economy in the 1840s engendered a parsimonious approach to public expenditure. That parsimony resulted in the prolonged delays in the asylum's construction and completion. It also contributed to identification and adoption of a site highly unsuitable due both to its lowland geographical situation and location near a large prison, confirming the status of the (predominantly Black) institutionalised insane as a deeply stigmatised group of people.

Even when the scandal of the old Kingston asylum rendered it imperative to open the new asylum in stages between 1860 and 1862, financial constraints restricted what could be made available. The consequence was an unprepared and inadequate institution, where its predecessor's practices were perpetuated. The political and medical establishments were forced to acknowledge that outside assistance was required, and that it had to be paid for. The appointment of the experienced Dr Thomas Allen proved crucial to transforming the Jamaica Lunatic Asylum into a site for the implementation of humane, enlightened principles and practices. The promotion of a 'moral management' system in the asylum accorded perfectly with Allen's intent to bring 'civilisation' to its patients. Despite his undoubted character flaws, including an increasingly racist perspective, Allen's achievements were considerable. It is interesting and ironic that his early reforming measures were strongly supported by Governor Edward Eyre. Eyre's evident progressive approach to the treatment of the insane during 1864 proved to be in stark contrast to the reactionary positions he adopted during Jamaica's political and social crisis of 1865.[127]

The Jamaica asylum, under Allen's leadership, became a model for emulation throughout the British Caribbean. However, some achievements gradually became hindrances. Allen's trumpeting of the considerable savings made in patient maintenance costs, as a consequence of earnings from employment projects, accorded well with the island government's interests in reducing expenditure.[128] As overcrowding increased, so conditions became increasingly Spartan. By the time Allen withdrew in 1886 the asylum was displaying many problems familiar in the large-scale institutions in England and elsewhere in the Empire.[129] When another experienced reforming medical superintendent, Dr Plaxton, took over, he was faced with formidable difficulties. Unlike his predecessor, however, Plaxton's expertise was overlooked in the crucial decisions on how to remedy the situation of a grossly overcrowded, counter-therapeutic institution. The decision to build a second asylum on the original site, rather than elsewhere on the island, was little short of disastrous. Based wholly on considerations of economy, with little apparent concern about implications for treatment and care, it confirmed that the asylum was now perceived primarily as a place to incarcerate deviant and challenging people rather than facilitate their recuperation and recovery.

The situation in 1900 paralleled that of 50 years before. Desperately needed new buildings stood empty due to the Jamaican government's inability or unwillingness to provide funds to furnish, equip and staff them. With patient numbers reaching 1,000 in 1904, the asylum's state of acute crisis was plain. When the new facilities finally opened, there was a period of relief for patients and staff, before natural disaster set the institution back again by several years. The repairs were eventually carried out, and measures implemented for further expansion. By 1914, with huge wards in place or projected, the Jamaica Lunatic Asylum had become little more than a vast, impersonal warehouse for the chronically insane. The therapeutic aspirations still receiving cursory expression in the medical superintendent's reports were now little more than a forlorn hope.

5
Colonial Asylums in Transition

The early development of institutions for insane people in the West Indies in the 1840s and 1850s had been piecemeal and patchy. In most territories the small lunatic asylum provided little more than incarceration in the crudest of facilities. The Colonial Office survey published in 1864 showed conditions in several Caribbean asylums differing little from exposures in the Kingston asylum scandal.[1] The report's recommendations, and the new requirement to submit annual returns, provided some impetus to consider improvements. The progressive reforms and practices that Thomas Allen put in place in Jamaica offered an example that could be emulated, given the right circumstances. During the 1870s and 1880s the model was adopted in the relatively prosperous colonies of British Guiana and Trinidad, where their governments opted to follow the Jamaican example and recruit a medical superintendent from England. Elsewhere, there was generally neither the preparedness nor the financial means to implement progressive reforms or even provide reasonable conditions. In some Caribbean colonies provision remained at the most basic level.

British Guiana – A glimpse of the vision

The British Guiana Lunatic Asylum that opened in 1866 at Fort Canje, near New Amsterdam in Berbice, was situated close to a river estuary, with ample adjoining land that offered considerable potential for the future. The initial stages were not promising, however, with problems in all aspects of its operation. Water supply, bathing arrangements, and sewerage were all inadequate. Patient numbers grew steadily, reaching 100 by 1869, and overcrowding necessitated provision of two additional wards. There were real shortcomings in the asylum regime, with a lack of occupation and diversionary amusements for patients. Governor Sir John Scott sought to account for this by highlighting British Guiana's diverse population – 'the difficulty is further increased here by the different races, and it becomes a matter of perplexity to amuse at once Chinese, East Indians, Portuguese, Africans, Creoles, and occasionally Americans and Europeans'.[2] This diversity, and particularly the large numbers of East Indian origin, became an important influence in the institution's development.

Following an inspection in 1872, Dr Gavin Milroy highlighted the asylum's considerable defects and severe overcrowding. He strongly advised that a medical officer be sent to view the 'many salutary improvements' introduced by Dr Allen in Jamaica, where the asylum was now 'the theme of admiration among both the inhabitants and strangers'.[3] Governor Scott took up the recommendations, and in 1873 the asylum's resident surgeon, Dr Cramer, was sent on an observation visit to asylums in other colonies. Unfortunately, he died suddenly following his return. The intended programme of reforms was subsequently adopted by Dr Robert Grieve, appointed by the Secretary of State in 1875.[4] A Scotsman, Grieve had previously worked under London's Metropolitan Asylums Board as medical superintendent of the Hampstead Fever Hospital. His tenure there, however, became mired in controversy. It seems likely that he accepted a move to the distant colonies as an opportunity to leave the limelight and forge a new career. Before taking up post he spent several months visiting and studying the practices of various lunatic asylums around Britain, imbibing the doctrines of non-restraint and moral management.[5] Despite his apparent failures in London, Grieve proved a highly influential figure in advancing provision for the insane in Guiana and the wider West Indies.

When Grieve commenced duties in 1875, patient numbers in the asylum had reached 170 and were increasing rapidly.[6] By December 1880 there were 330, rising to 456 at the end of 1885 and 640 in 1890.[7] This almost four-fold increase over a 15 year period placed tremendous pressure on buildings and facilities, necessitating an ongoing programme of alterations and additions. Overcrowding became a constant problem, only periodically relieved with the opening of a new ward or building.[8] The situation was greatly alleviated during 1885 with the long-awaited acquisition of the nearby Berbice general hospital's buildings, and their allocation to female insane patients, after the hospital's removal to a site in New Amsterdam.[9]

As in other West Indian asylums, water supply, sewerage, and sanitation presented major difficulties. Situated on low-lying land near sea level, the asylum was for some years dependent on rain water for drinking and domestic use; only salt water could be obtained from the River Canje. Drainage in the late 1860s was by means of open gutters and trenches. Excrement from the latrines was dealt with by the 'dry earth' system.[10] Under Grieve, this method was abandoned on account of the 'stinking mud' produced, and replaced by the use of tarred pails. These were removed daily and their contents either used as fertiliser for plantain roots or placed in pits around the asylum's ample grounds.[11] Despite drainage of much of the grounds during 1874, the area's swampy nature made it liable to outbreaks of malarial fevers, reinforcing the importance of personal hygiene and frequent bathing.[12] By 1881 water for bathing, domestic cleaning and laundry was being piped from the Calabash Creek near New Amsterdam, though drinking water was still accessed from rain storage tanks.[13]

Robert Grieve spent ten years as medical superintendent of the Berbice asylum.[14] During that relatively brief period he achieved a transformation comparable to that of Thomas Allen in Jamaica. Grieve's moral management approach informed refurbishment of the asylum buildings. Perturbed by their austere, custodial nature he had iron window bars, 'the last vestiges of the old prison style of asylum', replaced by jalousies or glass.[15] Ongoing renovations were carried out, and dormitories, day rooms and dining areas redecorated, all to create a more pleasant, comfortable and therapeutic environment for patients.[16] These developments accompanied a 'non-restraint' policy. Before 1875, many people were 'kept locked up all day in cells some for months at a time', and the asylum contained 'a fair collection of canvass jackets, leather straps and even iron handcuffs'. Grieve believed that the 'non-restraint regime' in English asylums represented one of the 'brightest ornaments' of civilisation. The practice of daytime seclusion in a locked room was also ended along with mechanical restraint.[17] He was unequivocal about these measures' effectiveness, whilst acknowledging their dependence on the availability of skilled attendants.[18]

Grieve's most notable achievement was the implementation of programmes of work and occupation. He was undoubtedly influenced by developments in both Jamaica and Britain. He considered a 'system of regulated labour and amusement' a necessary accompaniment to non-restraint, and arguably took things further than Thomas Allen.[19] Apart from the therapeutic benefits, in alleviating and treating mental disorders,[20] Grieve recognised employment to be a means of creating and maintaining order. This was important in a colony like British Guiana 'where the lower classes are totally unfitted to employ any leisure they may have in intellectual pursuits or amusements, even of the simplest kind', because 'deprived of the chance of working, they can only sleep, eat and quarrel'. Even people with incurable insanity, he suggested, when 'under the firm but kindly discipline of the institution' would 'work steadily and well'. Grieve considered the asylum one of the few places in the colony where 'to toil is looked upon as a privilege not as a punishment'.[21]

From 1876 onwards various enterprises were established. Agricultural and horticultural projects utilised the almost unlimited land available in the vicinity. Workshops were opened for craft trades such as carpentry, tailoring, painting, shoemaking, and sewing. A printing shop produced Robert Grieve's remarkable monthly publication, *The Asylum Journal*, between March 1881 and January 1886.[22] Domestic operations provided employment in cleaning wards and day rooms, as well as in the kitchens, the bakery, and the laundry. At the beginning of 1881, 226 patients were regularly employed out of a daily average of 326 in the asylum.[23] Agricultural production grew considerably, with large quantities of bananas and plantains, sweet potatoes, cassava, garden vegetables and coconuts harvested. Animal husbandry brought a substantial output of milk, eggs, beef and pork.[24] The asylum moved increasingly toward becoming a self-supporting enterprise.

As the patient population grew, so did numbers occupied. In April 1882, 282 people were employed out of 367 patients, including 78 in agricultural work, 36 in the laundry, and 18 in tailoring.[25] The percentage of working patients was now 80 per cent.[26] However, by March 1883, with 303 people employed out of 391, Grieve concluded that a limit had been reached if they were 'to have the benefit of active and interesting work'.[27] Although the proportion employed remained over 80 per cent for some time, he complained periodically of problems in finding sufficient productive work, observing in June 1883 that 'the difficulty of obtaining suitable and remunerative work for all who are able to do something is becoming greater from day to day'. The problem was accentuated by steady growth in the institution's size.[28] Some men were found employment in infrastructure projects like land reclamation, building works, decoration and maintenance.[29] However, the numbers engaged in trades or industrial activities gradually declined, blamed partly on a shortage of raw materials.[30] Only the agricultural sector remained relatively buoyant. A new piggery for fattening 150 pigs was completed in early 1883. By July 1884, 125 people were engaged in farm work. Even here, though, problems arose due to erratic weather conditions and soil deterioration.[31] By the end of 1884, with 418 patients, the proportion employed had fallen to 73 per cent.[32]

The employment programme was accompanied by measures to improve patients' quality of life through diversionary activities like entertainments, games, and exercise. Grieve adapted English asylum practices to local conditions. Weekly dances were established by 1881, attended by patients and staff. These brought increasing numbers of visitors from New Amsterdam. The dances' popularity among patients became so great that in May 1883 the large dining hall was 'completely filled'. Grieve had little doubt of their value in breaking 'the monotony of asylum existence' and 'promoting contentment in the institution'.[33] An asylum band played weekly in the grounds on Mondays, also attracting people from outside. There were occasional concerts and novelty shows, such as magic lantern displays. Games like bagatelle were provided. Large groups of patients, numbering over 100, were taken on walks in the neighbourhood on Sunday afternoons. In November 1884, a Sports Day comprising a wide range of races attracted over 200 patients and many visitors.[34]

Robert Grieve departed in late 1885 to become the colony's surgeon-general.[35] His assistant Dr George Snell took over,[36] maintaining the same overall approach. Published reports, however, became increasingly mundane, providing little of the rich information characteristic of Grieve's period. The regime remained based on employment and occupation. During 1895/96 an additional 15 acres of land were taken into cultivation. Despite fluctuating weather conditions, vegetable production continued on a large scale, as did rearing cows and pigs for their meat. Essential consumer items, like bread, clothing, and shoes were still manufactured in the asylum's workshops. Patients also participated in building repairs and maintenance.[37] Entertainments like the weekly dances continued. A large cricket field was laid out in the central square and used extensively by both patients and staff.[38] Due to steadily rising patient numbers, which reached 660 in 1895/96 and 689 in 1898/99, with about 200 new admissions annually, overcrowding remained a perennial problem. Notwithstanding building additions and enlargements, a dining hall and even attendants' accommodation had to be taken over and adapted as dormitories.[39] By 1900, the institution had lost much of its dynamism and was functioning in the ordered, routinised, statistically driven manner familiar in late Victorian lunatic asylums.

Trinidad – Toward the grand design

Developments in Trinidad showed resemblances to those in both Jamaica and British Guiana. Its population became as ethnically diverse as Guiana's, with the large-scale importation of indentured labourers, mainly from the Indian subcontinent, following the ending of African enslavement. That diversity became reflected within the lunatic asylum. The institution that had belatedly opened at Belmont in 1858 proved inadequate almost from the outset.[40] It was too small to cater for the needs of a growing population and its location on a confined site set among steep hills restricted space for patients' activities, as well as placing serious limitations on future expansion.

Dr Thomas Murray, the first medical superintendent, portrayed an optimistic picture of the new asylum, highlighting its cool and airy location, scenic views, and the 'large and handsome day room' provided (though left unfurnished). From the outset, however, he complained of inadequate staffing, the poor state of the grounds, insufficient water supply, and crude means of drainage and sewerage. With the asylum virtually full on opening he called immediately for construction of another wing and provision of staff accommodation.[41] Nevertheless, Murray was still projecting a positive perspective in January 1860. He claimed great success in achieving recoveries, attributed to the 'modern system of management', which comprised 'inspiring confidence and friendship', the absence of mechanical restraint, and strict adherence to 'wholesome rules of order and discipline'. However, with 45 patients in the asylum, he urged the governor to provide further accommodation 'with the least possible delay', in order to prevent some being returned to the Royal Gaol.[42]

Murray's entreaties drew a slow response and by late 1861 twelve men had indeed been transferred to the jail. Construction of a new two-storey building began a year later and a small convalescent ward was planned.[43] A bath and wash house was added during 1863. Following completion of the new wing for 26 people in February 1864, the patients were finally returned from the Royal Gaol. The asylum's capacity, including the convalescent ward, was now around 80 patients. Although Murray expressed gratitude to the Trinidad government, he continued to bemoan insufficient staff numbers.[44] The new building offered only temporary relief, for by 1870 nearly 100 people were crowded into an institution 'much too small for the number of patients'. A new ward costing over £2,000 was opened in 1873.[45] Nevertheless, overcrowding was again reported in 1875 and 1876, creating fears of night-time incidents. Wooden buildings were put up as a temporary expedient.[46] Governor Henry Irving, now taking a close interest, concluded that 'the building is insufficient for the growing wants of the Colony' and proposed erection of a new asylum on a different site, a view endorsed in principle by the Colonial Office.[47] Irving's condemnations became more forthright, arguing that 'in no case and by no management can the existing asylum be rendered satisfactory' and that it was 'not worthy of the Colony'.[48]

Recognising that a new asylum might be some way off, Irving's temporary successor Des Voeux initiated management changes, seeking transfer into 'younger hands'. The under-pressure Dr Murray retired in early 1877.[49] Hitherto, Murray had been required to work alongside a medically unqualified resident superintendent, William Pashley, who undertook the general management. Much of Murray's work was outside, and his direct involvement in the asylum limited.[50] Taking advice from Dr Crane, the surgeon-general, Des Voeux replaced Pashley's post with a 'chief warder', thereby enhancing the medical superintendent's power. Following the Jamaican model, a doctor with British asylum experience was recruited. Dr Alfred Martin had spent three years as medical officer at the Joint Counties Lunatic Asylum, Carmarthen in Wales.[51] Following Thomas Allen's example, he sought to appoint the head attendant from Carmarthen as his 'chief warder'.[52]

Governor Irving successfully demonstrated the case for a new asylum. In May 1878 a 63 acre site, two miles outside Port of Spain, was acquired for £850 and the colonial engineer was commissioned to prepare plans.[53] However, over 20 years elapsed before it actually opened. The urgency of the present situation was acutely apparent, with patient numbers rising rapidly from 172 in 1877 to 242 by the end of 1880.[54] Dr Martin was under no illusions about the task he faced. His reforming aspirations were severely hampered by 'want of accommodation' for patients, too few staff, and inadequate space for exercise or activities. He demanded urgent government action, referring to the 'very great liability to accidents' and the large numbers of 'unsafe' men required to sleep in dormitories due to lack of single rooms. He could only initiate limited changes himself, like improving sanitary arrangements, starting construction of additional accommodation, and altering the unfortunate title of 'chief warder' to 'head attendant'. He was somewhat reassured in early 1881 by steps to commence 'the sorely needed and long talked about' new asylum.[55] Martin, however, never got to witness this, for he died of fever in October 1881.[56]

Following Martin's death, after some debate, the Trinidad government opted to recruit another experienced British medical man and sought assistance from the Colonial Office and Commissioners in Lunacy.[57] The chosen candidate was Dr George Seccombe, Senior Assistant Medical Officer at Caterham Imbecile Asylum, Surrey. This huge institution, administered by the London Metropolitan Asylums Board, was opened in 1870 to accommodate 2,000 people with chronic conditions. Seccombe had worked there since 1875; his work was regularly commended, particularly his contribution to promoting patient welfare, recreation, and amusements.[58] He was appointed by the Secretary of State as medical superintendent of the Trinidad Lunatic Asylum, setting off for the colony in July 1882.[59]

In clear parallels with Thomas Allen, Seccombe arrived to discover conditions much worse than expected, and he employed similar means to expose them. The graphic depictions filling his initial report revealed serious deficiencies in both physical environment and management regime.[60] Seccombe described buildings irregularly scattered over a steep hill, having 'both internally and externally a prison-like appearance', badly constructed and insecure, offering opportunities for escapes and self-harm. Water supply was inadequate, sometimes needing to be carried half a mile from a mountain ravine. The same baths were used both for patients and laundry. Sanitation on the male side consisted only of two closets for 185 patients, based on the 'earth and charcoal system', and it was little better on the female side.[61] The therapeutic aspects were equally deficient. There were limited recreational facilities, with no day rooms for female patients, who spent much time in a 'small confined Gallery, where they while away the tedious monotony of their everyday existence in ceaseless quarrelling, the result of over-crowding'. The infirmary wards had no tables and the spectacle at meals 'is not such as one cares to witness'. In the absence of airing courts patients were 'confined to a certain position, on a certain bench' to keep them within view of the attendants. No patients were employed, apart from a few women in the laundry. There were neither amusements nor a suitable building to hold them in. Seccombe lamented that:

> One would have to go back forty years in the History of the Insane to find another such Asylum at Home...instead of the Asylum being a Hospital for the treatment of the Insane, it is rapidly becoming a Refuge for the chronic Insane, the patients lapsing, owing to the want of means for their cure, into hopeless forms of Insanity, in which state they will be life-long sources of expense to the Colony.[62]

Seccombe's scathing report was published in the *Journal of Mental Science* in Britain, ensuring the fullest exposure in influential circles outside the colony.[63] Over the next two decades Seccombe regularly utilised published reports to condemn the asylum's shortcomings, whilst describing steps taken toward addressing them.[64] His campaign to achieve fundamental changes comprised four main elements. Firstly, he highlighted persistent problems largely beyond his control, like ill-constructed, dilapidated buildings and defective sewage arrangements. He complained often about insufficient staff and the difficulties of securing people of suitable calibre. He particularly emphasised the worsening overcrowding that accompanied incessantly rising patient numbers, from 282 in 1880 to 360 in 1890, and 524 in 1900.[65] As he wrote in 1885, 'we are face to face with a very crowded Institution (composed of very defective buildings) on a cramped site, and an ever increasing resident population'.[66] In 1894 he protested that 'we are compelled to stow away at night 420 patients in buildings that should never contain more than 250 inmates'.[67] Seccombe's strictures and predictions became almost apocalyptic, as he regularly blamed the asylum's grossly overcrowded state for high mortality consequent on epidemic diseases and tuberculosis.[68] Congested wards were blamed also for the frequent injuries, and even danger to life, arising from quarrels and fights among patients.[69]

Secondly, notwithstanding the constraints of unsuitable location and extremely inadequate facilities, Seccombe managed to implement significant improvements in the asylum's material fabric. A new building with 20 female single rooms was erected. Adaptations and alterations were made to existing buildings, including conversion of single rooms into dormitories. Additional land was acquired and new buildings provided for a bath-house, kitchen and bake-house, infirmary, dispensary, and recreation room. Dormitories and day areas were painted and decorated. Improvements were made to sewerage and drainage, as well as arrangements for piping and storing water. In the grounds, steep slopes were evened out and pathways created between buildings. Ornamental gardens were laid out, and roads constructed up the hillsides to create exercise facilities.[70] By 1890, a marked transformation had been achieved and some prison-like features removed. Nevertheless, there was a limit to how far crowded old buildings could be upgraded. Proper laundry facilities were still lacking, and clothes had to be washed in 'very primitive fashion' in the female bath-house.[71]

Based on his previous experience and 'moral management' convictions, the key element of Seccombe's strategy was a programme to occupy as many patients as possible. His approach paralleled those of Thomas Allen and Robert Grieve. Within two years of taking up post, Seccombe claimed 172 out of 291 patients were constantly employed.[72] Where possible people worked on the limited surrounding lands, but the emphasis was more on light industrial tasks like tailoring and dress-making, and domestic tasks in the kitchen, laundry and on the wards. Workshops were established in the buildings' basements. As the system became more established, gangs of male patients were engaged in infras-tructure improvements, such as land reclamation, making roads, paths and drainage ditches, and erecting new buildings. Seccombe overcame financial restrictions, demonstrating that patient labour saved money by helping the institution become partially self-supporting.[73] In early 1888, he hailed the asylum's system of industrial employment as 'the greatest improvement of all', bearing favourable comparison with 'the more advanced Institutions at home'.[74] Over subsequent years arrange-ments were consolidated, despite limitations imposed by the buildings and surrounding terrain.[75]

These tangible improvements, in the face of adverse circumstances, greatly enhanced Seccombe's reputation and prestige.[76] The fourth element of his approach utilised this acquired status to criticise the authorities openly for dilatoriness in regard to the state of the asylum. He sought to pressurise or shame the governor and politicians into pro-viding money for immediate amelioration and, ultimately, to replace it. He several times denounced conditions in the institution as a disgrace to the colony, an assertion guaranteed to register with the governor and Colonial Office. He claimed in 1892 that 'in an English Asylum, this con-gested state of the buildings would never be allowed to exist'.[77] In 1894 he blamed severe overcrowding for sickness levels and high mortality rates, calling on the government to stop the 'Sacrifice of Life' entailed and get on with erecting a new asylum.[78] In 1895 he described the asy-lum as 'so far behind the age' that he felt ashamed to show visitors around. It was 'neither a credit to the Colony nor to those who are responsible for its management'.[79]

Seccombe agitated for years for a new asylum to replace that at Belmont.[80] The wheels of government moved painfully slowly. Although a site had been available since 1878, there were difficulties from the start. From 1880 onwards a whole series of plans were put for-ward, only to be rejected by the Colonial Office on various grounds.[81] One rejection was prompted by Governor Sir Sanford Freeling's concerns in 1881 that the estimated cost per head might be excessive for patients of 'low negro type' who composed the great majority.[82] Agreement was eventually achieved in 1893, but further delays occurred due to escalating costs partly associated with constantly rising patient numbers.[83] Joseph Chamberlain, when Secretary of State, intervened personally in 1895, describing the asylum's overcrowded state as 'a disgrace to the Colony' and calling for construction of a new asylum 'to be proceeded with as speedily as possible'.[84] The imposing new Trinidad Lunatic Asylum at St Ann's finally opened in 1900, having cost in excess of £50,000 (Figure 5.1).[85]

The patients moved from Belmont in March 1900 and by the year's end there were 524 people in the asylum.[86] Inevitably, St Ann's did not prove the complete panacea. There was a sharp initial increase in mortality rates compared to the old asylum, attributed largely to an influenza epidemic in April and May 1900.[87] Within a relatively short time most problems prevalent in English county asylums had become evident. Patient numbers climbed steadily, reaching 689 in 1914.[88] By 1904 Seccombe was renewing old complaints about overcrowding, and sought to transfer demented elderly and chronic 'imbecile' patients elsewhere.[89] A year later he claimed there was an excess of 117 patients and the asylum's 'congested state' made it difficult to maintain a 'quiet and orderly environment'.[90] Similar grievances were regularly restated.

Figure 5.1 Trinidad Lunatic Asylum, St Ann's (1908)
Source: Contemporary Postcard, in author's possession.

In 1908, he lamented that 'I have never known either the Old Asylum or the present Institution to be otherwise than overcrowded except for very short periods.' It was 'almost impossible' to maintain a regime conducive to treatment and recovery.[91] By now it was clear that Seccombe had had enough. The enthusiasm evident in the reforms of the 1880s had waned, reflected in increasingly cursory reports devoid of much interest. Exhausted and in poor health, he retired in March 1909.[92]

Seccombe, the reforming humanitarian, was succeeded as medical superintendent by his Edinburgh-trained assistant, Dr George Vincent, a man of firm eugenicist convictions.[93] Notwithstanding his pessimistic perspective, under Vincent's direction the asylum was quickly enlarged and improved to address overcrowding. New wards were built, including a large refractory ward and a detached 'villa' for convalescents, facilitating greater classification. New male and female head attendants were recruited from English county asylums. Patient employment and occupation were extended and once again accorded a high priority. The opening of a ward for private patients facilitated the social (and racial) segregation increasingly favoured by colonial authorities.[94] By 1914, the Trinidad Lunatic Asylum was operating as an orderly, model British colonial institution.

Barbados – A slow walk to Jenkinsville

Like much else in Barbados the development of mental health provision followed its own characteristic course.[95] The authorities consistently sought to provide no more than the bare minimum. The Barbados Lunatic Asylum that opened in 1846 directly adjoined District 'A' police station and was close to Glendairy Prison. Having its own high boundary walls, it visibly formed a part of the island's custodial framework. The asylum was deficient in almost every aspect. The Colonial Office report of 1864 showed small, inadequately ventilated rooms, some without windows. There was no proper sewerage or drainage, and the latrines consisted of pits. Little occupation was provided for patients. Mechanical restraint (in the form of manacles) and seclusion were used extensively.[96] Patient numbers continued to rise steadily, reaching 88 in 1868. Overcrowding persisted despite *ad hoc* building alterations, like conversion of the superintendent's house into patients' accommodation.[97] As in Trinidad, concerns about the deteriorating situation brought moves toward a replacement institution. Legislation was passed, and £1,900 spent in acquiring land on the nearby Codrington estate. Plans of English asylums were obtained to inform design of a

Figure 5.2 Barbados Lunatic Asylum (1872)
Source: British National Archives, CO 1069/243/48.

building.[98] However, nothing further happened at this point, initiating many years of prevarication. (Figure 5.2)

In 1871 Governor Rawson W. Rawson called the House of Assembly's attention to the 'pressing necessity' for a new asylum. Dr Francis Browne, the medical superintendent, had declared it impossible to admit any more lunatics and several were being held in jails for long periods. However, the reluctant assembly would only grant £300 to erect some wooden huts on the existing site as a temporary measure.[99] In early 1873, Rawson was optimistic that real progress was being made. Legislation to authorise erection of a new asylum had been passed, and expenditure of £25,000 agreed by the assembly. Architect's plans were drawn up based on the well-regarded Derby Lunatic Asylum in England. On the Secretary of State's recommendation, the asylum would also admit people from other islands in the Windward group, an idea previously mooted by Rawson in 1869. However, the project became mired in disputes about this, the numbers to be accommodated, and the likely need for more money than estimated.[100]

In May 1874 the Earl of Carnarvon (Secretary of State) issued his recommendation that Dr Thomas Allen be appointed Inspector of Lunatic Asylums in the West Indies. Governor Rawson showed interest and invited Allen to Barbados, primarily to provide expert advice on plans for the new asylum.[101] Allen visited in January 1875, though his final report only emerged a year later.[102] In characteristically bombastic fashion he criticised the plans mercilessly and in great detail, regarding even a modified version of the Derby Asylum design as quite inappropriate for the tropics. His criticisms covered arrangements for ventilation, lighting, water and sewage, the excessive amount of corridor space, the balance between dormitories and single rooms, and lack of separate dining accommodation. He insisted that likely numbers had been underestimated and that there was insufficient special provision for the most disturbed and disruptive people. Allen offered alternative proposals, based on detached pavilions rather than a single large building. Notably he argued for a location near the sea, between Bridgetown and Hole Town, rather than the inland site purchased at Codrington.[103] His belated report provided enough ammunition to delay a decision for several more years.

Thomas Allen, however, interpreted his commission wider than intended. Governor Rawson had asked him for suggestions on the best ways to improve the existing asylum pending erection of a new one. Allen approached this, as he had in Jamaica, by meticulously investigating all aspects of its operation and conditions. His lengthy report portrayed a dilapidated, dirty, overcrowded institution, defective in every conceivable aspect. Many of the 80 male and 61 female patients were accommodated in barely habitable rooms, some without windows or ventilation. There were insufficient beds, with some people sleeping on the floor. In many instances two or three patients occupied rooms intended for one. Infestations of bugs and cockroaches were common. Pigs, turkeys and chickens were running around yards and courts strewn with rubbish. The water closets and privies were primitive and foul. There were open drains and cesspits, one at the rear of the kitchen. The only bathing facilities were a tap or a mile-long walk to the sea.[104] Allen also castigated the staffing situation, regarding the head attendant Benoni Armstrong and chief nurse Mary Williamson as too old and incompetent for their duties. He considered the other staff illiterate, inexperienced, and of an inferior class, whilst acknowledging that they were lacking in numbers, overworked, poorly paid, and forced to live in conditions hardly better than the patients.[105]

Governor Rawson considered the 'minuteness' of Allen's investigation 'unnecessary & frivolous',[106] but the comprehensive findings confirmed the appalling state of the Barbados asylum in 1875. A report in August by Dr Charles Hutson, medical superintendent since Dr Browne's death in 1874,[107] gave dispassionate confirmation:

> The accommodation is in every way defective and insufficient, the rooms being hot, badly ventilated, and in some cases mere wooden sheds without opening of any kind save a slit in the leeward roof. Albeit in these single rooms we have often to lodge three and sometimes four patients at night and this too at great risk. The ventilation

throughout is wretched. There are no urinals, no lavatories, no baths, no day rooms, no worksheds. There is not even a verandah into which the patients might get during wet and rainy days and be dry. There is only one airing court on the Male side, and in it are congregated all the male patients whether violent, quiet, melancholic or idiotic. Assaults on the feeble and quiet by those who are violent are of almost daily occurrence.[108]

Hutson did his best to implement improvements based on Allen's recommendations. Notably he started compiling case records and a daily journal. These enable construction of a picture of the patients and the problems of managing the asylum in the late 1870s and 1880s.[109] They particularly illustrate significant levels of disorder, and the often vain attempts by medical officers and staff to provide a therapeutic experience for those confined there.

In the absence of progress toward a new asylum, the pressing overcrowding problem was addressed in piecemeal fashion. Two wooden dormitories were erected during 1877 after the authorities reluctantly contributed a meagre £200.[110] The situation reached crisis point in early 1879. Governor Strahan reported that, due to shortage of room, lunatics were 'now scattered about the Island at the Police Stations, the General Hospital, and elsewhere'.[111] A desperate plea was issued by Dr Albert Field, the recently appointed medical superintendent, describing the asylum's current state as 'dangerous to the health and lives of the unfortunate inmates'. He was particularly concerned about the lack of single rooms available for quarrelsome females. He cited the case of a vulnerable young woman forced to vacate a single room for a new patient; on removal to a dormitory she disturbed the others and received a fatal beating. His arguments were supported by his predecessor Dr Charles Hutson, now the island's Poor Law Inspector.[112] Following their urgent representations £650 was sanctioned to put up more wooden buildings to accommodate 40 patients and four attendants.[113]

With a daily average of 150 patients, more drastic measures were required. The use of jails, though universally condemned as inappropriate, was formalised by conversion of a cell block at District 'B' police station, several miles away at Boarded Hall, into a temporary asylum for 30 females. By the end of 1880, over 180 people were divided between the two sites.[114] Hutson contended that increasing numbers did not reflect a growth in the incidence of insanity in Barbados, but that the asylum's defects reduced recovery numbers. He cited its 'cage-like airing courts, the small and badly ventilated rooms, the enforced absence of system and discipline, the disorderly meals, the absence of amusement and employment, &c &c'. Most of the rooms were 'weak and dilapidated' and the medical superintendent was 'at his wits' end to know what to do with violent lunatics'.[115]

Hutson was highly critical of delays in developing a new asylum and maintained pressure on the government. His charges regarding the present asylum's insanitary state led to an enquiry in January 1883.[116] In July Governor Sir William Robinson declared it 'utterly unworthy of such an island as Barbados'.[117] Reports showed that, despite improvements in cleanliness and sanitary conditions, overcrowding persisted, sick and feeble patients were kept locked in single rooms for protection from rougher patients, and seclusion was used excessively.[118] By 1885 patient numbers had risen to 180, with a further 35 at District 'B'. According to Hutson they were 'crowded together without any possibility of classification, a very small percentage of them were employed; almost no amusements exist and it is only natural that frequent disagreements if not fights should frequently occur'.[119] A year later, though commenting oddly that 'Mad people can stand a good deal in the way of inconvenience', he claimed the noise and excitement of the overcrowded airing courts was retarding recovery, and that 'serious affrays' often occurred at night when patients needing isolation were placed with others. Naturally, Hutson was especially concerned about the many 'respectable or semi-respectable patients', whose 'peculiarities' made it more difficult to associate them with other patients.[120]

By the end of 1886, patient numbers had reached 208 and the overcrowding was 'unbearable', with only the 'dangerous and troublesome' considered for admission.[121] A government commission in 1887 acknowledged that the asylum belonged to a 'bygone age'.[122] The insanitary conditions impacted on death rates, as diseases like consumption, diarrhoea and dysentery became prevalent.[123] In an almost desperate measure 20 people were moved to cells in the abandoned prison at the adjoining District 'A' police station.[124] Numbers reached 230 during 1891.[125] The incidence of bowel complaints rose sharply and typhoid killed two patients and an attendant. The likely cause was identified when the drains were opened. According to Hutson:

> no one had any conception how bad things were... The main drain was found in one spot to be below the level of the rest of the drain and the accumulation of thick, foul-smelling ooze at the back of the kitchen, store-room, and Cross Ward was abominable.[126]

The doomed institution was now in a state of utter crisis. At the end of 1892, 258 patients were crammed in, an increase of 66 in two years. On an evening visit Hutson found 49 men in one dormitory, some 'huddled together on the floor', all under the charge of one attendant. It was proving difficult to maintain order; seclusion was being widely employed, particularly for disorderly females.[127] The agony was nearing its end, however, with a new asylum finally almost ready.

Almost 25 years had elapsed since the first steps toward a new asylum, and 18 years since Thomas Allen exposed appalling conditions in the existing one. The sequence of false starts and delays was largely attributable to the vagaries of Barbadian elite politics and the social perspectives of legislators constrained by indecision, parsimony and narrow, racially bound class attitudes.[128] In the year following Allen's visit, the asylum issue became enmeshed in heated arguments about creation of a Windward Islands Confederation, which escalated into the violent riots of April 1876. One of Governor Pope Hennessy's controversial six proposals was that the new Barbados asylum should be a facility for the whole Windward Islands, rendering it odious to opponents of Confederation.[129] In June 1878 a special commission was established to produce new plans and consider the relative merits of the present site and that acquired in 1869 at Codrington.[130] It concluded that an asylum for 225 people, including some from other islands, should be built at Codrington at a cost of £31,000.[131] The assembly was not convinced and in January 1879 appointed a committee to reconsider the commission's report.[132] After several months the plans were rejected, following objections by John Bourne, the Superintendent of Public Works.[133]

Nothing further occurred for two years. In April 1881, Bourne's successor Captain Monier Skinner reviewed all previous plans and produced new ones taking account of advice from Drs Charles Hutson and Albert Field.[134] By now the serious reservations expressed by Thomas Allen regarding the Codrington site were re-emerging, particularly because of water supply problems. Nevertheless, it was agreed in March 1882 to proceed with Codrington, mainly because £8,000 had already been spent on it.[135] In August another committee, including Field and Hutson, recommended abandonment of Codrington as barely large enough to contain the proposed buildings.[136] However, no suitable alternative was identified and in February 1884 the assembly resolved to persevere with the site and acquire additional land there.[137]

After two more years' inaction Governor Sir Charles Lees, facing serious criticism about the existing institution, issued assurances that the question was receiving renewed attention.[138] He enlisted the help of Dr Charles Manning, a Bridgetown doctor who had worked for six years in Japan and advised on the construction and organisation of its public asylum. Manning produced a measured report that took account of current literature and modern ideas. He advocated a curative hospital based on moral management principles, with an extensive programme of patient employment that required ample land. He endorsed Allen's recommendation of a location near the sea and Bridgetown, to ensure a ready market for its produce. He concluded that acquisition of Codrington had been a blunder, carried out by 'men who evidently knew nothing whatever about the special requirements of Lunatic Asylums'; to build there would be 'absolute folly'.[139] Even at the 11th hour, Charles Hutson submitted an alternative proposal for the new asylum to be on a site near the existing one at District 'A'.[140] Manning's intervention proved crucial, however, for he and Dr Field were placed on a commission that reported to the governor in April 1887. Having considered Codrington's attributes and defects, and the shortcomings of several other sites, they recommended the former Jenkinsville plantation, near the sea at Black Rock, a mile and a half west of Bridgetown. Within a few weeks the assembly agreed to purchase the estate, and the legislative council concurred in early 1888,[141] although two further years elapsed before £32,000 was granted in July 1890 to actually build the asylum.[142]

The third Barbados Lunatic Asylum was officially opened by Governor Sir James Hay on 5 April 1893, in a ceremony also attended by 58 patients and eight attendants. The new asylum had capacity for 434 people, including 20 private patients.[143] The first people were admitted on 10 July, when 32 females eagerly walked the four miles from the old asylum, arriving just after five o'clock in the morning. Over the next two days another 120 women arrived, some on foot and some by carriage. Early on 13 July, 80 male patients walked down, accompanied by ten attendants, the rest following in carriages. Altogether, 277 people were transferred, including 27 women from the outlying District 'B' asylum.[144]

The new asylum was not an unqualified success. Although conditions for patients had undoubtedly improved, there were considerable problems which proved more than mere teething troubles. Some building work had been shoddy, presumably because of restrictions on expenditure. Lessons had not been fully learned from the old asylum. The external walls were insecure, doors weak, locks not fitted properly, and spaces between window bars too wide, all facilitating escapes. There was insufficient ventilation, poor drainage, and defective sanitary arrangements, with raw sewage floating into baths from discharge pipes. Many patients and staff succumbed to diarrhoea and vomiting, and 22 people died within the first five months.[145] By 1897 the mortality rate was even higher; tuberculosis, dysentery and diarrhoea were the most common causes, with one death from typhoid. Contaminated water was identified as largely responsible.[146]

The worst defects were gradually addressed, though a high incidence of disease persisted.[147] From 1905 onwards the asylum was afflicted by outbreaks of pellagra and psilosis, which were prevalent in Barbados, accounting for large numbers of deaths. Pellagra was recognised to be associated with 'extreme poverty' and inadequate diet, and was often complicated by manifestations of insanity.[148] Despite a range of measures the diseases continued to have damaging consequences. In the first half of 1913, out of 51 deaths in the asylum, 35 (69 per cent) were attributed to psilosis or pellagra. Their ravages helped account for several years in which patient numbers barely increased or even decreased. In June 1913 there were 381 in the asylum, compared to 367 at the end of 1900. A year later the numbers had increased to 428, largely due to the repatriation of numerous insane men from Panama.[149]

There was scant early progress in most key aspects of the asylum regime. Dr Albert Field showed little commitment to introducing moral management practices. By the end of 1900 locked seclusion was still in frequent use, and mechanical restraint regularly employed. Patient occupation had not been significantly developed. Out of 367 inmates, an average of only 90 were working each day, at tasks including gardening and road-making for men, and sewing and laundry for women; the other 75 per cent were unoccupied. In regard to amusements, it was recognised that 'almost nothing' was being done. A new cricket patch was laid down in 1901, following intervention from the island's governor.[150]

After a period of poor health, Field died in December 1901.[151] Discussions followed around bringing in an experienced British asylum doctor 'who could introduce modern methods'. The Barbados government, true to form, would not agree to pay a sufficiently high salary. Dr Charles Manning was appointed, based on his local knowledge and earlier involvement in the asylum's planning.[152] He began fairly enthusiastically, having 'many plans for the improvement of the institution'. However, most were thwarted by the parsimonious government, like his proposal to employ a 'special agriculturist' to develop farm enterprises. He advocated provision of staff uniforms to distinguish attendants from patients but, although the executive agreed in principle, the expenditure was not sanctioned. There was similar resistance to paying for essential repairs to the asylum's defective, subsiding floors. Manning did manage, nevertheless, to extend the laundry and import pigs from Canada and two cows and a bull from Scotland for breeding purposes.[153]

Financial stringency brought greater efforts to promote work schemes. Building repairs and improvements were carried out by patients and staff during 1904. Nearly 100 sets of attendants' uniforms were manufactured in the asylum. Agricultural enterprises were gradually extended. By the end of 1904, 106 patients (29 per cent) were employed on average.[154] In 1906 seven acres of swamp land was reclaimed and planted with sugar canes.[155] However, overall progress remained relatively slow. In 1912 Manning had a fishing boat built; an old fisherman was appointed as attendant, some patients trained as crew members, and the fish caught was used to vary the institutional diet. By June 1913, employment reached a high point of 124 (31 per cent). Nevertheless, unfavourable comparisons were continually being drawn with asylums in England and elsewhere in the West Indies, one suggested reason being the 'large number of feeble, broken-down inmates'.[156]

Other areas of the asylum's operation continued to cause concern. In 1904, the lack of amusements, other than cricket for the men, remained a 'prominent defect'. The women had 'nothing of this kind to do, and they sit aimlessly during idle hours'.[157] Similar complaints were repeated over the following years. In 1905, lack of classification was deemed a 'serious blot' on the institution, with indiscriminate mixing of patients blamed for reducing the numbers of recoveries. Relatives were also criticised for not accepting people back home even when ready for discharge.[158] The use of seclusion remained high, with up to 40 instances per month in 1914, despite improvements in staffing and Manning's introduction of a training programme for attendants in 1912.[159] By 1914 the Barbados asylum had settled into a pattern of steady under-achievement.

Small islands, small aspirations

As outlined in Chapter 2, some limited asylum provision was made during the 1840s and 1850s in several eastern Caribbean islands, including Antigua, Grenada, St Lucia, St Kitts, and Montserrat. Conditions ranging from basic to extremely poor were illustrated in the Colonial Office survey of 1864.[160] The fundamental problem was identified by Governor Pine of Antigua in 1870, in accounting for its asylum's ineffectiveness:

All that these local Asylums can pretend to do, is to confine the lunatics with more or less comfort, and to feed them and attend to their bodily health. These institutions could not be rendered really effective for curative purposes except at a cost which the Islands individually could not bear.[161]

The conclusion was that groups of islands should combine to establish viable facilities. Although consideration had been given to creating a central lunatic asylum in Barbados, the Colonial Office had already concluded by 1870 that there would be 'some difficulty' in achieving this.[162] The formation of colonial federations in both the Leeward and Windward Islands in 1871 provided the necessary political frameworks for subsequent developments.

The Antigua Lunatic Asylum on Rat Island had opened in 1842.[163] Its unsuitable, isolated location and austere, dilapidated military buildings were widely condemned. According to its medical officer Dr Adam Nicholson in 1869, the asylum made a 'sorry figure' compared to institutions in Europe. The main part comprised a basement floor divided into 17 single cells for 'refractory' patients and an upper floor with nine rooms for the more 'orderly'. An adjoining wooden structure contained 12 rooms, and there was a recently erected day room with three open sides, in which meals were served. The asylum grounds consisted of one and a half acres of airing courts, although the inmates also had access to another six acres on the island for walks. The only occupation for the 40 patients appeared to be 'Music and Dancing'.[164]

In the absence of agreement on provision in Barbados, Governor Pine agreed with the Colonial Office that a central lunatic asylum be established for the Leeward Islands, and the necessary mechanisms were incorporated in the legislation forming the federation.[165] Notwithstanding reservations about Rat Island's great imperfections, within a few years it had become the central asylum and patients were admitted from other islands. However, St Kitts and Montserrat continued to retain their own small institutions.[166] In 1884 it was observed that, although the lunatics' accommodation behind St Kitts' Cunningham Hospital left much to be desired, it was preferable to the overcrowded Antigua asylum which was 'little more conducive to the cure of the insane than the very defective and unsuitable provision made in the other islands'.[167]

A novel solution was adopted to address the Leeward Islands' deficient asylum accommodation. By 1889 a second asylum had been opened in a former military barracks at The Ridge, near English Harbour, at the opposite end of Antigua to Rat Island. The Ridge catered for 'convalescent or harmless or quiet' people, whilst Rat Island continued to receive the newly admitted patients. Once settled and manageable they would be transferred to The Ridge. At the end of 1889 there were 69 patients at Rat Island and 36 at The Ridge. About half originated from Antigua, the remainder being from St Kitts, Nevis, Dominica, Montserrat and the Virgin Islands.[168] Despite the changes, concerns about conditions on Rat Island persisted, the building described in 1892 as 'quite unfitted for the purpose to which it is placed'.[169]

Rat Island's unsuitability became manifest during 1892 when mortality rates soared with the deaths of 29 people, well over one third of its patients. In an attempt to deal with the situation, an improved dietary was implemented and three trained nurses drafted in to attend on the sick.[170] These measures proved thoroughly inadequate, however, and another 29 people died during 1893. The problems appeared to lie in the asylum's poor sanitary state, and the Leeward Islands Legislative Council had to stump up funds for the buildings to be gutted and small confined cells converted into five spacious, well ventilated dormitories.[171] Staffing numbers were increased and, following recruitment in 1894 of a 'warder' and nurse trained in a European asylum, great improvement was reported in the patients' general treatment along with a marked fall in death rate.[172]

Rat Island asylum's shortcomings were, nevertheless, much too great to be addressed by cosmetic improvements. In 1907 a new Central Lunatic Asylum for the Leeward Islands was opened in a former juvenile reformatory at Skerretts, a mile outside Antigua's main town St John's. Its medical officer was shared with the hospital, the leper asylum, and the poor-house, whilst day to day management was provided by a resident superintendent. In May 1909 there were 113 patients in the asylum.[173] Contemporary photographs show four neat separate buildings, one each for refractory and for convalescent patients, male and female, all set in several acres of well laid-out grounds (Figure 5.3).[174] The vacated buildings on Rat Island were incorporated into the adjoining leper asylum.[175]

Parallel developments occurred in the Windward Islands, notably attempts to rationalise individual islands' inadequate provision by centralisation. With the continuing refusal of the Barbados authorities to adopt a leadership role, or even participate, in any pan-Windward ventures, Grenada became the main island for administrative purposes. When its asylum opened in 1855 there were ambitious proposals for it to become a 'General Lunatic Asylum for the West Indies'.[176] In reality the buildings, in an old fortress high on Richmond Hill, were far

The Central Lunatic Asylum, Antigua

Figure 5.3 Antigua Lunatic Asylum, Skerretts (early twentieth century)
Source: Contemporary Postcard, in author's possession.

too inadequate and unsuitable for such a function, comprising two two-storey blocks and a single yard. The three day rooms were formed from converted corridors. Sewerage was by means of inclined open drains and gutters. At the end of 1868 the asylum housed 24 patients, mostly in single cells ventilated and lighted through lattices.[177]

The possibility of a general lunatic asylum for the Windward Islands in Grenada, to include Barbados patients, was mooted in 1869. The site of a military sanatorium was considered, but rejected as too small.[178] Building alterations were made at Richmond Hill to facilitate 'the more entire separation of the sexes', and additional cells and dining areas were constructed to 'contribute to the comfort and well-being of the unfortunate inmates'. An 'apartment' was prepared for the medical officer to become resident.[179] Dr W.F. Newsam, who was also medical officer for St George's parish, took exception to being compelled to reside at the asylum and resigned in October 1871 to take up a post on St Vincent.[180] It proved difficult to find a doctor prepared to replace him.[181] The requirement to live on site was amended in new rules produced in 1872, with the medical attendant now expected to reside 'within or as near the Asylum as convenient'. There was to be a 'Keeper' and 'Matron', serving also as master and matron of the nearby poor-house. Basic standards were laid down, including separation of the 'convalescent and quiet' from the 'refractory, noisy, or dangerous' and 'clean' from 'dirty' patients.[182] The

Colonial Office were given assurances that the asylum building was now 'well adapted to the purpose' and the patients had grounds in which they could walk.[183]

It became increasingly evident, however, that the Grenada asylum was in a deplorable state, despite modest improvements like introduction of the 'dry earth' sewage disposal system on the female side during 1873.[184] An inspection by R.W. Harley, the island's administrator, in June 1877 revealed the true state of affairs:

> The Building requires a thorough cleansing. The Arched cellars on the ground floor female side are not fit places for the confinement of human beings, they are offensive in smell, damp and over run with rats, with but little light and no ventilation.... The Cells on the Male side are little better, the open drain which is intended for the reception of natural deposits and running through the whole range of them is most offensive...[185]

Harley oversaw some essential reforms, beginning with the appointment in May 1877 of Dr Charles Massiah as medical officer.[186] The Board of Guardians, previously responsible for the asylum's management, was abolished and its powers assumed by Grenada's lieutenant governor and executive council. Repairs and adaptations were made to the buildings, including extension of the dry earth system and cleansing of a cess pit near the kitchen.[187] With Massiah's participation, patients' comforts were improved by provision of new beds and bedding, furniture and tables. However, the asylum regime remained repressive, with mechanical restraint and seclusion used extensively. The supposedly liberal Massiah was unapologetic, pointing out the limitations imposed by a small staff of head keeper and male and female attendant.[188]

Improvements continued during 1878, with attempts to promote employment and recreation.[189] The Grenada asylum was now considered to be in a more acceptable state and the idea was revived of its taking on a wider role for the Windward Islands, following a visit by Lieutenant Governor Dundas of St Vincent. He was impressed with Massiah's achievements and proposed placing 16 lunatics from St Vincent and four from Tobago under his care.[190] The Secretary of State, Sir Michael Hicks Beach, approved the plan as an interim measure, with the Colonial Office still hoping for a Windward Islands' central asylum in Barbados.[191] Massiah was asked to propose essential alterations, but he was advised by a surveyor, Osbert Chadwick, that 'the present buildings are so old and dilapidated, that to attempt to produce a model asylum would almost amount to reconstruction'. Nevertheless, £280 was spent on measures to accommodate the additional people.[192] Ten male patients were transferred to Grenada in January 1880, before the necessary work had even been carried out.[193] The Richmond Hill institution effectively became the central lunatic asylum for the Windward Islands (Figure 5.4).[194] During 1883, 23 females from St Vincent and Tobago finally arrived, after building adaptations including erection of a new day room.[195]

At the end of December 1883, 32 of the Grenada asylum's 73 patients were from other islands. These included five women from St Lucia, which had recently started sending lunatics to Grenada.[196] St Lucia had its own limited asylum provision, dating back to 1855.[197] The much criticised facility was replaced in 1874 by a two-storey rented building on the harbour outside Castries. Men were housed on the ground floor and women upstairs, mostly in single cells with iron gratings on the windows. There were no washing facilities and patients bathed in the sea. Buckets were used for 'excrements', the 'ordure' being carried out to sea by patients at day-break. An enclosed area of one acre, surrounded by a seven feet high fence, served as the sole airing court, with men and women having to alternate. By 1877 there were 21 patients in this asylum, up to three or four in a single cell. Despite its manifest defects, little attempt was made at improvement because of the anticipated central

Figure 5.4 Grenada Lunatic Asylum (late nineteenth century)
Source: British National Archives, CO 1069/349/15.

asylum in Barbados.[198] Once it became evident this would not mate-
rialise the St Lucia authorities began sending people to Grenada, and
in early 1888 its asylum was closed.[199] In 1890 a temporary reception
house was set up in Soufriere to accommodate lunatics pending transfer
to Grenada, an arrangement that continued for several years.[200]

The growing pressure of demand for places from St Lucia, St Vincent
and Grenada necessitated a complete overhaul of the Grenada asylum buildings in the early 1890s.[201] By 1900, however, there were real
problems. Patient numbers had risen to 132 in early 1899, but only
120 could be comfortably accommodated.[202] Overcrowding rendered
proper classification impossible and 'convalescent' patients were associated with the 'noisy' and the 'excited', adversely affecting recovery
prospects.[203] The building's inadequacies were blamed for the increasing use of mechanical restraint and locked seclusion. Sanitary standards
deteriorated, dysentery and diarrhoea became prevalent, and death rates
rose alarmingly.[204] In 1904 the long-serving medical superintendent
Dr P.F. Macleod retired, to be replaced by Edwin Hatton. He appeared
powerless to bring about significant improvements and his reports catalogued similar hazards and defects each year.[205] During 1906 numbers
peaked at 148 and the ill-ventilated infirmaries had to be utilised as
general dormitories.[206] The asylum's unhealthy conditions, with its anti-
quated sewerage arrangements, precipitated further disease outbreaks.
In 1908 several people died from consumption; typhoid also gained
a hold, affecting staff as well as patients.[207] Water closets were finally
introduced in 1911, a roof was put over the 'washing trough' in 1912,
and more sinks provided in 1913. The old dining shed in the airing
court was repaired and turned into a day room. These and other minor
building alterations provided some work for otherwise unoccupied
patients.[208]

The overcrowding problem in Grenada was relieved during 1912, with
the removal of 27 St Lucia patients to its newly completed lunatic
asylum at La Toc, a mile outside Castries. The adapted military buildings could accommodate up to 100 patients.[209] Its facilities were basic,
with only a small amount of land for purposes of recreation, exercise
and employment. By the end of 1914, 78 people were confined there.
Despite its considerable limitations, however, the La Toc asylum was
superior to the one in Grenada, which remained in a parlous state.[210]

In British Honduras the defects in provision consistently surpassed
even those of Grenada and Antigua. The small asylum at Belize was the
subject of scathing condemnation in 1867.[211] In response the assembly spent £300 on re-covering the roof with galvanised iron, whilst

recognising that the asylum needed complete rebuilding along with the hospital and poor-house.[212] Nothing had happened by 1875, when a new asylum was deemed the most urgent public project.[213] The colonial administration agreed to pay for an inspection by Dr Thomas Allen of Jamaica.[214] He arrived in Belize in January 1876 and, characteristically, lambasted every aspect of the 'so-called Lunatic Asylum and its management', construing it as 'most improper and unsatisfactory'. The inmates were housed in 'some old dilapidated wooden buildings', which were 'so bad as to be scarcely fit for the occupation of patients'. The men and women 'associated together', and there were no safeguards against female patients being abused at night by the young male 'underkeeper'. The other people in charge were disparaged by Allen as 'a discharged black soldier – who was unable to read or write – and his wife'. Patients only washed if they pleased, apart from a weekly trip to the sea, and were dried by their own dirty clothes as there were no towels. There were no sheets or pillows and people slept in their day clothes. The only table was located in the kitchen, and patients ate anywhere they chose. No means were available to occupy or amuse the 18 patients.[215]

Allen's exposures had some effect, for the Honduras authorities agreed to provide a new asylum.[216] £1,000 was allocated to purchase a property near the public hospital, for conversion. Frederick Barlee, the lieutenant governor, anticipated that 'all modern conveniences and appliances will be available'. He promised closer supervision by the public medical officer and endeavours to 'amuse or employ' the patients 'in such manner as their peculiarities will admit'. Acknowledging the existing asylum's shortcomings, he explained a particular local obstacle to progress:

> The one great difficulty experienced is the means of communication with patients. Among the 15 patients Europeans, Indians, Spaniards, Chinese, Africans, Negroes, and West Indian Creoles may be seen. They cannot understand each other, and the overseers can only make themselves understood by signs. I mention the foregoing to show how difficult it is to treat these poor creatures or to work on their better feelings in the same manner that would be adopted with patients in a well regulated English asylum.[217]

The move to the new asylum finally took place in 1881, five years after Allen's visit. It claimed to offer 'every comfort' and 'ample care' to patients, and mechanical restraint was not permitted.[218]

The apparent complacency about the new asylum was short-lived. By 1896 average patient numbers had risen to 27. The 'total space available for all purposes' was only just over an acre, which the colonial surgeon described as 'palpably inadequate'.[219] The internal arrangements were deemed 'altogether unsuitable for the treatment of the insane', and a lack of funds was blamed for no remedial measures having been taken.[220] Overcrowding was partly addressed during 1898 by incorporating the buildings and grounds of the adjoining poor-house, which was converted into a female lunatic asylum.[221] This, however, provided only brief relief for, by late 1906, 45 patients were in buildings 'defective both as regards lighting and ventilation' and inadequate for the numbers. In the female asylum 17 people were crowded into eight cells, one containing four beds. Four temporary cells were created underneath the building. The situation on the male side was similar, with 27 inmates and 14 cells; one common cell was occupied by 13 men, which was 'inconvenient and dangerous, for at times they fight, which leads to a general melee, frequently resulting in injuries'. The medical officer was almost in despair, urging drastic improvements.[222] A year later, numbers had reached 57 and congestion was now 'far more serious'. Plans were at least in place for a new building.[223] There were inevitable delays before this finally materialised in 1913, by which time 69 people were crammed into the old asylum.[224]

Conclusion: Contrasting experiences

Great divergences emerged between different West Indian colonies in how mental health provision developed during the nineteenth century. This was related to factors political, economic and social. By and large, the economically stronger islands or territories gradually benefitted from more advanced public welfare facilities and medical institutions, including asylums for the insane. This was essentially true for Trinidad and British Guiana, as previously for Jamaica. In those places the governor and local politicians were able to co-operate and ensure provision to address an increasingly apparent social need. In Barbados the powerful White ruling elite had an evident distaste for providing welfare services to the Black and Brown lower classes. The Church of England minister Edward Pinder castigated them in 1858 for their indifference to 'the present pressing needs of our island'. The 'languishing state' of the hospitals, schools and charitable institutions constituted 'a standing reproach'.[225] They only responded after prolonged exposure of the scandalous situation in the old asylum.

In the small islands of the Leeward and Windward archipelago, and in mainland British Honduras, economic realities constrained the development of health care institutions. Low population numbers, with relatively few overtly insane people, restricted the options for providing localised facilities. Even when islands were federated the potential target group for asylum care was conceivably too small for development of an institution operating on progressive lines. The constraints existing in some places, however, provided a convenient cover for inaction and indecision. Attitudes prevalent in Barbados most likely applied also to the ruling classes of other West Indian colonies, ensuring that a low priority was accorded to any provision beyond the barest minimum.

Where real reform occurred and modern practices were implemented the role of highly motivated individuals was paramount. This was the case in Jamaica with Thomas Allen, and was replicated in British Guiana with Robert Grieve and Trinidad with George Seccombe. Their appointments signified the colonial governments' preparedness to invest in developing more effective asylum facilities, despite the risk that experienced British doctors could prove challenging. As relative outsiders, these men were not fully conversant with the ways of local colonial society and could tackle issues with energy and a degree of objectivity. Armed with their moral management principles and commitment to the centrality of occupation, they sought to ensure that their institutions could progress to be considered on a par with those 'at home'. For a period, at least, they appeared to achieve this.

6
Pathways to the Asylum

Over the last three decades, historians of mental health services in Britain and its former empire have increasingly tried to shift their focus from the development of institutions toward the people who entered and inhabited them. Since the ground-breaking work of John Walton on asylum admissions in mid-nineteenth-century Lancashire, scholars have sought to identify and analyse the circumstances that lay behind committal.[1] Examinations of case records regarding admissions to English public lunatic asylums, notably by Joseph Melling and his colleagues in relation to the county of Devon, have provided rich information regarding the types of people committed and the range of factors and situations that propelled them into the asylum.[2] In the imperial context similar approaches have been adopted, by historians such as James Mills, Catharine Coleborne, and Lynette Jackson. In each instance they have attempted mainly to identify the social, economic, and political circumstances that fostered mental distress and asylum admission, within a framework of colonial power and class relations.[3] Other historians have investigated the influence of more specific aspects such as gender, ethnicity, and migration in admissions.[4]

The various studies of asylum admissions in Britain and the colonies have generally concurred that the main determinant of referral for committal was the inability of the identified insane person's family to contain the manifestations of distressed or disturbed behaviour.[5] In most settings and cultures families were normally able and prepared to withstand a good deal of discomfort, inconvenience and hardship in order to maintain their suffering member at home. It tended to be only when circumstances became intolerable that he or she would be submitted to the authorities for institutionalisation. The evidence indicates that similar patterns prevailed in the West Indian colonies. Of course, many people who experienced acute mental distress did not have relatives or others actively interested in their welfare. Such individuals were more likely to figure among those whose admission to a lunatic asylum followed exhibition of violent, disorderly or disruptive behaviour in a public place.

Unfortunately, the vagaries of survival of asylum case records in the West Indies mean that a detailed analysis of factors leading to admissions, comparable to that undertaken by Melling and others, is not feasible.[6] In order to construct a picture of the people entering British Caribbean asylums it has been necessary to gather data from disparate sources. Some significant case material has survived from the Barbados Lunatic Asylum, mainly covering the period from 1875–80, and this provides an invaluable resource.[7] For British Guiana the printed journals compiled by Dr Robert Grieve included about 30 individual case histories from the 1880s, which in most instances gave information on circumstances that brought about their admissions.[8] One notable source for Jamaica is located within one of Thomas Allen's voluminous early reports, where he summarised some case histories primarily to demonstrate that the asylum was obligated to manage very difficult patients.[9] Other material, largely of a more general nature, has been found within official and semi-official reports, pamphlets, newspapers and journals.

Nineteenth-century published reports on lunatic asylums invariably presented statistics regarding people admitted, which included basic information on age, gender and occupation. In the West Indies, additional data regarding country of birth, and race or colour was often included. Reports would also normally provide, in tabular form, medical diagnoses and specific assigned 'causes' of the mental disorder. Medical superintendents were expected to submit this information, identifying a single cause for each patient. These causes might be of a designated physical origin, like disease, heredity or childbirth, or otherwise of a social or psychological nature, such as intemperance, grief, or domestic worry.[10] However, simplistic assignations of one 'cause' are of little value for the historian, for the circumstances contributing to an individual's descent into serious mental disorder were invariably more complex.[11] In this chapter the complexity has been addressed, at least in part, by differentiating what were pre-existing or underlying circumstances from those more immediate occurrences that brought about intervention from the authorities.

Background circumstances

The medical officer for the Antigua Lunatic Asylum reported in 1864 that the 'class of patients' admitted was 'for the most part composed of

Labourers, Artizans, and persons belonging to the lower orders'.[12] This concise statement would have been equally applicable for any of the West Indian asylums in the latter half of the nineteenth century. It highlighted the asylum's role as a place of refuge for some of the casualties of colonial society. The bulk of people received came from the labouring poor, often a reflection of the economic and social stresses to which they were constantly exposed. Recorded admission data repeatedly shows the largest single occupational group, for both men and women, to have been 'labourers' or 'agricultural labourers'. Annual admission figures from the mid-1880s provide illustration. During 1885, out of 101 people admitted to the British Guiana asylum, 68 per cent (48 men; 21 women) were agricultural labourers.[13] In Trinidad in the same year, of 114 people admitted 71 (52 men; 19 women), or 62 per cent, were described as 'labourers'.[14] In Jamaica, in 1885/86, with 139 people admitted, 71 (35 males; 36 females) were 'labourers', making up 51 per cent.[15] Across the three asylums, domestic servants formed the next largest group of females, followed by seamstresses and laundresses. Among the men, there were significant numbers of artisan tradesmen like shoemakers, tailors, and particularly carpenters.[16]

Barbados exhibited a somewhat different pattern. An analysis of 176 asylum admissions between 1875 and 1879 shows only 14 men (16 per cent) and 34 women (37 per cent) described as labourers.[17] The male patients came from a range of occupations, with urban-based workers prominent, including several shoemakers, carpenters, clerks, smiths, porters, coopers, boatmen, and seamen, as well as three fishermen and three soldiers. Interestingly, there was a significant middle class component which included three planters, two merchants, a druggist, a clergyman, a schoolmaster, and even a doctor. The female admissions included 12 (13 per cent) 'hucksters' and a similar number of domestic servants, as well as nine (10 per cent) seamstresses or needlewomen.[18]

The assigned social class background of people admitted was closely interwoven with race and colour. As Trinidad's Governor Sir Sanford Freeling crudely suggested in 1881, '9/10ths of our lunatics' were composed of 'the class of patient of low negro type'.[19] He was not alone in this perspective. Dr Thomas Allen drew similar characterisations on several occasions. For him, the 'uncivilized negro of the West Indies', and especially Jamaica, had a distinct predisposition toward insanity:

> The ferocity and ungovernable passion of the Jamaican negro is well known, and a trifling exciting cause, makes him lose all power of self control. Living in the mountains and leading a wild and savage life,

far removed from any civilizing influences, or European associations he has not only been uneducated, but has wanted that experience, example and discipline during childhood, which should fit him for the duties of society when he reached manhood. We have therefore present, deplorable ignorance and superstition, and besides the antagonism of race, that selfishness in all its brutality, with all its unsociability which has long been characteristic of an uncivilized people.[20]

Allen went further, contending that 'looking to the normal standard of intellect of the negro peasantry of the West Indies', many were 'so brutal, and low in their intellectual scale as to be little more than "thinking automata"'. He suggested contemptuously that their 'ungovernable feelings, passions and emotions' rendered many 'fit subjects for a lunatic asylum, when there is little or no decided intellectual aberration'.[21] Five years later he reiterated his outspoken contention of a close linkage between the intrinsic characteristics of many Black Jamaicans and insanity, referring to the asylum as a receptacle for 'a class of brutal negroes whose natural mental condition is so low, as to justify their being considered irresponsible and of unsound mind'.[22]

There was some grudging acceptance by Allen and others that the prevalent social conditions and living patterns of the West Indian Black population were still directly influenced by the legacies of enslavement. Commenting in 1882 on Jamaica's persistently high rates of illegitimate births, the island's Registrar General recognised that the 'widespread disregard of marriage' was the 'fruitage of the days before emancipation'. Governor Sir Anthony Musgrave suggested that the 'true present condition and character of the negro labouring population' was widely misunderstood. Those who condemned the levels of illegitimacy, he contended, had not properly considered that 'little more than 50 years ago it was not permitted to teach the fathers and mothers of these people to read; they were prevented, or at least discouraged, from marriage, and were encouraged to breed like cattle'. Acknowledging the validity of their arguments, Thomas Allen affirmed that 'a large proportion of the lunatics of the colony is manufactured from this class'.[23]

Information regarding the race or colour of people admitted to West Indian asylums was often recorded as a matter of course. The earliest available data relates to Antigua Lunatic Asylum between 1855 and 1863. Annual admissions were summarised according to patients' 'complexion'. Out of 184 people admitted during the period 142 (77 per cent) were described as 'Black', 21 (11.5 per cent) as 'Coloured', and

Table 6.1   Admissions by colour – Jamaica Lunatic Asylum, selected years 1863–98

	Black	Brown	'Coolie'	White	Total
1863/64	71	14	7	2	94
1875/76	82	9	9	3	101
1881/82	69	20	5	4	99
1887/78	111	27	9	2	149
1893/94	112	29	6	6	153
1897/98	153	38	10	2	203
	598	136	46	19	798
	[74.9%]	[16.9%]	[5.9%]	[2.3%]	

Sources: CO 137/388, 21 October 1864, fo.138; CO 137/484, 'Report of the Medical Super-intendent and Director of the Jamaica Lunatic Asylum for the Year 1875–76', fo.397; CO 140/86, Jamaica Lunatic Asylum, Annual Report, year ended 30 September 1882, p.158; CO 140/99, Jamaica Lunatic Asylum, Annual Report, year ended 30 September 1882, p.98; *Jamaica Gleaner*, 14 September 1894; *Jamaica: Annual Reports of the Superintending Medical Officers* (Kingston, Jamaica: General Printing Office, 1898), Lunatic Asylum, Annual Report, Year Ended 31 March 1898, p.4.

another 21 (11.5 per cent) as 'White'.[24] More detailed information is available for admissions to the larger asylums during certain periods. Data for admissions to the Jamaica Lunatic Asylum over several separate years between 1860 and 1900 is summarised in Table 6.1. The aggregated figures confirm that the great majority (three quarters) of people admitted in Jamaica were 'Black'. The 17 per cent referred to as 'Brown' corresponded to those deemed 'Coloured' in Antigua and comprised people of varying shades of complexion. The relatively small percentage of 'White' people admitted is roughly in proportion to the White population on the island, whilst the continuing steady flow of 'Coolies' into the asylum indicates the presence in Jamaica of several thousand Indian indentured servants.[25]

The picture presented by analysis of admission information in the Barbados case books reflects the relative uniqueness of Barbadian society. The careful stratifications of skin colour shades (shown in Table 6.2) were indicative of an acute consciousness, emanating from its White ruling class, that social distinction was largely determined by degree of darkness or lightness.[26] Table 6.2 illustrates that 'Black' people formed slightly less than half (46 per cent) of admissions to the Barbados asylum between 1875 and 1880, which differed noticeably from their predominance in the Jamaica asylum. People whose colour was somewhere between Black and White made up 35 per cent of those admitted. White

Table 6.2   Admissions by colour and gender – Barbados Lunatic Asylum 1875–80

Colour	Males	Males %	Females	Females %
Black	46	39.0	54	55.1
Dark Cob	5	4.2	1	1.0
Cob	13	11.0	21	21.4
Mulatto	24	20.4	8	8.3
Coloured	1	0.8	1	1.0
Fair Coloured			1	1.0
Light Brown			1	1.0
White	29	24.6	11	11.2
Unspecified	16		3	
TOTAL	134		101	

Notes: 1. The description 'Cob' normally referred to someone with one parent 'Black' and one 'Brown'.
2. Apart from three males, all of the instances where colour was unspecified occurred from October 1879 onwards. The calculation of percentages has been based only on those cases where colour was specified.
Sources: Barbados National Archives, Male Case Book, 1875–1916; Female Case Book, 1875–1916.

people accounted for 18.5 per cent, a high proportion compared to other West Indian asylums. When gender is considered the racial characteristics become yet more striking. Only 39 per cent of men admitted were deemed Black. White men (almost 25 per cent) formed the second largest group, and there were also significant numbers described as 'Mulatto' (20.5 per cent). Among the women those considered Black formed a clear majority (55 per cent), significantly so if taken together with those described as 'Cob' (22.5 per cent), whilst the proportions of White and 'Mulatto' women admitted were noticeably lower than for males. The admission data gives little indication of reasons for these marked gender differences. Poor White men and Black women were certainly among the groups subjected to the economic adversity and social deprivation afflicting the island in the 1870s and 1880s.[27] An alternative explanation might be around the relative likelihoods of disorderly dark-skinned women and light-skinned men to offend against social norms in mid-Victorian Barbados.

Information on admissions by racial category to the British Guiana Lunatic Asylum is also available for a five-year period, from 1881–85. The data, compiled by Dr Robert Grieve, is summarised in Table 6.3 and illustrates well the changing ethnic dynamics within the colony. The British Guiana population had been 97 per cent Black and Coloured in

Table 6.3 Admissions by colour – British Guiana Lunatic Asylum
1881–85

Colour	Numbers	%
Brown (East Indian)	224	45.1
Black	163	32.8
Coloured	53	10.7
White	35	7.0
Yellow (Chinese)	21	4.2
Brown (other)	1	0.2
Total	497	

Sources: The Asylum Journal, 16 January 1882, p.89; 15 January 1883, p.90;
15 January 1884, p.90; 15 January 1885, p.70; 15 January 1886, p.148.

1838, when its total numbers stood at 95,000. However, it was transformed by the large-scale importation of indentured labourers from the
Indian subcontinent. No fewer than 211,000 were brought in to the
colony by 1900.[28] As Table 6.3 shows, by the early 1880s people designated 'Brown, East Indian' constituted the largest single group admitted
into the asylum, outnumbering the combined figure of people who were
'Black' and 'Coloured'. As in Barbados there were gender differences.
Overall, male admissions were approximately double those of females,
but the differentials were even more pronounced among Indian immigrants, reflecting the much larger number of men than women brought
to work in Guiana.[29]

Inward migration constituted the most significant influence in the
changing nature of both Guiana's and Trinidad's populations.[30] The
dynamics were mirrored in the places of origin of people admitted
to their respective asylums. The figures shown for British Guiana in
Table 6.4 highlight that, between 1876 and 1885, people born outside
the colony made up almost 70 per cent of those committed to the asylum. The earlier data reflects the wide diversity of places from where
immigrants initially came, with not insignificant numbers of asylum
patients emanating from Africa, Madeira and China. Whilst admissions
from those areas then declined, in line with falling immigration, this
was largely offset by continually rising numbers from the Indian subcontinent. Migration from elsewhere in the Caribbean region in search
of work or land, especially from overcrowded Barbados, was a constant
factor in British Guiana, which asylum admission figures continued to
reflect.[31]

Table 6.4 Admissions by country/region of origin – British Guiana Lunatic Asylum

	1876–80	%	1881–85	%	1895/96–1899/1900*	%
B. Guiana	161	32.2	161	31.2	384	50.7
India	185	37.0	224	43.4	254	33.5
Barbados	48	9.6	54	10.5	65	8.6
Africa	39	7.8	17	3.3	14	1.8
Madeira	30	6.0	22	4.3	14	1.8
China	18	3.6	21	4.1	8	1.1
W.I. other	12	2.4	12	2.3	13	1.7
Europe	7	1.4	5	1.0	6	0.8
TOTAL	500		517		760*	

Note: *These figures exclude the year 1897/98, for which data is unavailable.
Sources: *The Asylum Journal*, 2 May 1881, p.18; 16 January 1882, p.89; 15 January 1883, p.90; 15 January 1884, p.89; 15 January 1885, p.69; 15 January 1886, p.148; *British Guiana: Reports of the Surgeon General* – 1895/96, p.44; 1896/97, p.39; 1898/99, p.44; 1899/1900, p.40.

Patterns of admission to the Trinidad Lunatic Asylum show some similarities to Guiana,[32] and are illustrated in Table 6.5. These data cover a slightly wider time period and show that, between 1883 and 1914, the proportion of immigrant patients admitted declined from 70 per cent to just over 50 per cent, whilst the numbers born in the colony rose substantially.[33] Table 6.5 reflects the falling numbers of people originating from India, as well as those born in Africa. This was partly offset by steadily rising numbers of people originating from Barbados and other islands in the Eastern Caribbean, particularly nearby St Vincent. As in British Guiana they had been attracted to Trinidad by better employment prospects and the possibilities of land acquisition. Barbadians became prominent in urban trades and in public services like the police force. Many, however, did not succeed in the ways they had anticipated and found themselves among colonial society's stigmatised social casualties.[34]

Migration has been recognised by historians as a significant factor in precipitating asylum admission, through the major social stresses generated for individuals and families.[35] Conversely, an unsettled mental state has been acknowledged as a factor that drives people to abandon previous life patterns and migrate. Robert Grieve took the opportunity presented by the British Guiana asylum's diverse population to investigate the links between migration, race and insanity. He calculated that, for the years 1876–80, the proportion of immigrants admitted compared

Table 6.5   Admissions by country/region of origin – Trinidad Lunatic Asylum

	1883–87	%	1900–14* (five separate years)	%
Trinidad	155	29.7	388	49.0
India	190	36.4	142	18.0
Barbados	51	9.8	120	15.2
W.I. other	62	13.0	100	12.6
Africa	20	3.8	2	0.2
China	3	0.6	6	0.7
G.B./Ireland	6	1.1	12	1.5
Madeira	4	0.8	2	0.2
Venezuela	8	1.5	14	1.8
Other	5	0.9	5	0.6
Unknown	12	2.3		
TOTAL	522		791	

Notes: 1. Data are only available for certain years – those included here are 1900; 1901/2; 1905/6; 1907/8; 1913/14.
2. The post-1900 figures for Trinidad also include Tobago. Previously some Tobago patients were included in 'W.I. other', though most were sent to Grenada.
Sources: CO 298/40, Report of Medical Superintendent for 1883, p.12; CO 298/41, Report of Medical Superintendent for 1884, p14; CO 298/42, Report of Medical Superintendent for 1885, p.27; CO 298/43, Report of Medical Superintendent for 1886, p.18; *Trinidad: Reports of the Surgeon General* – 31 March 1888, p.23; 1900, p.97; 1901/2, p.67; 1905/6, p.62; 1907/8, p.102; 1913/14, p.64.

to total population was more than double that for native Creoles. In his initial analysis, he suggested this was partially due to their different living situations. Immigrants were more likely to be alone and without family, whilst Creoles tended to live 'in the midst of relatives and friends' who would probably take care of them when unwell. Immigrants had experienced 'the change of circumstances, the separation from country and friends', which predisposed them to 'mental infirmities'. Grieve, however, considered that these factors did not offer sufficient explanation. From studying his patients he concluded that 'it is amongst those who possess a tendency to the insane neurosis, those who are mentally unstable that the emigration agent usually finds his recruits'. Once confronted with any difficulties 'he passes from the border land into the undoubted territory of insanity'. Many of the 'coolie immigrants', he noted, 'go mad on the voyage hither or within a very short period of their arrival in the colony'.[36]

Grieve returned to these themes several times, reiterating the wide disparities between immigrants and Creoles.[37] In a more detailed consideration he concluded that diverging admission rates were related both to 'the different conditions under which the people live' and to 'racial peculiarities of constitution'. He again observed that 'creole blacks' generally lived in stable circumstances among relatives and were unlikely to be sent to the asylum at an early stage of their illness, whereas immigrants had undergone a 'complete and sudden change in their condition'. Grieve singled out those born in Africa, concluding that insanity among them was 'more rife… than in any other section of our population'. Moreover, he thought that Africans were not only 'peculiarly liable to attacks of insanity but they suffer from it in a very incurable and fatal form', which confounded the 'popular belief that an unstable cerebral organisation is to a great extent the product of very advanced civilization'.[38] Dr James Donald, who worked briefly at the asylum, was more blunt about the nature of insanity among African patients, suggesting they were burdened with 'a mind uneducated' and consequently 'in many cases natural stupidity borders so closely on imbecility, dementia, and amentia, that it is difficult to say where a normal state ends and an abnormal one begins'.[39] Such views presaged what would later become part of psychiatric orthodoxy.[40]

For most asylum superintendents the liability to admission for vulnerable groups of people was greatly accentuated by excessive use of alcohol or drugs. Habitual use of cannabis or 'ganja' was considered directly causal in the onset of insanity, particularly among East Indian immigrants.[41] Both Robert Grieve in Guyana and George Seccombe in Trinidad latched on to evidence emanating from India regarding the allegedly undesirable physical and psychological effects of cannabis.[42] Grieve claimed in 1882 that over 75 per cent of the East Indians admitted were 'addicted' to ganja smoking and that the 'mental disease' arising from it was 'well known'. He had no doubt that 'Indian hemp' was 'a most prolific source of lunacy' in the colony; its effects on the mind were 'disastrous'.[43] He adduced the case histories of several Indian immigrants to support his contentions.[44] Seccombe came to similar conclusions soon after arriving in Trinidad. He noted that, of 29 male East Indians admitted during 1883, 'prolonged indulgence in this pernicious habit' of ganja smoking was the chief 'cause' in 12 cases.[45] He supported the campaign for a prohibition on growing cannabis. This had been largely achieved by 1887, and Seccombe attributed the sharp reduction in admissions of East Indians to the high prices now making its consumption a luxury.[46]

For other ethnic groups over-indulgence in alcohol was widely identified as a key causal factor leading to admission. This provided the authorities with a convenient opportunity to condemn the poorer classes' recreational habits. In Barbados, 'intemperance' was regularly blamed for the committal of many Black, Coloured and White patients, especially males. A calculation in 1892, based on ten years of evidence, showed it as the leading attributed 'cause', responsible for more than a quarter of admissions.[47] In 1885 the Poor Law Inspector claimed that intemperance accounted for 40 per cent of the previous half-year's admissions, where a cause was recorded, and that it was 'responsible in an indirect way for many more'.[48] However, by 1898 intemperance was regarded as a much less prevalent factor, suggesting either that drinking patterns in Barbados had altered or, more likely, that case recording had become more rigorous.[49]

In Trinidad alcohol was identified as the predominant underlying factor in many admissions, other than for East Indians. Thomas Murray, the asylum's long-serving first medical superintendent, did not doubt its significance:

> Amongst local causes I fear I must charge intemperance in the use of ardent spirits and its debasing results as predominant in the classes chiefly afflicted. Certain it is that this vice is to be treated in a large proportion of those committed to the Asylum, and there is little doubt its existence is to be apprehended in some others where the evidence cannot be so clearly apprehended.[50]

George Seccombe, with his recent British asylum experience, was equally convinced about alcohol's causative properties.[51] He suggested in 1890 that, as well as its direct effects, there were 'hereditary results' as drinkers' descendants were 'frequently born with unstable brains' and consequently 'liable to lose their balance from very slight causes'.[52] In British Guiana Grieve was also concerned about intemperance, claiming it as the main causal factor in Creole admissions. He considered the effects of alcohol on the brain more deleterious than cannabis, with a poorer treatment outlook.[53] However, alcohol abuse raised fewer concerns in Jamaica. Thomas Allen hardly referred to it, apart from suggesting in 1864 that drink was 'amongst the chief causes' of the mental disorders of people admitted.[54] In 1902 the acting medical superintendent Dr Williams observed that 'alcoholic intemperance' was practically absent as a cause in Jamaica, unlike in 'more civilised countries' where it was one of the 'fertile sources of insanity'. The *Gleaner* newspaper concluded complacently that 'it is just as well not to be too civilised'.[55]

The very poor physical state of many people admitted to West Indian asylums was constantly noticed by medical officers and public officials. The reasons advanced included old age, physical disease, and families' reluctance to refer a deranged member until there was no alternative or their condition became chronic. There were also the plainly evident effects of economic distress and deprivation, reflected in widespread poverty, neglect, and hunger. In Jamaica, as early as 1846, Doctors Joseph Magrath and James Scott observed that many of the lunatics in the Kingston asylum were in a 'very bad state of health' when admitted.[56] In 1860, at the height of the scandal, Scott argued that its high mortality rates were largely attributable to the 'wretched physical state' of people sent there.[57] Daniel Trench, the Inspector and Director, went further in reporting that many had been admitted in a 'broken down, debilitated, and not unfrequently mal-treated condition'.[58]

Little changed following the move to the new Jamaica Lunatic Asylum. In December 1863 Dr Charles Lake, accounting for a high death rate, claimed some patients were admitted 'either suffering from incurable organic disease, or, in a state of extreme debility and emaciation'.[59] Thomas Allen regularly drew attention to the poor physical health of people admitted.[60] In 1881 he repeated concerns about 'the deplorable class of patients admitted', incurable due to 'the prevalence of co-existing mortal disease' alongside their insanity. The situation had been worsened by a long drought in Jamaica, and a recent cyclone that destroyed many 'Peasant Residences'. Owing to inability to cultivate their 'Provision Fields', some families were compelled to seek the admission of 'afflicted relatives'. Allen lamented also that the asylum continued to act as 'a receptacle for criminals, aged and disgustingly infirm Patients', who in England would have been placed in workhouses.[61] His successors were equally concerned about the state of many of those received. The acting medical superintendent, Cormac MacCormack, referred in 1887 to the 'wretched condition of *bodily* health of a great number of patients when admitted'.[62] Dr Williams in 1907 spoke gloomily of the 'large number of physical and mental wrecks who were admitted in a moribund condition to swell the death-rate'.[63]

In British Guiana, Robert Grieve was extremely concerned about the 'miserable state of bodily health' in which patients were admitted. The majority appeared to be suffering from 'old standing disease or from prolonged starvation and illtreatment'.[64] His predecessor James Donald noted in 1876 that 'Coolies' in particular, on admission, were 'almost invariably very anaemic and half-starved', owing to lack of adequate nourishment. Some of them had indulged in dirt-eating, partly to assuage their hunger. They tended to make good recoveries in the asylum, their mental symptoms abating as a consequence of better nutrition.[65] In Trinidad, with its equally diverse population, there were similar issues. Dr Alfred Martin, commenting on the asylum's high death rate in 1880, attributed it to the unsatisfactory physical state of people admitted; a quarter were 'in such bad health and exhausted condition' that an early 'fatal termination' was almost inevitable.[66] George Seccombe was less graphic but frequently deplored the feeble state of the majority brought to the asylum.[67]

Similar problems were commonly reported in Barbados, particularly in the last years of the old asylum.[68] In 1890 Dr Albert Field complained that 'Many of our patients are sent up so feeble that there is very little to be done with them but to nurse them until they die.'[69] Indeed, 'feeble' or 'very feeble' were terms regularly used to describe the majority of patients admitted.[70] In 1901 it was noted that, in many cases, 'mental unsoundness comes on at a late stage of a chronic disease'.[71] The extremely poor physical condition of many people entering the Barbados asylum was represented in the surviving journals and case books. Dire economic and social circumstances often lay behind the narratives of individual breakdown, as with James C., a 44-year-old 'mulatto' shoemaker admitted in August 1878. One examining doctor described him as 'much emaciated' and the other observed that he 'recently had great trouble & has been in bad circumstances'. On admission he was 'thin & half starved looking', and appeared 'deeply poverty stricken – ragged & tattered'. James himself reported that he was 'left no work to do was starving & took to drinking'. He died of typhoid within a month.[72] Hezekiah A., a 17-year-old suicidal Black baker, was 'much emaciated and very weak' when examined in March 1878. On admission he was described as 'a very small shrunken looking youth eyes & cheeks being as great hollows & ribs & bones being in general prominent'; he was so feeble that he was 'scarcely able to stand'. In the asylum he gained a stone in weight within a fortnight, accompanied by a marked improvement in his mental state.[73]

Medical superintendents regularly referred to the poor physical and mental state of those admitted in accounting for low rates of recovery and discharge. This applied equally to the many cases where hereditary factors were deemed causative in the onset of insanity. By the early twentieth century, concerns about inherited constitutional deficiencies were leading to an increasingly eugenicist perspective. A doom laden headline in Jamaica's *Daily Gleaner* in September 1912 proclaimed 'HEREDITARY DANGER', and that 'Drastic Steps Should be Taken for the Good of the Race'. This referred to the latest annual report by Dr Williams, the asylum superintendent, in which he suggested that 'the time is approaching for preventing those who have suffered from insanity or who have inherited insanity on both sides from marrying'. He contended that 'For the sake of future generations' it was necessary to take 'drastic measures' for 'the suppression of the unfit'.[74] His counterpart in Trinidad, Dr Vincent, was almost equally apocalyptic, after pointing out that 'obvious congenital defect or a history of inherited mental instability' accounted for over half the admissions to the asylum. He went on to suggest 'the futility of expecting a reduction in insanity until some restriction is placed on the propagation of degenerate stock'.[75]

Whether or not heredity was directly influential, environmental and material deprivation were clearly prominent elements in the background circumstances of many people admitted to West Indian asylums. They interacted with, and were often largely determined by, considerations of social class and occupational status, race and colour, and transnational migration. Indeed, all these factors impacted on one another. It is important, though, not to minimise the ravages of mental illness itself. In some instances insanity developed in people who had not experienced marked deprivations or upheavals, but its baneful effects soon ensured that they suffered serious material and psychological losses. However, social and economic circumstances were normally not in themselves sufficient to provide the immediate rationale for asylum admission. Something had to occur that transformed a difficult but manageable situation into one beyond the ability of families and communities to contain.

Precipitants

Intervention by medical and legal authorities leading to committal to an asylum normally followed the exhibition of risky, disorderly, or deranged behaviour in the person's home or community, or in some public place. Within those parameters a range of incidents or situations could identify an individual as potentially insane. In most instances there were several contributory elements. In line with evidence from other British colonies, violence or threat toward people or property tended to be the key defining factors.[76] However, they were not the only ones. Offences against public decency or disruption in the streets could also trigger action. All these behaviours might otherwise come within the purview of the criminal justice system. What separated them out were accompanying symptomatic manifestations of mental disorder, in the form of elated or depressed mood, over- or under-activity, rapid or incoherent speech, and evidence of seemingly bizarre thoughts, delusional ideas, and hallucinations.

Most of the evidence here has been extracted from individual case histories.[77] In reading the narratives various factors have to be considered and caution exercised. In many instances information on medical certificates was provided by relatives extremely anxious to have a troublesome person removed from home and admitted to the asylum. It was, therefore, possible that some circumstances were exaggerated to emphasise the level of risk posed. Examining doctors had their own priorities, sympathies and prejudices which affected the content of their reports. Asylum medical superintendents were subject to similar influences, sometimes reflected in descriptions of individuals that were distinctly unflattering or racially stereotyped, whilst at other times demonstrating concern and compassion.[78]

The records provide numerous examples that confirm the frequency of violent or dangerous behaviours as determinants of the decision to commit. The partner of Thomas G. (45), a 'Dark Cob' tailor admitted to Barbados Lunatic Asylum in January 1877, reported that he suffered from 'fits' and had begun destroying things in the house. He talked incoherently, laughing and shouting, and had 'attempted to kill her several times'. On arrival at the asylum, being 'greatly alarmed', he 'fought & bit at the attendants with great maliciousness'.[79] A sense of strong grievance evidently lay behind the case of Joshua C. (36), a Black labourer admitted in June 1879. He believed that he had inherited a share of Bentley Estate and owned its cattle and horses. When evicted two years previously he refused to leave and began living in a tree on his former allotment, whilst continuing to graze his sheep in the estate's young sugar canes. He had become 'troublesome & violent', setting fire to his own furniture. He started 'menacing the managers with weapons' and threatened to burn down the estate and its buildings.[80]

Many women admitted had also exhibited violence, sometimes in conjunction with other concerning behaviours. According to the husband of Elizabeth G. (30), a 'stalwart' Black labourer sent to Barbados asylum in July 1878, her conduct had started to change during the 'crop season'. She became 'easily excited' and would sing late into the night. She became so violent that her husband tied her in bed with ropes. When examined at the police station she needed to be held by two men 'to prevent her from attacking the bystanders', whilst cursing and uttering 'pious ejaculations & broken prayers'.[81] Margaret S. (40), a 'lascivious' unmarried Black labourer, was admitted in September 1878 from St Mary's Alms House, having beaten other inmates and destroyed bedding and clothing. On examination she was reportedly indecent, exposing herself, filthy in habits, and talking 'incessantly & incoherently'. Within hours of arrival at the asylum she had 'engaged with several of the other inmates in mortal combat' and shown herself 'Strong at biting.'[82] Anna Maria P. (62), a 'robust' Black huckster, was admitted for the third time in April 1879 after her husband complained of her 'destroying his property, beating her neighbours & exposing her person to the public'. The authorities' greatest concern was that she walked around with a loaded gun 'to protect herself', which she 'loads & discharges...very often'.[83]

Several women admitted to the Jamaica Lunatic Asylum in the early 1860s had demonstrated extremely dangerous tendencies.[84] Eleanor R. had 'nearly killed' her infant and 'illused' all her other children. On examination she was 'very incoherent and unconnected', convinced that her children did not belong to her. She was also 'dangerous for anyone to go near'.[85] L.M.'s violent behaviour, reported by her sister, had included flying at people without provocation, biting them and tearing their clothes. She also 'wandered all about town & enter the houses of strangers & does mischief'. On the morning of her medical examination she threw a child down a flight of steps. Presenting to the doctor with a 'semi idiotic appearance', she would not respond to questions other than to keep repeating the last word.[86]

Suicidal or self-harming behaviour inevitably aroused alarm and figured quite frequently in the train of events. William D. (43), a 'Mulatto' carpenter admitted to the Barbados asylum in December 1876, had been experiencing violent delusions. He attempted to jump both into a well and out of a window. Committal followed him biting his wife and demanding an instrument to cut off his testicles.[87] Interesting circumstances surrounded the case of Elijah C. (30), a 'haggard' looking Black tailor admitted in May 1876. He had evidently never shown any sign of mental disorder before 21st April, when a 'large mob' of 'Confederation rioters' assembled near his home.[88] Elijah, a 'nervous, delicate & easily excited man', was extremely alarmed and for several days remained in a frightened and suspicious state, convinced that 'the soldiers are coming for me'. He became increasingly violent, destroyed his clothes, and then attempted to cut his throat with a razor.[89]

Confederation also figured in the case of Reverend N.H.G. (45), a White Church of England clergyman admitted in March 1877. He had a history of 'alternate paroxysms of excitement & depression', but despite his 'marked eccentricity' he was 'universally beloved by his people'. In April 1876 he attended a heated anti-Confederation meeting. Being 'much impressed' his mental state deteriorated into bouts of both depression and elation. In November his wife, to whom he was devoted, died 'under trying circumstances'. Reverend G. then became deeply depressed and 'could not be roused again'. As his condition worsened he became preoccupied with ideas of demonic possession and, in despair, he attempted suicide several times. The degree of violence toward himself and others caused him to be 'tied hand & foot' for some days before admission to the asylum.[90]

Circumstantial evidence from Barbados suggests that White people may have been more prone to suicide, sometimes adopting desperate measures.[91] The 1877 admission of Joseph E. (67), a White carpenter, followed 'a most determined attempt' to cut his throat and his wrists, whilst an inmate of St Philip's Alms House.[92] Arabella D. (30), a White seamstress, in 1875 attempted both to drown and set fire to herself.[93] Another White seamstress, Keturah B. (37), had endured the stillbirths of twin children. She developed alarming delusions and tried to drown both herself and one of her other children, precipitating admission in November 1877.[94] The case of Alexander G. (30), a White doctor admitted in December 1877 on a special order from the governor, was particularly graphic.[95] He had experienced two previous episodes of insanity before the latest in Demerara, where he was a district medical officer covering two districts.[96] Following prolonged overwork and sleepless nights he became preoccupied with delusional thoughts and was sent to Barbados for a change of environment. Evading his 'special attendant', Dr G. rushed out and 'took up a pantry knife & cut his throat resisting savagely before he would give up the weapon'. In the asylum he made further attempts and deliberately aggravated his injuries. He was dead just a week after admission.[97]

Sexually inappropriate behaviours were commonly reported in the circumstances leading to admission, particularly of female patients. Concerns centred both on the breaking of social norms and liability to exploitation. Mary C. (19), a 'Mulatto' needlewoman with a 'rather dull' countenance, was admitted to Barbados asylum in July 1875 after 'stripping & exposing herself' and spending entire days and nights 'in the open air'. Mary maintained that she took her clothes off to gain relief from the 'great heat'.[98] S.E.H. (28), an epileptic Black labourer, was 'always wandering about', often violent, and frequently 'strips off her clothes & exposes herself'. On admission in 1877 Dr Hutson described 'a dirty, repulsive looking black woman with a cross expression of countenance & a frightfully filthy smell'. Her vulnerability was plain when it was discovered that she was pregnant though unaware of having 'been with any man'.[99] A 16-year-old 'stout' Black girl, Theresa D., was admitted in July 1879 after 'running about the streets, tearing up her clothes and exposing herself' in Bridgetown. She was 'shameless & expresses herself in great need of a man'. This was reportedly her third episode of mania and, although normally 'very quiet & well behaved', when insane she became 'lewd' and wandered off from home.[100]

Insane men might also transgress social and sexual norms, as did J.H.H. (47), a 'tall semi respectable looking' White blacksmith. His first bout of mania followed the Confederation Riots in 1876. In his second episode in 1877 he was convinced he owned 5,000 acres of land and 8,000 head of cattle. He became 'very wild & unruly', breaking up his mother's furniture and destroying breadfruit trees. His admission followed reports of him 'wandering about day & night quite naked'.[101] The case of George E. (25), a White 'freighter', was more outrageous for his community. A patient in the asylum six years before, George suffered a relapse and began drinking heavily 'contrary to his usual practice'. He started stripping himself and wandering about. At night he was 'noisy', singing hymns and 'making use of obscene language'. Matters became unacceptable when he was observed making a 'deliberate attempt ... to have connexion with a cow'. He was charged with bestiality but, rather than being imprisoned, he was committed to the asylum in October 1879.[102]

Other forms of socially disruptive behaviour could lead to committal, particularly where they posed risks to property or challenges to public order. Amelia S. (34), a 'Cob' labourer with a 'repulsive' countenance, was admitted in 1879 on account of a 'growing propensity to burn things and her threats to burn cars'. She had been living rough in a gully for four years and had burnt down her hut, thrown away her food, and refused her weekly pension from the parish.[103] The admission in January 1876 of Susannah E. (44), a Black labourer, followed pulling up potatoes and cutting sugar canes on her husband's land. He described her as 'strange in her habits' for several years, sometimes 'tearing clothes, getting out of bed & sleeping on floor, breaking up furniture & crockery & talking idly'. After five days of cutting his canes he 'had her removed'.[104] In November 1876 Mary Louisa Y. (16), a 'Cob' bottle seller, attracted a crowd by 'throwing all of the things out the house' and tearing up her clothes. After days of 'cursing' and flinging things at people, her mother 'felt compelled to take her to the Station house'. An examining doctor observed that she 'raves incoherently, shouting & gesticulating incessantly' and her language was 'disgusting in the extreme'.[105]

In some instances particular alarm was aroused by an incident's location. In early 1878 Venus C. (47), a Black washer woman, 'began to be peculiar walking about all night from house to house, talking & making a noise'. She was apprehended at Government House, in the bedroom of Governor Strahan's wife, having gone there 'to take possession' in the belief that God had given it to her.[106] In December 1877 Joshua C. (20), a 'depressed' Black labourer, was found 'kneeling down' in the Council Chamber in Bridgetown. He was 'put out in the yard', but later returned and was arrested and sent to the asylum.[107] Romulus S. (60), a Black sugar boiler admitted in late 1875, suffered from 'melancholia' and believed that he was to be severely punished or even hanged. At night he roamed around 'disturbing the neighbours by knocking at their doors', making frequent threats to kill himself. He went each night to the gate of Glendairy prison, 'clamouring for admission'.[108]

The most notorious incident in a public place resulting in asylum admission involved the Jamaican messianic preacher and faith healer Alexander Bedward.[109] Bedward has subsequently gained almost mythical status, deriving partly from the politically and racially defiant content of the 1895 sermon that led to his first committal. Born in 1848, he was ordained as an elder in the Free Baptist Church in 1876.[110] Questions were raised about his sanity as early as 1883.[111] He subsequently travelled to Colon, Panama, where he 'received his vision' and returned to Jamaica, establishing himself as the 'prophet' of August Town.[112] His large-scale meetings and mass baptisms in the early 1890s excited the political and religious authorities' alarm. The *Gleaner* newspaper sought to undermine the influence of the 'prophet', citing a Dr Bronstorph who examined him in 1892 and concluded that he suffered from 'religious monomania' and ought to be in a lunatic asylum (Figure 6.1).[113]

In 1894 Bedward established his own church at August Town, later known as the 'Jamaica Free Baptist Church', adopting the title of 'Shepherd'.[114] In January 1895 he addressed a large revival meeting at Hope River. Governor Henry Blake had despatched police to observe for, as he told Secretary of State the Marquess of Ripon, 'in such movements among a black and coloured population there is always an element of possible danger'.[115] The *Gleaner* sent a reporter to cover the event. Bedward spoke in apocalyptic language with apparent revolutionary intent. He advised his followers that 'We are the true people. The white

Figure 6.1   Alexander Bedward
Source: Martha Warren Beckwith, *Black Roadways: A Study of Jamaica Folk-Life* (Chapel Hill, North Carolina: University of North Carolina Press, 1929), adjoining p. 170.

men are hypocrites, robbers and thieves. They are all liars'. He allegedly urged the crowd to action:

> Brethren Hell will be your portion if you do not rise up and crush the white man. The time is coming. I tell you the time is coming. There is a white wall and a black wall and the white wall has been closing around the black wall, but now the black wall is becoming bigger than the white, and they must knock the white wall down. The white wall has oppressed us for years. Now we must oppress the white wall.

The Government passes laws that oppress the black people. They take their money out of their pockets. They rob them of their bread, and they do nothing for them.

Bedward's implication was clear enough when he added 'Let them remember the Morant war'. He declared the governor 'a scoundrel and robber', along with other politicians, constables, and ministers of religion, whilst declaring himself 'the prophet of Jesus'.[116] At his ensuing trial for sedition Bedward was acquitted on grounds of insanity, and he was committed to the asylum by order of the governor. However, he remained barely a fortnight, his lawyer securing his release after the verdict was overturned on a point of law.[117]

The Bedward movement, and the great enthusiasm aroused among the Black and Coloured masses, provided ammunition for those in Jamaica who construed a close connection between religious revivalism and the onset of insanity. By the end of the nineteenth century it was becoming almost axiomatic.[118] One correspondent to the *Daily Gleaner* from Savannah-la-Mar, in May 1899, did not think 'that there is any doubt that it is the principal feeder of the Lunatic Asylum'.[119] In a survey the paper quoted a stipendiary magistrate who 'does not recollect a single case of insanity brought before him which had not been directly attributable to revivalism'. The Rev. James Chapman of St James's parochial board claimed to express 'what is generally believed' in saying that 'these night orgies produced evils and excitement among the people, which finally ended in madness'. Several parish boards called for revivalist meetings to be suppressed because they were 'responsible for an increase in lunacy'. However, the asylum's medical superintendent Dr Plaxton cast doubt on the claims, suggesting that 'religious mania' was more effect than cause of insanity. His view was that 'The masses of Jamaica have nothing to occupy their leisure except religious exercises'; they were 'saturated' with religion. Therefore, when 'the mind gives way' they would 'indulge in that with which they are familiar – religion'. In the face of Plaxton's 'authority' the *Gleaner* conceded that 'the belief that revivalism is fast producing lunatics is a fallacy'.[120] The debate was nevertheless far from closed and a common belief persisted that religious revivals were directly causative of mental disorder.[121]

Religious preoccupations predominated in many recorded cases of people admitted. R.R., committed to the Jamaica asylum in January 1864, had declared himself 'a prophet'. He thought that a Reverend Mayhew 'wishes to witch him in order to sacrifice him in the church at Halfway Tree'. He told a policeman that 'Christ had been on earth & had two bastard children in St Andrews'. Rather more alarmingly, he claimed to have 'a long sword with which he intends to cut off heads'.[122] Selina K.'s admission to the Jamaica asylum followed reports of her being 'violent in the extreme'. She had previously experienced a 'revival' and was talking 'wildly and incoherently' on religious matters. She claimed to be acting under the direction of the Holy Spirit when 'breaking into her neighbours houses & threatening to assault parties with an axe and a knife'.[123] There were also highly risky aspects with Anne W. (32), a Black huckster, admitted to Barbados asylum in December 1875. She was a Sunday school teacher, prone to singing hymns and praying. However, she had exhibited a sudden change, 'becoming excited, repeating scriptural expressions & praying incoherently'. She gathered her children round the bed and started 'praying & singing very loudly', continuing for three days without food or sleep. As Anne's condition worsened she tried to kill the children, and was then tied 'hand & feet' for several days before finally being admitted.[124]

Case histories like these confirm that admissions of people deemed insane normally came about when others could no longer put up with their behaviours. When incidents occurred in a public arena the police were likely to arrest the offending person, to be followed by medical examination. If the disruptions happened within the person's home it would fall on relatives or a wider community network to decide when the time had come to seek intervention from the authorities. As Thomas Allen suggested, 'it has not been until the friends were unable to manage them, or that they manifested dangerous propensities, or were troublesome, offensive, destructive and filthy, in their habits and a "nuisance" – that they were finally sent to the Lunatic Asylum'.[125] Even the most supportive of families would have their tipping point, when the decision had to be taken to seek the removal of a loved one who had become unmanageable or too unsafe to remain in a setting without adequate care and supervision.

Becoming a patient

The legal requirements for committal to a lunatic asylum tended to develop separately in each British colony, though with some commonalities. The need for certification by at least one medical officer and a magistrate's order emanated from accepted practice in England, as consolidated in legislation of 1853.[126] An act of parliament passed in 1858 for the Indian states gradually became the standard for much of the British Empire.[127] There were, nevertheless, marked differences between West Indian colonies as regards specific requirements, due both to independently minded governments and local expediency. In some instances lunacy was dealt with as a criminal matter, whereas elsewhere mechanisms of civil law were applied. A key factor was the availability of medical practitioners, which largely determined whether one or two certificates were required to effect committal. What was common throughout, however, was the involvement of police and magistrates in the admission process. As Catharine Coleborne has shown for Australia, this inevitably confirmed perceptions that the lunatic was deemed a criminal and that the asylum formed an integral part of state systems for the control of disorderly and deviant behaviour.[128]

In Barbados, legislation of 1840 and 1842 provided that any person 'so furiously mad, or so far disordered in his or her senses' that it might be dangerous for them to 'go abroad' could be committed by two justices of the peace, based on certificates from two physicians or surgeons.[129] A similar law was enacted in Jamaica in 1861, and Thomas Allen helped draft an amending Act in 1873 providing for detention of people 'wandering at large' even if not dangerous.[130] In some instances the requirement for a second medical certificate was overlooked.[131] Two medical certificates were required in British Guiana following legislation in 1867, based on Colonial Office recommendations. However, the arrangements differed from elsewhere. Initial admission was to a general hospital for a probationary period, based on one medical certificate. A second certificate would be provided by the hospital's medical officer if the patient was subsequently transferred to the Berbice asylum.[132]

From 1844 only one medical certificate was required in Trinidad, due to a shortage of practitioners in the colony. The legislation provided only for admission of people indicted for a criminal offence and acquitted on grounds of insanity, and for those apparently deranged and liable to commit a crime, with cases adjudicated by a stipendiary magistrate. There was no provision whereby 'dangerous lunatics at large' could be 'apprehended and kept in safe custody'. Following advice from the Commissioners in Lunacy in London an Act of 1877 extended the parameters for committals in Trinidad, in conformity with legislation elsewhere in the empire.[133] In the smaller, less populous islands local legislation normally provided for committal based on one medical certificate and a magistrate's order. This was the case in Antigua from 1843 and remained so when its legislation was updated and codified in 1863.[134] Similar legal arrangements prevailed in Grenada.[135]

For legal proceedings to begin, the mentally disordered person had first to be submitted for medical examination. This normally meant either a police arrest or desperate relatives taking them to a police station. In Barbados people in the outlying parts would be brought to a district police station, whilst those from around Bridgetown were generally brought to the central police station or 'Main Guard'.[136] Delays were common, either due to awaiting the attendance of two doctors or because the asylum was full. This could result in people being held in a police cell for several days, sometimes sharing with a 'common felon'.[137] The problems worsened during the 1880s as the asylum's overcrowding crisis became acute. In May 1883, H.K. was kept in Speights Town police station for 12 days and Margaret W. at District 'C' police station for 23 days, before they could be admitted.[138] Arrangements in Jamaica were similar, with most patients coming via the police and being held in constabulary 'lock-ups' before transfer to the asylum.[139]

Conveyance to the asylum was potentially problematic, particularly where considerable distances might be involved. In a relatively small, compact island like Barbados most parts were accessible by foot or horse within a day. However, in larger colonies such as Jamaica or British Guiana a sea voyage could be necessary, as also in the Leeward and Windward Islands where transfer to Antigua or Grenada was entailed. Whichever the mode of transport, there was scope for coercion, maltreatment and neglect, particularly on longer journeys. In some instances conveyance was sympathetically handled, as Ann Pratt confirmed in regard to her four day boat journey from Lucea to Kingston.[140] However, Jamaica posed particular difficulties, both due to the island's mountainous interior and the distances involved. When Emma Steele was admitted to the Kingston asylum in early 1860, her voyage from Falmouth took three weeks.[141]

The problems that arose en route to the new Jamaica asylum at Rae's Town were highlighted by Thomas Allen soon after taking up post. He deplored the 'unsatisfactory manner' in which patients had been sent:

> In some instances they have been dressed in rags, had dirt or vermin about their bodies, or were suffering from foul or contagious disease – Some have been escorted with prisoners on their way to the General Penitentiary, or had walked long and fatiguing distances – others had their hands confined in wrist locks or tied with cords.[142]

He presented examples of particularly rough or inappropriate treatment, like the male patient R.M. who was required to walk over 30 miles through mountains from Port Maria, accompanied by a policeman on a horse. Another man (D.) arrived exhausted, having been tied to a cart, wearing only filthy trousers, his hair 'long and filthy' and his hands 'much swollen', with deep cord marks around his wrists. B.I. arrived with his shirt 'filthily dirty' and his pantaloons 'nothing but rags, and covered with excrement'.[143]

Allen was particularly concerned about the experiences of vulnerable women sent by boat. He lamented that 'On board the Droghers, the safety, comfort and social decency of the female patients particularly, seems to have been quite disregarded and no responsible person appears to have accompanied them.' He cited several cases, including that of A.L. who was put on board a drogher with her baby but no female to look after her. She was placed in the cargo hold, 'to which the sailors had access'. On admission she had 'vermin about her body'. G.S. had been looked after by the cook and a 'lad of about 17', who had 'insulted her' during the voyage and made 'indecent proposals'. M.I. had been 'very bad' on board, 'stripping herself naked before every body'. She was 'very dirty in her habits' and had to be washed by the sailors. Sailors also washed J.A., who had 'filthy habits' and stayed naked on deck whilst her petticoat was drying.[144]

Allen initiated reforms in the conveyance arrangements and gained agreement from the governor and executive for use of the 'Jamaica Packet', in order to speed up the voyage from the 'outports'.[145] New rules were brought in and the use of handcuffs and ropes largely eliminated, the aim being the exercise of 'kindness and tact' to gain the patient's confidence and to encourage them to 'come much more quietly' to the asylum.[146] By the mid-1890s patients were being transported by various means, including carts, steamships and train.[147] The railway between Kingston and Spanish Town presented both opportunities and challenges. In 1893 'considerable indignation' was expressed by passengers on the morning train to Kingston, due to the 'conduct and language' of a woman being conveyed to the asylum. Demands to have her removed to the baggage compartment were refused.[148] Such complaints led the railway company in 1899 to refuse to carry lunatics bound for the asylum, the ban lasting more than two years.[149] Even after it was lifted the company only agreed to convey them on Wednesdays and Saturdays.[150]

Similar problems were replicated elsewhere. In Trinidad George Seccombe noted the 'unpleasant scenes' occurring when people were removed to the asylum, and quickly introduced new regulations.[151] However, these were widely ignored and, during 1883, 17 male patients were brought in handcuffs and 19 women without any female escort. Seccombe sought to ensure that 'patients brought to a Hospital for treatment should not be impressed with the idea that they are going to a

Gaol for detention'.[152] Significant improvements followed and it was noted in 1886 that greater care was being taken, with few men arriving handcuffed and most women coming either with a female escort or relatives.[153] However, standards proved hard to maintain and in 1895, despite the regulations, eight female patients were brought without female escort.[154]

In Barbados, by 1875, patients were all brought to the asylum in a 'carriage', 'cab' or cart, accompanied by the police. There were no arrangements for women to be attended by another female, until new police rules were implemented in 1882.[155] Restraint and excessive force were clearly used on occasions. In February 1883 a man arrived with bruises on his hands from 'handcuffs used in bringing him here'. Dr Field, the medical superintendent, contended that such restraint was almost never necessary.[156] In 1892 a woman brought in handcuffs was found to have a broken arm. Hutson, the Poor Law Inspector, protested that *'handcuffs should never be used for restraining the violence of a lunatic'*.[157] He and Field arranged for their replacement by special belts with wristlets.[158]

The final stage in acquisition of patient status occurred with the rituals of admission. The key elements were succinctly laid out in Antigua as early as 1842: 'All patients on their admission to have their hair cut, be well washed, and clad with the clothing of the Establishment.'[159] In the Kingston lunatic asylum, hair cutting and bathing became opportunities for abuse and humiliation, before its practices became discredited.[160] The procedures of being bathed and clothed in asylum uniform persisted generally. Gradually medical superintendents took an increasingly key role in the post-admission processes. They oversaw physical examinations and, where possible, interviewed patients to gather information on their circumstances and the likely 'causes' of their insanity. The evidence of the Barbados case books and Grieve's British Guiana case histories suggests that these interviews were often conducted sympathetically. Patients could be afforded opportunities to tell their own stories and ventilate feelings and emotions. Crucially, however, these early interactions confirmed the asylum's power differentials, albeit within the parameters of a relatively benign institution. The transition to the role of 'patient' was now complete.

Conclusion: Benevolent intervention or social control?

Although the lunatic asylums of the British West Indies did admit some people from the 'respectable', mainly White sector of the population, the great majority came from the poorer Black and Coloured classes. The descent of people into insanity largely occurred against a background of poverty, deprivation and poor physical health. Behind these key determinants were the effects of economic dislocation, under-employment, labour surplus, racial oppression, and migration. However, although there were numerous casualties of economic and social conditions in the West Indies, only a relatively small number developed a mental disorder of a degree necessitating removal to an asylum. Indeed, as in other British colonies, it was consistently acknowledged that the incidence of insanity reflected in asylum admissions fell well below that in the metropole.[161]

The low overall numbers admitted largely negates a contention that committal to the asylum represented a mechanism of social control on a 'macro' level. There is, however, a stronger argument on the 'micro' level. People sent to the asylum were those whose mental distress was manifested in ways that caused or threatened harm to themselves or others and their property, offended against public decency, or otherwise disrupted the lives of family, community or society at large. Their deviation from social norms had become unacceptable, and removal provided relief to those who could no longer live with them or contain their behaviour. The direct involvement of police and magistracy in the process served only to confirm that control was at its heart, even if there was also an element of humane concern for suffering individuals.

7
The Patient Challenge

As the previous chapter has shown, most people entering West Indian lunatic asylums did so after arousing anxieties and concerns, owing to the perceived degree of risk, disturbance, or hopelessness they had exhibited. As a consequence of some extreme situation or series of events they were now in an institution providing, in varying measures, custodial care and recuperative treatment. Patients all arrived with their individual backgrounds, experiences, symptomatologies, and presenting behaviours. Once in the asylum they reacted in different ways to the environment and expectations placed upon them. They responded also to the people encountered – the medical superintendent, the staff tasked with their management and care and, not least, fellow patients. As this chapter will illustrate, asylum patient populations were highly diverse, both in terms of personal characteristics and in their adaptations to the regime. It becomes evident that, far from being passive recipients, many patients demonstrated clear agency in their responses to the institution's routines and practices.

Of class, colour and race

When enumerating the Barbados Lunatic Asylum's many deficiencies in 1875, Dr Thomas Allen singled out its lack of separate provision for 'respectable persons'. He was, of course, referring primarily to White people who made up about one fifth of the asylum's population.[1] He considered it highly inappropriate that 'governesses, daughters of clergymen, decayed gentlewomen, and men who have been engaged in various pursuits, in the fierce struggle for existence' should be 'brought into relationship with negroes or others of such savage habits'.[2] Perspectives that linked 'negroes' to savagery and barbarism were far from unusual, having become part of White ruling class belief systems in Britain as well as the West Indies, especially since the Morant Bay rebellion in Jamaica.[3] Allen had made these linkages almost since taking up post in Jamaica in 1864, and his views remained consistent.[4] He referred in 1872 to the 'ungovernable passion' that typified the Jamaica asylum's inmates. They were 'vindictive and savage', and any opposition would throw them into 'ungovernable rage'.[5] He reiterated similar ideas many times through the years. By 1881 he was seeking to raise money to construct a separate ward for 'respectable' patients, to prevent them having to associate with 'debased pauper negroes'.[6] A year later he had broadened out his stigmatised group, lamenting that patients 'related to persons of the highest respectability in the island' were still required to mix with others who included 'criminals, coolies, and a low type of the insane poor'.[7]

Thomas Allen's pronouncements were made against a background of striking figures for relative proportions of patients by racial background in Jamaica's asylums. In the Kingston asylum in September 1858, as it was becoming engulfed in scandal, there were 116 patients, of whom 95 were Black, 17 Coloured, and only four White.[8] Five years later, just prior to Allen's arrival, there were 159 people in the new asylum, of whom 110 were Black, 34 Coloured, and five White, in addition to ten 'Coolies'.[9] By the time Allen retired in late 1886 the overall breakdown remained largely similar; of a total 389 patients, 266 were Black, 68 'Brown', and 13 White. One significant change, however, had been among the 'Coolies', whose numbers had climbed to 42.[10] At almost 11 per cent of the asylum's patients their relative numbers far exceeded their presence in Jamaica's wider population and, as in Trinidad and British Guiana, the overwhelming majority were male. Their over-representation in both the asylum and the penitentiary did not go unnoticed by the authorities.[11] Dr Blumer, an American asylum superintendent visiting Jamaica in 1899, was particularly struck by the large number of 'Coolies' in its asylum. His host, Dr Plaxton, in acknowledging the disproportion in their numbers, observed that they 'were probably men of unstable brain when they left India to engage in work here'.[12]

The issues raised by Thomas Allen in relation to the Barbados asylum in 1875 had clearly resonated with the island's authorities. The problems emanating from the mixing of social classes and races were exacerbated by the asylum's increasingly overcrowded state during the 1880s. Representations were invariably made in careful language, with social class and degree of respectability acting as coded inferences for race and colour. In 1886 Dr Charles Manning, in making proposals for the new asylum, regretted that 'patients of the upper and middle class' were at present compelled to mix with 'persons in a lower sphere of life'.[13] As it became more difficult to allocate single rooms to 'respectable or semi-respectable' patients, with their 'peculiarities', there were heightened tensions.[14] A year later protests were made about the absence of rooms for private patients, which meant that 'ladies' who became insane were 'mixed up with the vilest and most filthy of women'.[15] In 1888 Gaskin, the Acting Poor Law Inspector, expressed 'the horror' of a situation where 'unfortunate lunatics of a better class' were mixed with those of a 'low class', for whom even the asylum was an improvement on their own homes.[16] The attitudes he reflected, however, rather paled next to the comments of the American Dr Blumer after visiting Barbados in 1899. He had been delighted, on arriving at the asylum, to find Dr Field reading Plato in classical Greek. Blumer praised his management, noted that the patients 'looked well and contented', but commented acerbically that many were 'no higher than the anthropoid apes'.[17]

Perceptions of race and class permeate the surviving case records and journals for the Barbados asylum in the 1870s and 1880s, often in unspoken assumptions but occasionally in overt comments and actions. As the Barbados historian Pedro Welch has highlighted, the medical superintendents would sometimes describe Black and Coloured patients in derogatory or stereotyping language whilst White patients were regarded more sympathetically.[18] Richard L. (18) was described in July 1875 as 'a dark, cob boy', with 'a large mouth and stupid countenance'.[19] Charles G. (22), an epileptic Black cooper, was referred to in January 1876 as 'a stalwart large limbed man with coarse heavy features & very thick neck'. He had a 'dull & heavy' expression, 'becoming silly when he speaks'.[20] Descriptions of female patients tended to be more thorough. In July 1875 Mary Frances G. (21) was referred to as 'a stalwart black woman with enormous mammary development & decidedly vulgar look & manner'. It was observed, however, that she was 'clean and respectful in her conversation' and had arrived at the asylum 'dressed in the finery of which the negroes are so fond & evidently considers herself above the normal run of her class'.[21] Another with ideas beyond her station was Mary E. (26), 'a respectable looking, fairly well nourished black woman'. She appeared to be 'a little conceited & boastful about her bringing up, respectable training &c', which was 'by no means an uncommon trait in the character of the negro races'.[22]

Entries in the Barbados medical superintendents' journals show that, although the asylum was not segregated, there were differences in how the races were managed and treated.[23] The limited number of single rooms would be allocated either to particularly disturbed patients or, where possible, to 'respectable' White patients for whom adjoining rooms were retained on both the male and female sides.[24] The manner of addressing or naming patients consistently differed according to their race and colour. Black and Coloured patients would be referred to by their full names or, frequently, just their surnames. White patients were usually recorded as Mr, Mrs or Miss followed by their surname, unless their behaviours became sufficiently challenging or antisocial to justify a lowering of class status. Their cases were more likely to arouse the medical superintendent's interest and he would spend additional time with them, sometimes allowing special privileges. Where their previous status in society had been relatively eminent they might on occasions be taken into his confidence or offered a role in the workings of the institution.[25]

Patients were clearly conscious of racial and class disparities within the Barbados asylum, and these frequently impinged directly on relationships and interactions. Theodore L. (29) was a pugnacious Irish seaman who had previously spent two years in a lunatic asylum in England, in addition to spells in both the Barbados asylum and St Michael's (Bridgetown) alms house. Despite his own lowly social status he clearly retained perceptions of superiority toward some groups. He was observed in October 1878 to be 'very irritable specially if one of the black inmates troubles him', provocatively calling them 'dog of a nigger'.[26] John L., a White Portuguese seaman admitted in August 1891, had not expressed his prejudices through violence or threat. However, he was reportedly 'so terribly afraid of "niggers" that he becomes quite nervous and prostrate when brought in contact with them'. His mental condition improved greatly when he was moved to a single room.[27]

The asylum also presented opportunities for projection of strong negative feelings in the opposite direction. Two female patients, Nancy B. and Mary F., were recorded in October 1883 as being 'very violent still against white people'.[28] In May 1889 Nancy B. was sent to the lower ward (for refractory patients) after she beat the housekeeper's daughter, 'simply because she dislikes white people'.[29] Evidently Nancy B. and Mary F. nursed some very strong grievances, for in September 1889 both were being kept in the lower ward as they were 'still much enraged against white people in general'.[30] In August of the following year Nancy B. was physically unwell, but nevertheless refused to be seen by the medical superintendent, steward and head attendant, 'nor any white person'.[31] Joseph W. in August 1891 kicked two White patients due to 'his rage being directed against white people'.[32] In these cases insanity provided an apparent shield for expressions of defiance against perceived representatives of the hated White ruling elite.

The multi-ethnic composition of British Guiana's asylum population afforded opportunities both for comparative studies and generalisations about racial and cultural characteristics. James Donald, who spent a period as resident surgeon in the mid-1870s, drew a number of conclusions from his experiences. He acknowledged the difficulties associated with 'an inability to converse personally with some of the patients', particularly the Chinese, and shrewdly conceded the importance of avoiding 'national peculiarity' being 'mistaken for mental derangement'.[33] Among the Black Creoles from Guiana and Barbados, who made up one quarter of the asylum's inmates, he considered the predominating form of insanity to be mania, in most cases 'complicated with delusions of a religious character'. Donald connected this with the frequent incidence of revivals and the 'strong religious sentiment (that) exists in the mind of the negro'. Some suffered from distressing 'religious melancholia', which he likened to the experiences of African patients preoccupied by 'the terrors of the mysterious Obeah'.[34] 'Coolies', who formed over half the patient population, suffered principally from mania and 'dementia'. Their presentation was 'generally characterised by great destructiveness and impulsiveness', frequently with 'homicidal and suicidal propensities'.[35] The few Chinese patients were generally 'quite docile' and amenable, whilst there was 'nothing particular to note' in the insanity of the Portuguese.[36]

During his ten years at the British Guiana asylum Robert Grieve gave much consideration to how racial differences influenced the expression of insanity. Though generally sympathetic, his approach betrayed evidence of cultural stereotypes that later became familiar. Observing in 1882 that the asylum contained many 'maniacal' patients of 'African descent', Grieve suggested that their 'mental manifestations' lacked sophistication, with an absence of 'the imaginative phase'. There were not the 'well known characters', with their grand schemes, as found in European asylums. Claims of possessing great wealth were relatively simple, 'only to be associated in idea with the power of unlimited eating and drinking'. Persecutory delusions, although common, almost invariably consisted of 'having been obeahed' by the person particularly distrusted. Mania among 'Africans' was 'characterized more by emotional than intellectual perversion', whilst depression (or 'melancholia') was an unusual feature among Black patients.[37]

Grieve questioned whether the differing proportions of asylum inmates were due to 'racial peculiarities of constitution' or the conditions under which people lived, and decided that both sets of factors applied. Based on comparative recovery rates, he studied whether racial differences also affected outcomes. He calculated that only 11 per cent of Black Creoles recovered, compared to 27 per cent of East Indians, attributing these disparities partly to the circumstances leading them to the asylum, and partly to the comparative treatability of conditions related to excessive use of alcohol or 'ganja'. However, having computed that 25 per cent of Coloured Creoles recovered, Grieve concluded that 'race does influence the result in insanity', a contention that received further support from comparing the low recovery rates of pure blood Africans with much higher rates for White Portuguese people. The issue was not straightforward, for he noted that Black natives of other West Indian colonies showed a relatively high recovery rate. Taking all the figures together, Grieve ascertained that only 16 per cent of all patients of African descent had recovered, compared to 30 per cent of the rest.[38]

Grieve's researches also demonstrated that recovery rates inversely mirrored mortality rates. Whilst 51 per cent of Black patients died in the asylum, only 30 per cent of Indian and White patients succumbed.[39] In attempting to identify probable causes he compiled detailed data regarding 229 deaths over a five-year period. The largest single cause, accounting for 69, was chronic kidney disease (Bright's Disease). The condition had proved fatal particularly for Africans and, to a lesser extent, Black natives of British Guiana and the West Indian Islands. Grieve inevitably drew a close linkage between kidney disease and excessive indulgence in 'alcoholic stimulants', regarded as more prevalent among Black people than East Indians. The latter, however, showed a higher death rate from phthisis (tuberculosis), related both to hereditary factors and to their 'previous mode of life in its sanitary aspects'.[40] Ultimately, however, Grieve's tentative explanations could not provide adequate interpretation of the discrepancies in outcomes between ethnic groups in British Guiana or elsewhere.

Categorisations and presentations

In this strange caravanserai are gathered a great number of insane folk, mostly Negroes. In the centre of the quadrangle a grey-headed mulatto is kneeling in the sun and praying with breathless eagerness. He is a religious monomaniac. A comparatively young man, sweating with excitement, and puffing out his cheeks, is calling out that he is

Lord Nelson, and wants boots. Lying senseless in the shade is a man recovering from a fit. Drooping on benches are listless melancholics, while among them is a man who sits bolt upright and for ever pats his hand to the moaning of some fragment of a song. A very cheerful being, squatted on the ground, is professing to make a hat out of grass roots collected with infinite assiduity. There are, besides, idiots and dotards and the absolutely mindless.[41]

An impressionistic picture of the asylum patient population can be constructed from surviving case notes, contemporary journals and reports. In compiling their records, medical superintendents distinguished people according to diagnoses, symptoms, and behaviours. The Colonial Office provided guidance on patient categorisation in its 1863 circular and in the subsequent call for regular returns regarding all hospitals and lunatic asylums. The information required became standardised, with insane patients grouped into four classes:[42]

(I)   Maniacal and Dangerous
(II) Quiet Chronic
(III) Melancholy and Suicidal
(IV) Idiotic, Paralytic, Epileptic

By and large the statistical data compiled by asylum superintendents conformed to these guidelines over the next half-century. The problem for historians is that the categories left plenty of leeway for discretion in interpretation. Significant differences emerged between colonies in the proportions of patients considered to fall within each class. These differences could reflect local circumstances, but they could also be affected by individual doctors' approaches to assessment and diagnosis. The inconsistencies within individual colonies over relatively short time periods support a conclusion that the criteria for determining illness categories were neither standardised nor static.

Table 7.1 illustrates both similarities and disparities between colonies. It is evident that 'quiet chronic' patients generally constituted the majority group within the West Indian asylums of whatever size, from Antigua in 1868 with its 40 patients to the huge Jamaica asylum in 1909 with its 1,160 patients. The notable exception was Barbados, where figures for both 1872 and 1889 show that 'maniacal and dangerous' people far exceeded the other categories. This may have been partly related to higher admission thresholds in the island. However, a significant contributory factor was the poor physical state of the asylum,

Table 7.1   Classifications of patient symptomatologies

	Maniacal and dangerous	Quiet chronic	Melancholy and suicidal	Idiotic, paralytic, epileptic	Total numbers
Antigua 1868	4	23	8	5	40
Barbados 1872	64	28	6	22	120
Barbados 1889	129	78	12	37	256
B. Guiana 1889	163	358	56	57	634
B. Guiana 1891	70	486	32	34	622
Jamaica 1898/99	54	577	34	101	766
Jamaica 1908/09	321	671	61	101	1160

Sources: CO 7/137, 27 September 1869, Antigua, Lunatic Asylum Return, p.4; CO 28/216, Barbados, Lunatic Asylum Return, 1872; BPP 1890, Vol. XLVIII, Papers Relating to Her Majesty's Colonial Possessions, 72, British Guiana, 1889, p.19, 107, Barbados, 1889, p.16; BPP 1893–94, Vol. LIX, Annual Series of Colonial Reports, 55, British Guiana, 1891, p.34; *Jamaica: Asylums and Hospitals (Returns)*, 1898–99, Lunatic Asylum Returns, p.8, 1908–9, Lunatic Asylum Returns, p.8 – [Copies in LSHTM].

with its severe overcrowding. These conditions fostered discontent and violent behaviour among patients, amply evidenced in case books and medical superintendents' journals.[43] The British Guiana figures are interesting for the significant changes over a two-year period, reflecting either a marked lessening in numbers of disturbed patients or, more likely, alterations in diagnostic criteria. Shifts in the opposite direction apparently occurred in the Jamaica asylum. Whilst in 1899 less than one tenth of patients were designated 'maniacal and dangerous', by 1909 their numbers had climbed steeply to more than one quarter. It seems likely that, rather than signifying changes in composition of the asylum's population, assessments of patient presentations had become more rigorous.

Despite 'quiet chronic' patients forming the largest group in most asylums they often went relatively unnoticed, apart from when becoming the focus of complaints about the presence of too many incurables without prospect. The large numbers in this category were related in part to the poor state of many admissions.[44] After a year at the Jamaica asylum, Thomas Allen concluded that a lot of patients were 'hopelessly incurable', mainly due to physical health problems, and would be better managed in a poor-house or by families with financial assistance from the parish.[45] He was still making these complaints in 1880,[46] and in 1881 claimed that the asylum was partly a home for 'the aged, imbecile, idiots and incurable insane'.[47] The problem persisted, and in 1903 Dr Williams again argued for parish authorities to be responsible for the care of 'chronic and harmless' people.[48] It was also an issue in Trinidad where, within months of his arrival, George Seccombe complained that the asylum was rapidly becoming a 'Refuge for the chronic Insane', with many patients lapsing into 'hopeless forms of Insanity'.[49] Chronic patients were periodically blamed for contributing to overcrowding. A campaign for their removal achieved some success in 1895 when 50 people were transferred to new 'Insane Wards' at the St Clair House of Refuge.[50]

Those falling within the broad category of 'idiotic, paralytic, epileptic' were often lumped together with the chronic patients. However, they became noticed when their behaviours occasioned alarm or impacted adversely on others. One such case was S.B., a Black native of British Guiana admitted to Berbice asylum in July 1880 when aged 19, after violence toward children and others. On admission he was 'completely idiotic, being unable to understand the simplest thing said to him'. He could not 'articulate a single intelligent word' and the only sound emitted was a low growl. When settled he smiled vacantly, but if 'thwarted' he became extremely angry, striking and biting anybody within reach. He ate voraciously, did not attend to the 'calls of nature', and also displayed 'marked and irregular erotic tendencies'. S.B.'s condition remained unchanged in the asylum and he died in March 1883 from Bright's Disease.[51]

One patient deemed paralytic was J.D. (46), a Black cooper from Barbados, admitted to the British Guiana asylum in July 1883 and later diagnosed by Grieve with 'general paralysis of the insane'. Prior to admission he had been violent and threatening, with grandiose delusions that he was the son of an English duke. In the asylum he behaved 'in a very childish manner', rambling and at times incoherent in conversation. His memory for both recent and distant events was poor. He trembled when standing and his gait was unsteady. Within a few weeks his childishness and incoherence had worsened and he had 'no idea where he is'. By early 1884 he was increasingly demented and at times dirty in his habits. His physical condition became more feeble and unsteady, and his conversation almost unintelligible. He died at the beginning of April.[52]

The Barbados asylum records include several case histories of people who suffered badly from epilepsy. Mary S. (19), a Black servant admitted in early 1876, had experienced regular epileptic fits for the previous nine years, which became worse in 1874 after childbirth. Following an attack she would become 'stupid', violent, abusive and 'indecent'. Six weeks after admission she had seven fits in two days and was 'dreadfully excited & impulsive' as well as 'extremely filthy & indecent'. Over the following months there were frequent fits and episodes of excitement. Nevertheless, in October 1878 Mary was considered generally 'One of the best workers in the place'. She died in the asylum in July 1879, aged only 22.[53] A similar fate awaited Alonzo W. (21), a Black carpenter, admitted in July 1878.[54] He was at first restless and overactive but then remained mostly settled, apart from after periodic fits when he would become 'much excited'. Despite this he too proved himself a 'good worker'. However, in late June 1880 the fits became suddenly more frequent, with 16 in one day and 15 the next. Alonzo died shortly after, aged 23.[55]

The group of patients presenting as 'melancholy and suicidal' in Barbados was relatively small, but they could occasion great concern both because of the risks posed and the anguish projected. This was certainly the case with Reverend N.H.G., the White clergyman admitted in March 1877.[56] He was deeply depressed and preoccupied with delusions of 'Satanic possession', telling Dr Charles Hutson on admission 'I'm a miserable man Sir – I have sold myself eternally to the Devil – made a bargain with him – & he has taken complete possession of me'. Whilst describing his torment he shouted 'pitifully', contorting his mouth and body. A few nights later he had to be prevented from choking himself with his sheet. At the end of May he was sometimes unable to speak, 'only going through the most horrible grimaces'. He would howl loudly and constantly walked up and down until his feet were swollen. His habits became dirty and on occasions he soiled his room and daubed the walls. In late June he remained 'very depressed, crying a great deal', claiming that only 'the dread of Hell' prevented him ending his 'miserable life'. Gradually his mood improved, but the delusions remained fixed and Reverend G. was still in the asylum at the end of 1878.[57]

The large numbers comprised within the description 'maniacal and dangerous' occupied most of the lunatic asylums' resources and had great influence in shaping internal dynamics. Their disorderly behaviours included quarrelling and fighting, violence toward staff, and damage to property. However, as became clear in the Kingston asylum even before 1840, the violent nature of patients' behaviour was as much related to the conditions and regime in the institution as to their mental states or inherent dangerousness.[58] The levels of apparently unmanageable disorder in the Kingston asylum, in its successor Jamaica asylum prior to Thomas Allen's reforms, and in the Barbados asylum before Jenkinsville opened in 1893, underlined the close relationship between environment and patient behaviour.

The voluminous evidence regarding the Kingston asylum in the late 1850s revealed a body of ostensibly violent and unruly patients, mostly female. Their conduct was greatly exacerbated by the overcrowded, inhuman conditions into which they were forced. According to the former warden Richard Rouse, they would fight violently among themselves until 'maimed and covered with blood', making use of 'broomsticks, brick bats, and any and every thing they could lay their hands on'. Their attitudes toward one another were 'harsh and unfeeling'.[59] He claimed that violent incidents occurred daily, as well as during the night in the crowded cells, and that frequently women fought naked.[60] William Harrison, for several years the 'dresser' at the public hospital until dismissed in 1858, confirmed that latterly there had been 'a great deal more wounds to dress' than previously, because of the 'more unmanageable class of lunatics' being admitted.[61]

The patients at the new Jamaica asylum proved to be similar in character. Dr Charles Lake claimed that many 'evince a refractory and destructive disposition', the women especially displaying a 'mischievous propensity' to destroy their clothing.[62] On his first visit in October 1863 Thomas Allen found many patients noisy, disorderly and violent to the point of being out of control. His initial impressions were confirmed over the following weeks as he catalogued numerous incidents of violence, destruction, menacing demeanour, indecency, foul language, soiling, shouting, screaming, and so on.[63] He lost few opportunities to liken the patients, collectively and individually, to 'savages'.[64]

As part of a drive to obtain more single rooms Allen compiled descriptive summaries of about 100 patients who had exhibited dangerous or unpredictable behaviours.[65] These cameos conveyed a strong impression of a challenging body of people. W.H., in the asylum since 1850, was 'Very dangerous and violent, very passionate, and easily excited'. He would strike violently 'without the least provocation', and was also guilty of 'attempted Sodomy'.[66] J.L.C., a patient since 1855, was 'Dangerous, passionate, and easily Excited, will assault'. On one occasion he had bitten off another patient's ear and also nearly bit off an attendant's thumb.[67] D.D., admitted in 1861, was 'Dangerous' and liable to strike out 'if interfered with'. He tore his clothes after they were put on and remained in one position all day in the sun 'almost in a state of nudity'. Attempts to encourage him to keep on his clothes or give him his medicine could result in attendants being struck 'violently'.[68] J.F., who entered the asylum in March 1863, was considered 'Very dangerous', for he assaulted fellow patients and was 'very destructive'; he attempted to strangle another man with his bedding. He also had 'nasty habits' like 'daubing his room and smearing his head with his own excrement', painting pictures on the walls of his room with faeces.[69]

A greater number of female patients were included in Allen's summaries, reflecting notions that they were more troublesome than the men. Rebecca F. typified many women in the asylum:

> very violent dangerous, destroys her clothing, obscene in her language, and conduct, assaults patients, and nurses, very unmanageable and of dirty habits she attacks the nurses when they are engaged in controlling another patient, exposes her person and she is very passionate.[70]

J.K. was 'Very dangerous exceedingly noisy day & night', talking constantly on religious subjects whilst 'putting herself in different postures'. At night she would beat the floor of her room with her feet or bedding and 'sings very loud'. She was 'unmanageable', owing to a liability to 'assault any person without provocation', whether patients, staff or visitors.[71] M.B. was another deemed 'unmanageable', being 'Dangerous at times & violent', destructive of clothing, dirty in habits and 'indecent'. She had attacked patients and nurses, broke windows, and exposed herself. She constantly sang 'revival hymns' loudly and 'dances & beats her feet on the floor' during the night.[72] To Thomas Allen the turbulence of so many justified assertions that their 'inclinations, habits, and uncontrollable passions' stemmed from their 'naturally savage and uncivilized condition'.[73] These perceptions persisted even after his reforms stimulated significant improvements in patients' demeanour.[74]

The records of the troubled Barbados asylum from the late 1870s and 1880s provide numerous instances of patients whose dangerous presentations were perpetuated within the institution. One example was Justina M., a 'tall finely made cob' woman aged 23, admitted in January 1877. On the day of admission she assaulted the sempstress, and the next day was 'extremely restless jumping up fr. chair & wandering about room'. Over the following week she continued 'very restless and troublesome' and, having taken a 'special dislike' to the sempstress, severely assaulted her twice. She became 'destructive to clothing & very filthy in her habits' and in early May appeared 'decidedly worse than she was on admission'. Justina then had a long relatively settled period when she was seen as merely very troublesome, but not dangerous. Her mental condition, however, remained unimproved. In early 1880 she once again became violent and for several months was regularly fighting with other patients, causing serious injuries more than once. The violence abated during 1881, but the last entry in March 1882 reported her mental state as 'not improved'.[75]

Many patients experienced fluctuating mental conditions, like Thomas H. (37), a Black boatman admitted for the third time in January 1877 with acute mania. He came through the gate 'shouting & singing merrily', having stripped off most of his clothes. He was tall, very muscular and agile, and shortly after admission managed to scale the airing court wall. Three days later he exhibited 'Constant & violent excitement' and walked up and down 'shouting & threatening in rather alarming manner'. Over the next few weeks, episodes of great excitement alternated with more settled phases. By late March he was again 'Very threatening & violent'. At night he frequently broke down partition walls or pulled up the floor of his room. Dr Hutson lamented that he was 'very powerful' and there was 'no room able to restrain him'. Following a quieter spell, another period of great excitement followed in June; Thomas spent several days in seclusion and was 'intensely noisy & violent'. At the end of January 1878 he was again 'excited & noisy' and 'obscene & abusive', though not dangerous. By December he was working 'cheerfully' and was 'less noisy' than before.[76]

Elizabeth (or Betty) W., a Black huckster, was another whose violent periods interspersed with more tranquil interludes. She was first admitted in December 1854, aged 19, remaining until July 1855, and again from July 1856 to June 1857.[77] She then stayed well for 20 years before readmission in July 1879, diagnosed with mania. Betty proved very difficult from the outset. In October it was recorded that she had been 'occasionally violent and raving almost ever since her admission, fighting with patients, striking attendants and being often naked and dirty', requiring frequent seclusion. Following a settled period in 1880 she was sent out on leave but readmitted a few weeks later. Between September and November she was constantly engaged in fights. A pattern of episodic violent, unmanageable behaviour succeeded by calmer periods continued for the next two years and beyond.[78] In June 1883 she beat pregnant nurse Elizabeth Holder so badly that she suffered a miscarriage.[79] By the end of 1884 Betty's presentation had altered, with the violent outbreaks channelled primarily into systematic destruction of the rooms in which she was placed.[80] In April 1888 she 'tore up several boards in the floor of her room, got under the cellar & tried to dig through the wall', and during November and December she destroyed at least four more rooms.[81] Over the next three years there were numerous similar incidents, alternating with periods when she was settled and sometimes contentedly working in the laundry.[82] The case was particularly striking in several ways. It exemplified the changeability and unpredictability characterising many asylum patients. It also showed the difficulty in distinguishing between 'maniacal' behaviours and actions that represented a clear, direct provocation to the institution itself.

Protest and confrontation

The case examples already cited demonstrate that many West Indian asylum patients did not passively accept their situations. Their various forms of disturbed and antisocial conduct were usually interpreted by doctors and officials as symptomatic of insanity. To some degree behaviours like tearing clothes, stripping, shouting, quarrelling, and even low-level violence were indeed manifestations of distress or despair. However, overt physical attacks or menacing threats toward staff, serious damage to surroundings, soiling and smearing of faeces, and active escape attempts all carried strong elements of challenge to the institution, its regime, and those responsible for it. There is ample evidence that individuals and groups of patients were at times either defiantly refusing to co-operate or openly provoking the asylum and its representatives.

The singularly oppressive practices of the Kingston lunatic asylum between 1855 and 1860 precipitated acts of resistance, even before Ann Pratt and others laid their experiences before the public. Some female patients stood up to Mrs Ryan and the nurses attempting forcibly to 'tank' them. Elizabeth Scott 'knocked them away' before extra assistance was summoned to subdue her. John Williams, the 'labourer' brought in to assist with tanking females, confirmed that many struggled violently against it.[83] The public exposures of the asylum's widespread abuses, and the removal of Mrs Ryan and some nurses for criminal trial, sparked an 'ebullition of feeling' among the patients. A complete breakdown of order had occurred on the female side of the asylum by September 1860, with patients 'violently and systematically' obstructing the staff.[84] Dr James Scott decried the 'manifest tendency' among the female patients to ignore all the rules and defy the matron's orders. Patients were intimidating staff and 'the discipline has been subverted'. On his daily visits to the female asylum Scott found the patients 'boisterous, obstinate and rude', where previously they had been orderly, quiet, and 'respectful and submissive to authority'.[85]

This spirit of resistance carried over into the new Jamaica asylum. The general disorder and mayhem confronting Thomas Allen as he familiarised himself with the asylum in 1863 was partly a manifestation of patients' continuing reluctance to accept structure or discipline. Allen was left in no doubt about the obstacles faced in implementing unpopular reforms. He could stand firm against recalcitrant staff and secure their dismissal, but patients determined to protect perceived entitlements proved another matter. He concluded that the patients' diet scale was 'unnecessarily, and mischievously high', with females allowed the same as males and no distinction made between those employed and the 'indolent'. His proposed solution was a reduction in the meat allowance. However, Allen abandoned the measure 'on account of the violence of some of the patients, and a spirit of insubordination manifested'. In the face of patient power he only managed to reduce the bread ration.[86]

It took at least a year for Allen's reforms to take effect properly. In the meantime many female patients continued as defiant as ever. Elizabeth Y. was a 'Violent & dangerous patient passionate easily provoked', who believed that she had been 'obeahed'. She was prone to assaulting both other patients and nurses. She adamantly refused to wear the blue asylum uniform dress, and was prepared to fight if staff attempted to make her do so.[87] R.A. was 'very refractory' and 'exceedingly noisy', to the point of being 'unmanageable'. She was fond of exciting 'a spirit of insubordination' among the patients; if there was any disturbance she 'dances & sings as if it were a delight to her', in order to encourage them.[88] One female patient broke through the doors of her single room and escaped into the airing court, where she tore up some sheets and pushed them down the water closets. She was then put into another room and quickly proceeded to smash its door.[89] Another made several threats to 'burn down the Asylum', but her dramatic protest went badly wrong. Having acquired some lucifer matches through the airing court fence, she set fire to her bedding but died as a result of burns.[90]

Once the moral management regime became established in Jamaica, reports of insubordination by patients declined steeply. Whilst this may have reflected reality, the exposure of such behaviours was not convenient for the triumphant message that Thomas Allen and his successors wanted to convey. Occasionally evidence filtered through to confirm that some patients remained as truculent as ever. An unfortunate victim was the nurse Sophia Scotland, pensioned off in 1884 due to incapacity stemming from injuries dating back to the late 1860s. She had been violently attacked with a heavy iron corridor key by Elizabeth F., described as a 'big strong woman with a full beard like a man' who frequently assaulted both staff and patients. After a gradual recovery from her injuries Sophia was again badly hurt by a bite on her right arm from another patient, Rebecca V. The actual circumstances surrounding these attacks were not reported.[91] Assaults on staff remained a straightforward, easy means of angry protest. In 1911 'criminal lunatics' were blamed by the medical superintendent as a 'constant source of irritation' on the wards, because of their attempts to escape and for 'conspiring to injure those in charge of them'.[92]

The seething discontent evident among some patients was liable to be aroused by their living conditions. Soon after arriving in Trinidad in 1882, George Seccombe was struck by the disorder prevailing in the asylum, attributing it largely to the extreme discomfort consequent on overcrowding. Tensions were heightened by the lack of facilities for exercise, recreation and amusement. The insufficient airing court space restricted free movement and patients were required to remain for long periods on benches without moving. Their resentments found expression in frequent 'unseemly struggles' with the attendants.[93] Likewise, the Barbados asylum in the late 1870s and 1880s was characterised by frequent violent commotions, at times to the point of virtual anarchy.[94] Its confined space and overcrowded, insanitary conditions only exacerbated the disruptive and defiant behaviours of its growing numbers of patients.

In the turbulent circumstances of the Barbados asylum, physical attacks by patients on staff and even on the medical superintendents were not uncommon. Sarah Elizabeth K. (19), a Black domestic servant suffering from mania, was being held in seclusion in the refractory lower ward during a period of 'General excitement' in June 1876. When Dr Albert Field arrived to see her she attacked and kicked a nurse before biting Field on his forearm.[95] Even patients from privileged backgrounds might aggressively express their discontents. Joseph T. (62), a White planter suffering from mania, proved quite a handful after his admission in 1876. He had been threatened with force-feeding, and placed in seclusion because of the annoyance he caused to other inmates. In his anger he spat in the face of Dr Charles Hutson, the medical superintendent.[96] A fortnight later a female patient was secluded on the lower ward after swearing and trying to kick Hutson in the face.[97]

Hutson was replaced in 1878 by Albert Field, who was subjected to patients' violent wrath on numerous occasions. Arabella S., a 'Mulatto' seamstress and former prostitute, went in and out of the asylum many times before being continuously detained for four years. One day in September 1882 she rushed to get out of the upper ward as Dr Field entered. When he tried to push her back she bit him on the arm and pulled away his watch chain.[98] A few months later Field intervened when he found patient W. beating an old woman on the upper ward, 'whereupon she pulled my moustache and threatened me most violently'. He commented that she was a 'very dangerous character'.[99] This was a difficult period for Field, for in March 1883 he was targeted by M., a male patient, who gesticulated 'furiously' and 'swears to strike me dead' for keeping him in the asylum. Four days later M. 'came at me' in the airing court and had to be restrained until Field got out of the way.[100]

The attendants were often the most readily accessible targets for patients seeking to vent their resentments. Theodore L., the White seaman whose racist behaviour has been noted, was readmitted to the Barbados asylum in 1878. Field regarded him as a 'perfect pest'. In November 1880 he 'cuffed' and kicked Campbell, the head attendant, as well as kicking another patient.[101] In September 1883 he stoned and 'cuffed' attendant Ford.[102] Thomasin S., a 'Cob' domestic servant admitted in 1878 aged 22, gained notoriety for attacking and biting other patients and staff.[103] In January 1883 she was described as a 'furious biter', after biting 'a piece out of the arm' of an attendant.[104] It seems that teeth proved useful offensive weapons for several disgruntled inmates. A male patient, M., became extremely problematic in the late 1880s, due both to his violence and a tendency to destroy furniture and fittings. In early 1889 he 'went in search' of the medical superintendent (Field). Attendant Chandler managed to stop M. but was himself struck and bitten on the forearm.[105] A year later M. badly injured another attendant, Samuel Cox.[106] Chandler had a few months previously received 'severe blows' on the chest and face in an attack by Obadiah B., a 'very powerful man'.[107] Johanna W., a 'respectably dressed white person' admitted in October 1888, remained in an excited and 'very troublesome' state for several weeks, during which she attacked the experienced nurse Julia Paul, giving her a 'severe blow' in the back.[108] Social standing was clearly no bar to violent expressions of discontent.

Certain patients demonstrated their defiance through direct confrontation, like James W., admitted to the Barbados asylum from Glendairy prison in 1889. He attracted little notice before his campaign of resistance began on 11th June, when he climbed through a trap door in the bathroom, got into the roof and did some damage. On the 25th he set his room on fire and two days later broke up the room. He was 'very excited' on the 28th, 'resisting the Attendants in order to have his own way', and destroying whatever he could. Over the next few days he broke up several more rooms. On July 9th he got into the ceiling, remaining there for the night before escaping through the roof. He was recaptured four days later in Hole Town and put in handcuffs after his return. By late July he was again smashing doors and masonry, and on August 4th Dr Field arrived to find him 'keeping the attendants at bay' whilst tearing up the ceiling of the lower ward. In the early hours of the 9th he pulled out the iron bars in a new door and escaped again. Over the next six weeks he continued to break doors, destroy rooms and make escapes.[109] His openly rebellious activities aroused fears of contagion, after an 'epidemic of escapes' was sparked.[110] Fortunately for the authorities James's fluctuating mental condition eventually settled down. However, by the end of November 1890 he had relapsed and resumed breaking ceilings, climbing into the roof space, and making escapes, as well as fighting with other patients.[111] He settled again within a few months and was even well enough to be allowed periods of leave. By November 1891 he was back in the asylum and the pattern restarted.[112]

Another determined patient who posed a serious challenge to the Barbados asylum's fabric and functioning was Mary Anne C., transferred from the overflow asylum at District 'B' police station in February 1882. She immediately began a process of destruction, ripping up flooring and forcing her way into the roof of her room. By mid-April she was habitually digging up the walls and a mason had to be 'constantly employed' to keep them in repair.[113] Despite being restrained with wrist straps, over the following weeks Mary managed to kick down doors, pull down railings, dig a large hole through the wall of the ward and do other damage, as well as getting into fights and participating in the general disorder on the lower ward.[114] She eventually calmed down but was active again by June 1883, when she pulled down a fence, got into the piggery and climbed over the asylum wall. She began demolishing her strong concrete room and Dr Field expected it soon to be 'down & useless'.[115] During 1888 Mary was involved in several violent incidents with other patients.[116] In July she pulled up flooring and caused more damage in the lower ward, and in November she broke out of her room and destroyed fence palings.[117] She evidently hated change, for within two days of arriving at Jenkinsville in 1893 she had resumed her old ways, breaking down a door and smashing the thick glass contained within it. She then set about digging up the wall of her room and tearing the galvanised iron roof. Dr Field once again had to demand provision of a concrete room.[118]

A particularly graphic illustration of a rebellious patient emerged from the small Grenada Lunatic Asylum. The revelation that a man named Amand had been in 'close confinement' for 18 years caused consternation at the Colonial Office in 1870. According to Robert Mundy, the colony's lieutenant governor, he was a criminal lunatic 'of a most violent and ferocious disposition', too dangerous to be permitted even the 'slightest liberty'. He was, however, allegedly shown 'every attention and kindness' in the asylum.[119] Amand's initial committal in 1852 followed indictment for the murders of two men, 'whom he had chopped to pieces with his cutlass' whilst 'labouring under a fit of insanity'. His resentment against incarceration was exhibited by 'numerous acts of outrage and violence'. On one occasion he seized the medical attendant by the throat and would have killed him but for the intervention of the 'keepers', and in another instance he almost strangled the head keeper. As well as several serious attacks on staff, Amand showed defiance in every possible way. He tore mattresses into shreds, rarely slept, and was heard 'singing and screaming at the pitch of his voice day and night'. Despite many years of captivity, Amand remained 'a man of immense physical strength and great muscular development', his capacity for violence and intimidation undiminished.[120]

Conclusions

The dynamics of race, colour, and class status were of considerable significance in the West Indian lunatic asylums, fully reflecting the social and power relationships of wider colonial society. Within the institution those dynamics influenced perceptions of people's liability to particular forms of insanity and how they were manifested. They influenced how the all-powerful medical superintendents conceived their patients, whether as brutal savages to be tamed or as interesting subjects whose ethnically determined characteristics merited serious study and sympathetic consideration. The dynamics also directly affected the content and character of relationships between medical officers, staff, and patients, as well as the nature of contacts and interactions between the different groups of patients. The races and classes were forced to mix with one another in the asylum whether they liked it or not, for insanity could prove a great leveller.

The diagnosis and categorisation of patients' mental conditions and associated symptoms or behaviours remained a problematic and contested area. The classification system recommended by the Colonial Office provided a convenient framework, although its simple, general headings proved unhelpful for comparative analysis. The returns from the various asylums largely concurred that the numerically dominant group were the generally quiescent, chronic, often physically deteriorated, 'incurable' patients. Other than in Barbados the violent, noisy, and confrontational formed no more than a vociferous minority. Whatever the overall numbers, the prevailing culture of the institutions was at least partly determined by the relative dominance within them of the 'quiet, chronic' or the 'maniacal and dangerous' patients.

The ways in which patients presented in the asylum were determined by a number of factors. Predominant among these was the nature of their individual mental health conditions, as mediated through external social and cultural influences. However, the experience of the institution itself was of crucial significance, whether in regard to physical and environmental conditions or the nature of its systems of control, care and treatment. The degree to which behavioural presentation was related to individual and collective pathology or prevalent circumstances within the asylum remained an unresolved issue. The pervasive nature of active dissent or outright defiance shows that many people were unprepared to acquiesce with a regime conceived as alien and oppressive. In this aspect the asylum arguably constituted a mirror of West Indian colonial society.

8
The Colonial Asylum Regime

> The order and quiet of the Asylum, and the absence of all
> irritating causes, oftentimes seem to transform him who was
> violent and dangerous outside its walls into a very harmless
> and tractable patient.[1]

The main ostensible purposes of colonial lunatic asylums were to provide effective curative treatment for mentally disordered people and
to ensure the safety and security of the public at large. In some cases
patients achieved recovery, followed by a resumption of social and economic roles in family and community. In other less promising instances
the goals were more modest, comprising behavioural improvement,
risk reduction, and a reasonably comfortable existence in controlled
surroundings. In effect the asylum regime aimed to change the agitated, distressed, or threatening person into a settled and calm, even
almost rational member of the patient community. The primary means
employed were the techniques associated with moral management,
which were known about but hardly implemented before British doctors
brought them out to the West Indies. These were augmented, to some
extent, by medically based remedies that involved medicines or physical
interventions. Key to the whole enterprise were the asylum attendants,
expected to provide the direct care and management of the patients.

The moral management system – civilising the lunatics

As was suggested in Chapter 1 the principles and practices of moral
management were particularly suited to colonial settings. They resonated with the methods employed by the British to bring order,
control and 'civilisation' to diverse peoples who were reluctant subjects

of an unwelcome external colonising power. Moral management was essentially an over-arching psychological approach that combined organisation, structure and discipline, whilst displaying a degree of benevolence and sympathetic regard for the needs of the troubled individual. It encompassed aspects of positive role modelling, along with the use of reward and reinforcement of what were deemed appropriate behaviours. The key practical elements of a moral management system were classification of patients according to symptomatology and behaviours, an absence of mechanical restraint, and programmes of work or other occupation. These would normally be augmented by organised recreational activities and religious observance.

Dr Lewis Bowerbank advised the public enquiry in Jamaica in 1861 that classification was 'a measure of vital importance' to the welfare of patients; it was 'the essential requisite of all treatment, especially of that termed the "Modern, or moral system"'.[2] However, it had long been acknowledged that anything more than gender separation was impossible in the Kingston asylum, due to the nature of its buildings and facilities.[3] The lack of scope for classification proved a perennial problem throughout the region, whether buildings were adapted or newly constructed. In smaller asylums, as in Grenada and Antigua, statements in their rules that there should be separation of the 'convalescent and quiet' from 'those who are refractory, noisy and dangerous', and 'the clean from the dirty' were hardly more than pious aspirations.[4] As late as 1900 Grenada's Colonial Surgeon lamented that 'convalescing cases are associated with the noisy, excited and talkative', thus retarding recovery and even causing them to relapse. The asylum building did not 'admit of a classification of patients' and it was impossible to provide curative treatment, 'more especially a moral treatment'.[5]

The recently erected Trinidad asylum was criticised in 1863 by Thomas Murray, the medical superintendent, for its inability to classify patients 'in accordance with their special characteristics'.[6] George Seccombe's attempts to remedy the situation in the 1880s were constrained by the building's great limitations.[7] The new asylum at St Ann's presented opportunities, but these were overtaken by the rapid growth in patient numbers before a large new refractory ward and a villa for quiet, convalescent patients were opened in 1913.[8] In Barbados in 1863 the asylum claimed to provide some classification, with quiet, well-behaved people in the eastern wards, and the 'noisy, violent or obscene' in the western wards.[9] This overstated what was available, and in 1881 the lack of suitable classification was largely blamed for so many patients having to be placed in seclusion.[10] These problems worsened over the ensuing years due to the severe overcrowding.[11] The only semblance of classification offered was the frequent removal of violent, noisy or otherwise troublesome patients to the lower ward or the 'railed gallery'.[12] The opening of the new Barbados asylum in 1893 was expected to produce significant improvements, but these had hardly materialised a decade later.[13]

Despite early recognition of its importance, it remained constantly problematic to achieve an effective system of classification in the Jamaica asylum. Thomas Allen's predecessor Charles Lake observed in late 1863 that classification, 'so essential to successful treatment', was impossible; inoffensive, quiet lunatics were 'unavoidably associated' with the dangerous and refractory.[14] Allen himself quickly recognised its 'necessity' in the circumstances prevailing at the asylum, the absence of classification being 'the most serious and fatal barrier' to its being a curative institution. With only one ward for males and one for females, 'Rest and the preservation of order, and control' were 'nearly impossible'. The provision of separate wards and rooms for 'the tranquil, for the sick, for the helpless, for the unruly or violent, and the dirty' was 'indispensable'.[15] Allen continued to campaign strongly for building modifications and additions.[16] Although some of his initial demands were met, he was still complaining in 1886 that the layout of the asylum buildings prevented proper classification.[17] His replacement Joseph Plaxton seemed to have more success. For example, by 1891 Ward B for female refractory patients comprised four divisions, the women being separated according to mental condition with the 'worst' in the west division and the 'third worse' in the east.[18] Notwithstanding such developments, however, classification of patients was never really accorded the prominence sought by its advocates in the West Indies.

The situation regarding implementation of the 'non-restraint' system was rather more complex. By the mid-1840s the abolition of mechanical restraint had become the accepted standard for best practice in British public lunatic asylums.[19] The principle was disseminated throughout the empire but the degree of adherence varied greatly. An awareness of 'non-restraint' even permeated through to the Kingston lunatic asylum and the paraphernalia of chains, locks and stocks were virtually dispensed with in 1842–43, whilst all its other abuses continued.[20] However, similar measures were not taken in the other functioning asylums at that time. When it opened in 1842 the Antigua asylum was well equipped with leather gloves and straps, manacles, and strait-waistcoats.[21] By 1864 violent or suicidal patients were being confined in a 'Canvass Sack, which extends from the chin to the feet, laced behind, so as to prevent their doing mischief'.[22] Five years later Colonial

Office officials condemned the 'very frequent recourse' to restraint in the Antigua asylum, which was 'ill built and therefore ill-managed' and caused the lunatics to be 'very unruly'.[23] Mechanical restraint was used extensively in the first two decades of the Barbados asylum's operation. Seven patients were reportedly restrained by leather straps and wrist locks during 1862, including three women for periods exceeding ten weeks.[24] By 1869 the practice was said to have largely ended, though Thomas Allen found leather straps being occasionally used in 1875.[25] The female casebook and medical superintendent's journals confirm the periodic employment of leather straps, wrist locks, strong canvas suits, locked gloves, and even handcuffs, in the years between 1878 and 1891.[26]

The asylum doctors recruited from Britain brought with them their non-restraint convictions. This was one aspect where Thomas Allen did not have to introduce major change in the Jamaica asylum, though he remained keen to point out its complete absence of mechanical restraint at every opportunity.[27] The British Secretary of State for War, Hugh Childers, was suitably impressed when visiting in 1879. Complimenting Dr Allen on having apparently succeeded where 'almost every Asylum in Europe has failed', he wondered if 'the circumstance that you have to deal with an inferior race' may have contributed.[28] George Seccombe proclaimed his non-restraint credentials soon after taking up post in Trinidad, though it appears that he too inherited an institution where the steps had already been taken.[29] In British Guiana Robert Grieve was confronted in 1875 by a highly custodial institution, with an 'armamentum of manacles and straight waistcoats'. He dispensed with them rapidly and claimed a progressive increase in the 'quiet and order of the wards' as a consequence. He maintained in 1881 that the asylum provided 'strong and clear evidence' of the success of the non-restraint system.[30]

Non-restraint was still practised in the Guiana asylum long after Grieve departed in 1885.[31] However, the principle was increasingly being breached elsewhere. Even during Thomas Allen's tenure in Jamaica there were instances in 1883 where locked leather gloves were employed for extended periods, ostensibly to prevent female patients from interfering with surgical treatments.[32] As the asylum became increasingly overcrowded and order difficult to maintain, the need for drastic measures became more pressing. Returns for the year 1897–98 show that five female patients were restrained on several occasions, by means of 'canvas camisoles with sleeves stitched to the sides'.[33] In the new Barbados asylum (opened 1893) an initial non-restraint policy did not survive long, as practical considerations of managing difficult patients took over. In the latter half of 1900 eight men were subjected to mechanical restraint, either in a strait-waistcoat or by having their wrists and ankles secured to the bedsteads.[34] In small island asylums, like those of Grenada and St Lucia, there was a marked upsurge in the use of straitjackets, manacles, and leather straps in the early twentieth century. Their constricted buildings and surrounds offered little scope for diversionary activities and medical officers and staff felt forced to adopt repressive measures in response to violent or disturbed behavioural manifestations.[35]

A less overt form of restraint was enforced seclusion in a locked room. Practitioners like Robert Grieve in British Guiana sought to avoid the use of seclusion, though accepting that it could sometimes be justified.[36] Like some others he regarded its selective, occasional use as consistent with a non-restraint policy. However, locked seclusion was open to considerable abuse, if used as punishment rather than for separating and calming agitated people. Dr Charles Massiah justified its use in the Grenada asylum on both grounds in 1878:

> I find that this treatment judiciously combined with bread and water remarkably useful in subduing and bringing to a proper state of mind refractory patients of whom we have several and I am also of opinion that the quiet cell is beneficial to any mania with or without violence which may coexist.[37]

Such an approach inevitably linked seclusion to imprisonment in the minds of many asylum patients. The Grenada asylum employed seclusion to a very great extent over a long period. In 1877, 19 people out of 31 in the asylum were subjected to periods of locked seclusion lasting up to four weeks.[38] In the early twentieth century, an increasing use of seclusion accompanied the rise in mechanical restraint. During 1914 there were almost 150 instances of 'seclusion under lock and key', mostly following misdemeanours like fighting or removal of clothes.[39]

Locked seclusion was imposed frequently in the old Barbados asylum, where the options for dealing with recalcitrant patients were severely limited by inadequate facilities and overcrowded conditions. In 1869, although there were no instances of mechanical restraint, 33 people out of an average 88 in the asylum were subjected at some time to seclusion 'under Lock and Key'.[40] Dr Charles Hutson, its former medical superintendent and now Poor Law Inspector, reiterated in 1884 that there was an 'excessive' amount of seclusion in the asylum, with at least 56 patients having been subjected to it over the previous year.[41] A year later, in highlighting the asylum's desperate state, he protested that during a six-month period, 32 people had been involved in 285 instances of seclusion, totalling 2,027 hours. He compared this to large asylums in England and the West Indies 'conducted absolutely without seclusion', proclaiming that he could 'never be silent' about the situation.[42] The eventual move to the new asylum initially brought a sharp reduction in the use of locked seclusion. However, within a few years the trend was reversed and the numbers of people subjected to it rose significantly. In the first half of 1907 there was an average 25 seclusions per month, each lasting between ten hours and 29 days. By 1912 the average monthly number had reached 35.[43] An unusually custodial asylum regime was being perpetuated.

Productive occupation was at the heart of the moral management system, as transported to the West Indies by Allen, Grieve and Seccombe.[44] Their key practical measures were outlined in chapters 4 and 5. Each articulated ambitious aspirations for what could be achieved by work schemes. Thomas Allen made clear from the outset his belief that there was 'nothing more important in the moral treatment of the Insane' than bodily and mental employment, which would divert attention from disordered thoughts and feelings.[45] Robert Grieve adopted an almost evangelical perspective, suggesting there were no 'medicaments' available 'whose efficacy exceeds that of labour'. He placed work firmly in the arena of medical treatment:

> Conducted, as medicine in hospitals always is, or ought to be administered, that is, under the direction and supervision of the medical authorities, labour holds in the Lunatic Asylum a foremost place in its pharmacopeia. As in the Asylums of England so in British Guiana, from the workshops and farms proceed the largest number of our recoveries.[46]

For George Seccombe the work schemes in Trinidad were central to his wider therapeutic vision: 'It is by employment, and by employment chiefly, that we may hope to render these patients useful members of society.' Artisan trades like tailoring, boot-making and carpentry, along with the cultivation of land and stock rearing, were 'potent agents in the cure of Insanity'.[47] He maintained consistently that 'industrial' employment increased the number of recoveries.[48] Leonard Crane, Trinidad's energetic surgeon-general, advised the island's governor that Seccombe's views on employment as a 'curative agent' were 'in strict accordance with the modern treatment of insanity'.[49]

As the employment schemes progressed, their therapeutic objectives were increasingly accompanied by assertions regarding contribution to the economics of asylum management. Medical superintendents would annually submit detailed returns of the different agricultural, industrial, and domestic occupations pursued in their asylums, including how many patients of each sex were engaged in them, the quantities of articles or foodstuffs produced, and their respective financial value.[50] Even small asylums, with limited employment opportunities, were expected to produce something and make inroads into operating costs. In larger institutions the intentions were more ambitious. As George Seccombe explained in 1886, he had endeavoured 'to increase, as much as possible, the number of industries, always having in mind those which tend to render the Institution self-supporting and decrease the cost of maintenance of the inmates'.[51] Thomas Allen had stressed early on that he anticipated considerable savings to be achieved through patient labour.[52] In 1872 he proudly claimed that it accounted for £650 in value during the preceding year, as well as a further sum added to the 'Patients Fund' which paid for amusements and treats. He believed that the value of the 'Industrial Occupations' would 'bear favourable comparison with any asylum in Europe'.[53]

The highlighting of savings achieved through patient labour inevitably drew criticism that asylum work schemes were exploitative. This was not a new complaint, for it had surfaced in relation to the old Kingston asylum when Mrs Ryan and others were accused of profiting from patient labour and resisting the discharge of people who were valuable workers.[54] Allegations were made in 1886 of exploitation occurring under Thomas Allen, in relation to skilled tradesmen like carpenters. An outspoken critic suggested that patients were being treated 'as slaves, for nothing save food and clothing'.[55] In the context of recent Caribbean history, the apparent reluctance of some patients to labour in fields or workshops may well have been related to unwillingness to work for no pay. In Barbados in 1896 it was argued that farm produce could be increased 'but the inmates as a rule are disinclined to work'.[56] In 1904 there was reported to be 'solid inertia' among the inmates, with many of the men being characterised as 'incorrigible loafers'.[57] The numbers of patients employed at the Barbados asylum hardly rose above 30 per cent (Figure 8.1).[58] Even in the model British Guiana asylum, where the figures reached nearer 80 per cent, Robert Grieve acknowledged that some were unwilling to work.[59]

Figure 8.1   Inside Barbados Lunatic Asylum (early twentieth century)
Source: Contemporary Postcard – Courtesy of the Barbados Museum and Historical Society.

One of the chief benefits of organised employment was its contribution to the imposition of order and decorum in the institution. Thomas Allen articulated how 'useful occupation' in the Jamaica asylum had valuable effects in 'diminishing a Tendency to violence, improving the bodily health, and increasing tranquillity and cheerfulness amongst the Patients'.[60] George Seccombe drew similar conclusions in Trinidad within two years of the introduction of 'various industries' employing a large proportion of people from both sexes. Not only was there individual gain, but the 'greater quietness' pervading the asylum increased its efficacy as a 'hospital for the cure of the insane'.[61] He had similar expectations regarding organised entertainments, such as asylum band concerts and dances, which patients looked forward to eagerly. Attendance could be subject to good behaviour and 'a line of conduct conducive to their recovery'. It might also act as a 'wholesome restraint' in 'chronic and troublesome cases', assisting in maintaining the necessary 'order and regularity'.[62]

Similar motivations lay behind the development of recreational and sporting activities at all the larger asylums. The most extensive and diverse range was provided at the British Guiana asylum under Grieve's stewardship. The concerts, dances, and sports days attracted high levels of patient participation, and outside visitors were encouraged to attend.[63] In Jamaica by the time Allen retired in 1886 organised leisure activities were in decline. His successor Joseph Plaxton oversaw a partial revival, establishing theatrical and musical entertainments as well as organised cricket matches against teams from the outside community.[64] A makeshift form of cricket had also been encouraged at the old Barbados asylum, and other sporadic amusements included music and dancing, parlour games, walks, sea bathing, and annual trips to the Garrison races.[65] A proper cricket pitch was provided at Jenkinsville in 1901. However, the asylum's lack of recreation options was frequently criticised.[66] The small island asylums never offered much scope for amusements for patients, apart from some games, and perhaps a reading room as in the Bahamas asylum on New Providence, adding to difficulties in maintaining a settled environment.[67]

The inculcation of Christian values through religious observance had an acknowledged place in a moral management system. The reverence, quiet and order encouraged in services could serve as a calming influence in the institution. Most West Indian asylums provided some form of weekly worship, normally based on the Church of England model.[68] However, despite the entreaties of people like Thomas Allen, none were equipped with a proper chapel as in British public asylums.[69] Services were usually held either in rooms intended for other purposes or in the open air. Only in the British Guiana asylum did significant numbers go to Sunday services, in the dining room. In late 1884 almost 300 people were attending each week, out of just over 400 patients.[70] Around the same period only about a quarter of patients at the Jamaica asylum were going to services.[71] In Barbados the numbers also remained relatively low.[72] Attendances may well have corresponded less to levels of religious observance than adherence to the Church of England as distinct from Methodist, Baptist, Moravian or other dissenting Christian sects.[73] The absence of provision for alternative forms of Christianity was doubtless related to concerns about the alleged links between African-influenced religious practices, revivalism, fundamentalism and insanity, particularly in Jamaica.[74] Asylum authorities wanted to ensure that religion was a tranquillising element in the institution, rather than risk the opposite.

Medical interventions

In Britain and its empire, from the mid-nineteenth century, lunatic asylums were increasingly being promoted as 'hospitals for the treatment of the insane', a description readily adopted by George Seccombe in 1884.[75] Treatment that was strictly medical or surgical in character, however, played a relatively minor part in the asylum regime. The available information for West Indian asylums emanates primarily from Barbados and British Guiana, and indicates that only a limited range of medicines were in regular use. Some were primarily intended to treat physical conditions, but had concomitant effects on mental symptoms, whilst a few others were specifically aimed to address mental phenomena. Their main therapeutic effects were generally exhibited more in behavioural control than curative treatment.

One of the more commonly utilised drugs was potassium bromide, which was primarily a treatment for epileptic fits. People who suffered from epilepsy were frequently admitted to lunatic asylums because of the unpredictable or violent outbursts that could attend the condition. Potassium bromide proved effective in reducing the numbers of fits as well as helping to control the behaviour. Charlotte C. was admitted to the Barbados asylum in July 1879, her family having become unable to cope with her 'furious' behaviour after epileptic attacks. Dr Albert Field reported having witnessed her fits three months earlier and took action: 'I gave her large doses of Bromide of Potash, & she has not had any fits since.'[76] The treatment proved similarly effective with William C., admitted in 1878 after regular fits accompanied by florid delusions and violence. Following treatment with potassium bromide he recovered sufficiently to return home, though was subsequently readmitted and discharged again.[77]

Potassium bromide became increasingly used in British asylums as a tranquillising agent.[78] Dr Charles Massiah in Grenada recognised its efficacy in 1878, having found it of 'marked benefit in several instances of Acute Mania'.[79] Grieve employed it in the British Guiana asylum in the case of J.C., a Coloured man from Barbados suffering from 'syphilitic insanity' who, on admission in 1881, was restless, excited, incoherent, and experiencing auditory and visual hallucinations.[80] He utilised it, along with various other drugs, in several cases of 'general paralysis of the insane' featuring highly excited, overactive, destructive behaviour.[81] Bromide of potash was used similarly by Dr Field in Barbados, as in the case of an acutely manic White woman, Miss N., in February 1891.[82] In one instance in 1889 the medicine was disguised in some rum to ensure it was taken by Patrick M., a newly admitted large, aggressive and excited seaman from Newfoundland.[83]

Chloral hydrate, though nominally a narcotic to induce sleep, was another drug widely employed as a tranquilliser in British asylums.[84] It was in regular use for both purposes in the Barbados and Guiana asylums, and presumably throughout the West Indies. Grieve utilised it either on its own or in combination with other drugs.[85] The Barbados asylum records provide numerous examples of its use for quelling restless, excited, manic behaviour accompanied by sleeplessness. Sarah

Elizabeth K., a young Black woman aged 19, was admitted in 1876 raving and shouting both day and night. Several doses of chloral had little effect before Dr Hutson discovered she had been spitting it out.[86] Samuel A., a thin 'diminutive mulatto' of about 16, was admitted in a manic state in 1878. Three days following admission he was, according to Hutson, 'noisy & wandering about incessantly – saying nothing rational – ordered dose of chloral at bedtime'. After a few days of the treatment Samuel was 'wonderfully better', but relapsed a week later. He recovered after more chloral, remaining settled until sent out on leave at the end of August.[87] The treatment did not always prove quickly effective, however. In late August 1883 three male patients were simultaneously 'reported sleepless & noisy, altho' they had Chloral'.[88] Other medicines employed occasionally included digitalis, hyoscyamus, cannabis indica, sulphonal, strychnia, morphia, quinine, and potassium iodide, all utilised primarily to tranquillise over-excited patients.[89]

Evidence for the use of other forms of physical treatment is limited, though they were certainly employed. Various water treatments like cold applications to the head, warm or cold baths, plunge baths, shower baths, and douches were in common usage in English public asylums by 1850. Although all were claimed to have therapeutic benefits, there were clear elements of punishment and threat.[90] These aspects became prominent during the Kingston asylum scandal, where Dr Scott sought to explain 'tanking' in a context of the therapeutic use of cold baths and showers. Even after his public disgrace he continued to maintain that the 'judicious employment of a general douche, jet or any other description of Bath' was 'an important agent in effecting an amelioration of their condition'.[91] Scott's arguments were echoed in Trinidad in 1863 by Thomas Murray, delighted with his asylum's new shower baths, which constituted 'the most powerful means of controlling violent paroxysms'.[92] Adam Nicholson, medical officer of the Antigua asylum, was equally convinced of the value of frequent shower baths for those 'labouring under Acute mania'. He found that 'baths of surprise' had a 'very salutary effect', though insisting they were never used as a form of punishment.[93]

The employment of a stomach tube or pump to force-feed unwilling patients was frequent practice in the Barbados asylum, for refusal of food on account of severe depressive inertia,[94] delusional beliefs about being poisoned,[95] or gross overactivity associated with acute mania.[96] Dr Field was clearly an enthusiast for the treatment, to judge from the increasing regularity of its utilisation. In some cases force-feeding was employed at least daily over an extended period,[97] none more so than for Elvira S. who was subjected to it continuously for more than two years.[98] For other patients the mere threat or sight of the stomach tube proved sufficient to induce them to resume eating.[99] Increasingly, newly admitted people who showed any reluctance to eat were fed by tube almost immediately, with little prior attempt at persuasion.[100] It is difficult not to conclude that force-feeding and its associated paraphernalia were employed to represent the power of the institution and its doctor, and to encourage docile acquiescence in the regime.

Attendants and nurses[101]

> The difficulty of obtaining steady and efficient attendants is one that has been mentioned frequently ... Comparing asylum work here with that in England, the difference lies in the kind of attendants more than in the patients. Asylum inmates are as manageable in British Guiana as in Europe, Asylum attendants are much less so.[102]

Whilst it was the medical superintendent who determined policies and oversaw practices in the asylum, the people who exerted the greatest influence on patients' lived experiences were the staff with whom they interacted on a daily basis. For John Conolly, the foremost British champion of non-restraint and moral management, the attendants were 'the most essential instruments' for the effective operation of a lunatic asylum.[103] However, reform-minded medical men frequently lamented the educational and social limitations and the alleged unreliability of their subordinate staff. Similar complaints were commonly made by British-trained medical superintendents working in colonial lunatic asylums.[104] In the West Indies, perceptions about staff inadequacies were further complicated by distinctions connected to race and class.

Nurses and attendants provided the direct care and management of patients. The work was arduous, exacting, and at times dangerous. The requirements placed on them were considerable and comprised menial domestic tasks, direct personal care, supervision, encouragement, and risk prevention. Their working day was normally ordered from sunrise until sunset. At the Antigua asylum in 1842 the routine included cleaning rooms, washing the patients, checking and reporting on their 'evacuations', giving medicines, supervising meals and exercise, and then locking them back in their rooms at night. Even at this early date certain standards of conduct were anticipated. The attendant's 'whole time' was supposed to be 'devoted to the Patients'. He or she had to 'pay constant attention to their food, dress, occupation, exercise, amusement, and general conduct', preventing 'impropriety of manners and language'. They were 'never to punish the insane for their misconduct, but they are at all times to endeavour to put a stop to it, and to prevent quarrels and violence'. They were not permitted to use 'violent or intemperate language' or speak to patients in a 'loud or scolding manner', but rather 'endeavour to be calm and forbearing'.[105]

The standards laid down in Antigua certainly had not applied in Jamaica at that time. The unrestrained verbal, physical, and even sexual abuse of patients carried out by some nurses and 'labourers' contributed to the appalling situation that developed in the Kingston asylum.[106] In the aftermath of the crisis and the public enquiry, detailed regulations were produced for use in the new asylum and approved by Governor Edward Eyre.[107] Day attendants and 'servants' had to 'punctually and cheerfully obey' the orders of the head male attendant (or warden) and head female attendant (matron). Their duties were closely prescribed, as was their conduct; breaches of discipline or 'unfaithfulness' were punishable by dismissal.[108] Interactions with patients were expected to combine control with consideration and respect:

> They shall on no account at any time attempt to deceive or terrify the patients or attempt to irritate them by mockery, mimicry or wanton allusions to any thing in their personal appearance or past conduct. They shall exercise the greatest vigilance over the Patients and always watch carefully their behaviour at all times during the day. They shall never manifest vindictive feelings towards a Patient, but forgive all petulances, abusive language, or sarcasm, treating with equal kindness those who give the most, and those who give the least trouble . . .[109]

These regulations preceded Thomas Allen's arrival by a year, but his reforms were needed for the principles to be properly implemented and staff encouraged to become role models.[110]

By the early twentieth century expectations on staff were becoming rather more sophisticated, to judge from printed regulations for the Barbados asylum.[111] All had to sign firm undertakings regarding misconduct and its consequences:

> I acknowledge the right of the Superintendent to suspend me without warning for acts of unkindness, harshness or insolence, violence to patients, disobedience to orders, transgression of rules of negligence, also for intemperance within the precincts of the Asylum.

There was a new emphasis on the 'most important character' of the attendants' duties and their 'very great responsibilities', acknowledged as 'often irksome and disagreeable' but demanding 'unceasing kindness, caution, activity, and vigilance'.[112] Although domestic tasks still formed an important part of the work, the therapeutic elements had come more to the fore. They were expected to 'study carefully the temper and character' of their patients, and be 'gentle, firm and persevering' in encouraging them to eat, work, and join in recreation. They had to watch them carefully, 'endeavour to win their confidence' and gently correct 'slight acts of insolence or mischief'. They were cautioned not to respond roughly to 'petulance, irritating complaints, abusive language, objectionable habits or threatening or irritating demeanour', although they were expected to prevent violence and fights.[113] Whilst strict discipline and a rigid daily timetable still prevailed, the humane, consoling aspects of the role were now more prominent.[114]

The development of moral management systems brought important role changes. The work programmes required people able to supervise patients in various trades or tasks. In Jamaica, Allen established early on that the attendants should be 'Artizans', participating in the asylum's 'industrial occupation' and providing instruction.[115] In British Guiana, by 1881 the attendants were divided into 'first class' and 'second class', with commensurate salary levels. The first class included men with a skilled trade such as printer, tailor, shoemaker, or carpenter, whilst farm workers and gardeners were placed in the second class.[116] George Seccombe in Trinidad sought attendants and nurses who could 'direct or assist in the industrial departments', or promote recreational and musical activities.[117] He argued forcefully that the development of 'industries' required 'attendants for their exclusive direction', but he still had to justify the expenditure to retain a mason attendant in 1886.[118] The surgeon-general recognised that 'artificers' provided the 'skilled superintendence' needed for patients renovating the asylum's buildings and drainage system.[119]

Extensions to the roles of attendant and nurse brought a degree of professionalisation. A visible representation was the growing expectation that staff should wear smart uniforms. In British Guiana, male and female attendants wore a brown linen uniform until 1881, after which the men's dress was changed to a white military-style jacket with red facings.[120] In Trinidad, at this time, female staff had uniforms but not the males. Seccombe considered this unhelpful in managing patients who needed to be 'orderly clad' as part of their treatment. It was impossible if the staff in charge of them are 'models of anything but neatness and order', their clothing always liable to spoiling by 'an unruly patient'. He arranged for the men to receive a uniform.[121] As usual the Barbados authorities were much slower in addressing the issue. Strong representations about the need for uniforms were made in 1903 by the Poor Law Inspector:

At present on entering an airing-court it is not always possible to distinguish the attendants from the inmates unless from personal knowledge, and the influence of the attendants over the patients would be greatly increased by this simple expedient.[122]

After a lengthy delay, 95 'suits of uniform for the attendants' were manufactured by patients' labour.[123]

The recruitment of staff regarded as suitable by medical superintendents was a perennial difficulty. The senior positions normally went to White people, often from England. Thomas Allen brought the Jamaican asylum's warden and matron from the Lincolnshire county asylum, in 1863.[124] A subsequent matron, Ann Blake, was selected by the Colonial Office in 1888 from a similar post at St Pancras workhouse in London; she remained until 1906 when succeeded by Lilla Parnther, an American trained nurse.[125] An English steward and matron, Mr and Mrs Stevenson, began at the British Guiana asylum in early 1881.[126] In Trinidad Charles Bizzell was recruited from Britain as head male attendant in 1878, and Miss Flood was appointed female head attendant in 1887 after six years of 'training' in various English asylums. They remained until 1904 and 1912 respectively, to be replaced by other experienced people from England.[127]

Most of the subordinate staff in West Indian asylums were Black or Coloured. Governor Darling of Jamaica in 1861 noted the 'African temperament' of the 'subordinate Servants' at the Kingston asylum and their consequent difficulties in managing violent lunatics.[128] The serious misconduct of some staff, albeit sanctioned by the matron Mrs Ryan, raised important questions about recruitment and working conditions. Dr James Scott thought the nurses ought to be 'a more respectable and intelligent class of persons', paid more than their current nine shillings per week.[129] The commission of enquiry agreed that poor remuneration discouraged 'all but persons of inferior caste and qualifications'. The nurses and servants were characterised as 'few in number, inferior in intelligence, and greatly more devoted to their own interests than the welfare of their charge'.[130]

The move to the new Jamaica asylum brought little improvement in pay and conditions apart from the inclusion of cooked meals.[131] It also did not resolve other fundamental problems. When Thomas Allen began implementing reforms he faced direct opposition from resentful attendants; even some appointed by him proved 'quite unworthy of trust'.[132] However, within a few months he claimed marked improvements. Many had become positive role models for patients, and he praised their 'good conduct, creditable demeanour, and willingness to learn'.[133] Allen's optimistic assessment was short-lived. By 1870 he was complaining about the 'independence of servants and their utter indifference to all the obligations which usually attach to such a class', and their defiance when threatened with dismissal for 'an irregularity'. He reiterated these views again in 1883.[134] Allen adopted the novel solution of dealing with 'serious breaches of discipline' by either suspending people or dismissing and later re-engaging them, regarding this as preferable to recruiting new, untrained staff.[135] His successor Joseph Plaxton found great difficulty in employing people 'likely to become efficient' nurses and attendants. It seemed a 'hopeless task endeavouring to raise and retain a skilled staff', and he feared that the asylum could consequently never reach a 'high standard of efficiency'.[136]

Medical officers were rarely satisfied with the quality of people recruited as attendants and made their opinions clear. Robert Grieve complained in 1881 that it was impossible to get attendants prepared to learn their duties well enough, attributing it to the 'indolent habits' of British Guiana's people.[137] The 'labouring classes', he claimed, disliked regular, steady employment and believed that 'government work is synonymous with perfunctory work'. A 'moral code' that legitimised theft 'when there is want' brought 'special obstacles' in maintaining discipline.[138] He returned regularly to these themes, noting in June 1883 that, although well-trained attendants were an 'essential element' in enabling patients to enjoy a degree of freedom, 'everyone connected with this colony' knew the 'special difficulty ... of creating such a class'.[139] In 1885 he again bemoaned 'the difficulty of obtaining intelligent and trustworthy attendants'.[140] Grieve was greatly exercised about the influence of obeah, particularly among Creoles.[141] He had no doubt that it was prevalent and 'interferes with discipline'. He thought every attendant believed in 'this horrid Obeahism'. Many were 'intelligent and respectable members of their class' who attended Christian churches, but none were 'free from the dread of it'. It affected all aspects of their conduct, preventing 'one of their own class' being able to exercise authority over others and making them 'deceitful to their superiors'.[142]

In Trinidad George Seccombe was equally critical of the calibre of staff. He observed in 1885 that, of the many applicants for posts as attendants and nurses, 'few are found worth a trial, and the services of fewer still are found worth retaining'.[143] Things were no better two years later, for 'little dependence' could be placed on the 'excellent testimonials' that accompanied the numerous applications. Although the men worked well for a few months, having once 'settled down comfortably in a Government appointment' they became 'careless and indifferent'.[144] In 1892 he condemned their lack of *'esprit de corps'*, for only eight attendants or nurses out of 46 had completed more than eight years' service.[145] He observed that most were found unsuitable within their probationary period, and protested regularly that low salary and poor conditions were not attracting the 'right class of persons' for the 'often arduous and unpleasant' work.[146] He anticipated that provision of accommodation and board in the new asylum might attract a 'better class' of staff.[147] However, in 1906 Seccombe was again complaining that, of the 'shoals of applicants' for employment, few were suitable and many did not complete their probation.[148]

Dr Charles Hutson in Barbados acknowledged in 1889 that it was hard to attract 'decent officers' when the work was 'not sufficiently remunerative'.[149] He was acutely aware of the steady stream of dismissals of asylum staff for offences including neglect of duty, violence toward patients, threatening behaviour, drunkenness, and theft. At least seven attendants (male and female) were dismissed during 1883, and six in 1889.[150] Ill-treatment of patients accounted for significant numbers of dismissals from the asylums of both Trinidad and British Guiana, although more were ascribed to negligence or general unsuitability for the attendant role.[151] A succession of dismissals at the Grenada asylum in the early twentieth century, for violence to patients as well as neglect of duties, confirms that perceived problems of staff suitability were quite widespread.[152] Poor pay and conditions reflected the low status of asylum workers, ensuring many were not equipped to deal with difficult and demanding situations. An anonymous female patient exposed some serious problems in the Jamaica asylum in 1905, alleging that Dr Williams (the medical superintendent) was wrongly blaming the 'poorly paid' nurses, who became 'irritable and discouraged'. She sympathetically alleged that there were instances of patients not getting their proper food because 'some nurses are hungry'.[153]

A proportion of the attendants dismissed were victims of rough justice. Some of the violence shown toward patients arose in circumstances where staff members had been physically attacked. In the Jamaica asylum in 1876 a nurse was summarily dismissed for not having shown 'sufficient forbearance' during an incident where she wrestled with a patient, who received an injury on the forehead from the nurse's keys.[154] In 1893 two nurses, Jane Mendes and Sarah Ann Wynn, were dismissed after an incident involving a patient named Dennis, known for being violent and a 'bad biter'. They had retaliated after Dennis struck Jane Mendes over the back with a broomstick and tried to bite her.[155] In Barbados in 1878 an attendant named Brookes was discharged after a fight in which Alfred N., a White seaman, received a black eye, even though the patient admitted that he struck the first blow and it was his fault.[156] The experienced nurse Julia Paul was almost dismissed in 1888 after she hit Mary N., 'to make her let go of her hold of her breast'; she was only reprieved by the governor on condition of accepting a reduction in yearly wages from £12 to £10.[157] As shown in Chapter 7, staff could be injured by agitated, disturbed patients who might be expressing their discontent against the asylum authorities. Although violent responses by staff were often explicable they tended not to be viewed sympathetically, for dismissed attendants and nurses could soon be replaced.

Lunatic asylum attendants in the Caribbean, as elsewhere, had much to contend with. The nature of historical records is that inadequacies and malpractices receive attention, whilst routine good, humane practice passes relatively unnoticed. However, even in the turmoil of the Kingston asylum around 1860, the conduct of certain staff received praise. Ann Pratt observed that some were 'good and humane', not participating in abuses.[158] Henrietta Dawson named several nurses and 'labourers' who treated her kindly.[159] Dr Thomas Allen occasionally spoke positively about staff in the new asylum. He acknowledged in 1864 that 'willing Attendants' did their best 'to cheer the sad, to give a turn to moroseness, violence, and destructiveness, to correct filthy and disgusting habits, and to minister to the wants of those sunken to the lowest state of bodily and mental decrepitude'.[160] In 1876 he cautiously reported their conduct as 'upon the whole good'.[161] Dr Williams in 1905 was more positive, referring to their cheerful performance of 'arduous and irksome duties...notwithstanding great risks to life and limb'.[162] Elsewhere, there were occasional references to the good or satisfactory conduct of staff, as in Grenada in 1883,[163] and British Guiana in 1881 and 1897.[164] George Seccombe in Trinidad wrote in 1884 of their 'general good conduct', despite the 'necessarily uncongenial' duties.[165] In Barbados, Charles Manning in 1908 described the 'present staff of assistants' as the best he ever had. They had 'well and cheerfully' upheld the 'discipline of the institution', and he commended their willingness to undertake extra work in addition to their regular duties.[166] However, such relatively rare praise hardly did justice to the significant contribution of under-paid, overworked, poorly trained attendants and nurses.

Conclusion: Unfulfilled ambitions

The principles and ideas of moral management provided a paramount influence on the regimes of West Indian asylums, but their translation into actual practice proved problematic. In the smaller islands low patient numbers were seen not to justify expenditure on an institution providing the necessary infrastructure. In the larger, more advanced colonies attempts to implement and sustain the main elements were usually impeded by some combination of defective buildings, insufficient surrounding land, lack of facilities, and inadequate staffing. British-trained asylum doctors, however well intentioned, were unlikely to achieve their aspirations with the restricted resources that colonial governments were able to allocate. The key elements of moral management could only be partially established. The classification of patients was never properly effected, due mainly to building inadequacies. Although non-restraint was upheld to a large extent, it was compromised by the extensive use of locked seclusion. Organised religious worship was little more than makeshift, there being no asylum chapels as routinely provided in English asylums. The only aspect where apparent success was achieved was in the quite extensive employment of patients.

The anticipations of medical superintendents that patient employment would of itself bring about many recoveries were always over-ambitious. In practice the main therapeutic benefit of the work programmes was their contribution to instilling order and tranquillity in the asylums. The more medically orientated aspects of treatment, with their punitive elements, served a similar function whilst confirming the power and control exercised by the medical superintendent and the institution. The key instruments for enforcing that control were the staff of attendants and nurses. The high expectations placed upon them were invariably difficult to meet, bearing in mind their poor remuneration, low status, and insufficient numbers. Subjected to constant criticism for their limitations, they became a convenient focus of blame for various failures within the asylum regime.

Conclusion

The crisis that descended upon the Kingston Lunatic Asylum between 1858 and 1861 marked a watershed in the development of provision for mentally disordered people in the British colonies of the West Indies, and arguably beyond. The exposure of prolonged, entrenched squalor, abuse, and disregard for basic human decency in Jamaica demonstrated the existence of a huge gulf between aspirations for treatment of the pauper insane in enlightened post-reform English public asylums and the reality to be expected by individuals in the unfortunate colonies. The inadequate buildings, poor facilities, and deplorable conditions characteristic of early Caribbean institutions reflected their particular societies' slow and painful emergence from the era of enslavement, as well as endemic elite perceptions of the lowly status and entitlements of the Black and Coloured poor. The situation in Jamaica was even exacerbated by the unhelpful inheritance of an early established public hospital and lunatic asylum in Kingston, for its confined and outdated buildings were in no respect adaptable to the altered, increased demands of a post-emancipation society.

The scandal of the Kingston asylum brought into sharp relief the nature of any conception of imperial 'civilising mission'. The outrage expressed in the British press about the situation reflected a strong current of public opinion which held that colonial officials and institutions should not be permitted to treat formerly enslaved, vulnerable, and impoverished people in this manner.[1] For *The Anti-Slavery Reporter* it was 'a state of things disgraceful to any civilized community'.[2] *The Lancet* called for 'justice to the suffering poor'.[3] The active engagement of the English Commissioners in Lunacy, operating well beyond their normal sphere of responsibility, showed how unacceptable the situation in Kingston was perceived to be. Some British government credibility was reasserted with the enforcement of a public enquiry in Jamaica. The later empire-wide investigations ordered by the Secretary of State, at the behest of Colonial Office officials, and the resultant published report on hospitals and lunatic asylums in 1864, were at least a practical attempt to reinforce the responsibilities of colonial governors and their staffs in regard to health and welfare provision in their territories.[4] The annual submission to London of endless reports from all round the empire, however, may have had little real positive effect beyond ensuring the continuous accumulation of data about the defective provision and conditions remaining in many colonial places and spaces.

The vestigial paternalistic inclinations of British colonial stewardship in the Caribbean were constrained in many instances by an alleged paucity of available resources, though in some cases there was a distinct lack of preparedness to expend the funds available. The unremittingly meagre lunatic asylum provision in the Leeward and Windward Islands, as in British Honduras, meant the perpetuation into the twentieth century of institutional conditions acknowledged by colonial officials and medical men as little short of disgraceful. Despite its long history as an important colony, and close ties with Britain, the Barbados legislature successfully resisted providing any more than the absolute minimum for as long as it possibly could. It required the exposure in 1875 of scandalous conditions, not dissimilar to those in Jamaica more than a decade earlier, to escalate demands for a new lunatic asylum more in keeping with acceptable modern standards and the status of the colony. The critical situation prevailing throughout the 1880s was met by a level of official prevarication that confirmed the indifference of Barbados's White elite to the plight of disadvantaged people, particularly if Black or Coloured. The eventual provision of the new asylum at Jenkinsville, with all its limitations, only followed a reluctant acceptance that matters had become intolerable by any contemporary standards.

Ideas of social responsibility, encouraged by reformist elements in the Colonial Office, proved rather more influential in Jamaica, Trinidad, and British Guiana. In Jamaica, the shock of the Kingston scandal had ensured the opening of the new asylum at Rae's Town, even though it was hardly in a fit state of readiness. In Trinidad and Guiana, institutions for the insane continued to demonstrate for some time many of the defective features and shortcomings revealed in Jamaica and Barbados. In each case the necessary modernisation of both fabric and practices came only after advice was accepted from London to recruit a medical man with sound experience in the English asylum system. The arrival of Thomas Allen, Robert Grieve, and George Seccombe represented the assertion of British hegemony in colonial mental medicine, thereby setting aside the discredited efforts of locally based practitioners. Allen, Grieve and Seccombe brought sufficient knowledge and experience to enable quick identification of the problems in the asylums and of the steps required toward implementation of remedies. Their demonstrable expertise, along with varying degrees of determination, dynamism, personal ambition and charisma, were utilised to pressurise their respective colonial governments into providing significant material resources. Eventually, the energies expended over a prolonged period led to their exhaustion and disillusion, partly contributing to the subsequent decline that became apparent in each institution.[5]

The British doctors took up their posts armed with the principles of 'moral management'. The accordance of its practices with conceptions of the 'civilising mission' was almost uncanny. The intent to bring civilisation to the colonial insane was rarely expressed directly, apart from Thomas Allen's repeated crude assertions regarding the 'savagery' of Black Jamaicans. Nevertheless, the overt attempts to create order, discipline, structure and system in the large asylums amounted to a comparable process of bringing about tranquillisation of those perceived to be disadvantaged and inferior. Moral management, however, was shown to have serious limitations, especially if some of its key essentials were not implemented, due to the constraints of inappropriate location, unadaptable buildings, inadequate staffing, or insufficient funds. In particular, suitable arrangements for the classification and separation of patients, according to behavioural presentations or types of disorder, remained largely elusive. With notable exceptions, mechanical restraint continued in extensive use, and locked solitary confinement even more so. The result was the perpetuation of colonial asylum regimes of a particularly custodial and repressive nature.

The element of the moral management regime implemented with most success in the British Caribbean asylums was organised labour. The establishment of a range of employments, both inside and outside the asylums, proved to be the crucial factor in turning around the fortunes of the institutions. The respective medical superintendents demonstrated a remarkable degree of entrepreneurial ability in devising the various arrangements, some of which were quite sophisticated. The system exhibited a number of manifest advantages. It could be directly therapeutic for individual patients. It helped to promote discipline, order and tranquillity in the asylum. It assisted in preparing people to resume their assigned places in colonial societies where their role almost invariably was to labour. Finally and importantly, it contributed materially to the economy of the institution by reducing running costs. In the West Indian context, it can hardly have been coincidental that the arrangements bore a distinct resemblance to a well-managed plantation. Of course, these supposed advantages of the moral management system were confined only to the main Caribbean colonies. The asylums established in the lesser islands barely progressed beyond the basic, inadequate facilities so strongly criticised in 1864.[6] Even by the early twentieth century the central lunatic asylums in Grenada and Antigua could provide hardly more than the necessaries of life.

Although the smaller colonies suffered most from resource constraints, inadequacies in the extent and quality of provision for the insane were apparent throughout the region. This in turn reflected the general state of public services, where piecemeal measures designed to bring small, cosmetic improvements were normally the most that could be achieved. It appeared increasingly that the imperial government in London, represented through its colonial governors, was becoming ever less committed to the West Indian colonies. The region no longer represented the source of prosperity, prestige and strategic significance that had formerly been the case. If anything, it had become an economic, social and political liability, at a time when many of the larger and newer colonies elsewhere offered untold potential. In these circumstances, it was unlikely that much contribution would be forthcoming from a reluctant imperial centre toward public and health services in the Caribbean, including those for mentally disordered people, and local resources were needed to address more urgent priorities. Benign neglect was now the most that could be expected from the imperial government and its representatives.

The new reality was most apparent in measures taken to address lack of capacity in the lunatic asylums. As in Britain and in most of its other colonies, the problems associated with overcrowding became perennial in all the Caribbean asylums, large and small. Demand for asylum places constantly outstripped supply. The consequences went well beyond the implications for individuals' comfort, health, and safety. As the larger asylums' populations grew and their buildings multiplied, the difficulties of maintaining any semblance of a therapeutic regime were exacerbated. The ideals of moral management were being steadily eroded well before 1914. In Jamaica these trends were reinforced in the 1890s by the cynically pragmatic decision to expand on the Rae's Town site rather than construct a new asylum in another part of the island. By the time the new buildings opened in the early twentieth century, it was clear that the asylum was now regarded as an institution for cheap, large-scale incarceration and containment, rather than a facility geared toward conceptions of therapeutic treatment and promoting recovery.[7] In Trinidad, despite its fine buildings, the new asylum rapidly became overcrowded and began displaying similar phenomena. Ultimately, the moral management regime deteriorated into little more than a set of mechanisms for controlling a large inchoate group of dysfunctional people.

For much of the period between 1838 and 1914 the incidence of admission to a lunatic asylum within the West Indies, in comparison to overall population, remained well below that in Britain, indicating a relatively high threshold for committal. In general, taking account of the limited accommodation in asylums, the evidence suggests that most of those admitted had exhibited symptoms or behaviours suggestive of a high level of distress, distraction, or disturbed functioning. Violence or threat to other people or property, serious acts of self-harm, or severe self-neglect, formed the frequent precipitants of committal to the asylum. Economic marginality, material deprivation, and crude, insanitary housing all operated as underlying constant elements in the chain of circumstances contributing to collapse of individuals' mental health in the attenuated economic and social state of the nineteenth and early twentieth century British Caribbean. As was the case in Britain and elsewhere, complaints from medical superintendents about the deteriorated physical state of the patients they had to admit became increasingly regular. Not without reason, doctors would lament that some people had been sent in to die or to be maintained in a state of basic existence, adding to the accumulating numbers of chronic patients for whom 'modern treatment' had not succeeded.

The whole argument about public lunatic asylums, and particularly colonial asylums, as instruments of social control remains contentious and contested. In the West Indian context, the notorious Alexander Bedward case has provided material to fuel the debate. There was certainly evidence of committal being used as a means to remove an awkward vocal critic of the political and religious status quo, especially with his second admission to the Jamaica asylum in 1921.[8] On a wider level, significant numbers of people were admitted whose behaviours in public had offended against standards of propriety or otherwise caused distress or discomfort to 'respectable' citizens. The asylums provided the means both to remove them and to attempt to inculcate more quiescent and socially acceptable forms of conduct. Indeed, with the various factors that had directed people into them, the asylums became populated with an amorphous and challenging range of people. For the most part they were doubly stigmatised, the ravages of mental disorder compounding the manifestations of enforced racial disadvantage. Yet, the stories of the Kingston asylum crisis and of the Barbados asylum in its years of turmoil showed many examples where ordinary Black and Coloured people, despite their alleged insanity, demonstrated personhood, 'agency', and a capacity to resist. Such people were far from prepared to accept the subservient and marginal roles assigned to them within the colonial dispensation.

Notes

Introduction

1. For example – L. Bell, *Social and Mental Disorder in Sub-Saharan Africa: The Case of Sierra Leone, 1787–1990* (New York, London: Greenwood Press, 1991); C. Coleborne, *Madness in the Family: Insanity and Institutions in the Australasian Colonial World, 1860–1914* (Basingstoke: Macmillan, 2010); W. Ernst, *Mad Tales From the Raj: The European Insane in British India, 1800–1858* (London and New York: Routeldge, 1991); L.A. Jackson, *Surfacing Up: Psychiatry and Social Order in Colonial Zimbabwe, 1908–1968* (Ithaca and London: Cornell University Press, 2005); R.C. Keller, *Colonial Madness: Psychiatry in French North Africa* (Chicago and London: University of Chicago Press, 2007); J. McCulloch, *Colonial Psychiatry and the 'African Mind'* (Cambridge: Cambridge University Press, 1995); S. Mahone and M. Vaughan (eds.), *Psychiatry and Empire* (London: Palgrave Macmillan, 2007); J.H. Mills, *Madness, Cannabis and Colonialism: The 'Native-Only' Lunatic Asylums of British India, 1857–1900* (Basingstoke: Palgrave Macmillan, 2000); L. Monk, *Attending Madness: At Work in the Australian Colonial Lunatic Asylum* (Amsterdam and New York: Rodopi, 2008); J.E. Moran, *Committed to the State Asylum: Insanity and Society in Nineteenth-Century Quebec and Ontario* (McGill: Queen's University Press, 2000); J. Parle, *States of Mind: Searching for Mental Health in Natal and Zululand, 1868–1918* (Scottsville, South Africa: University of KwaZulu Natal Press, 2007); J. Sadowsky, *Imperial Bedlam: Institutions of Madness in Colonial Southwest Nigeria* (Berkeley and London: University of California Press, 1999); N. Beng Yeong, *Till the Break of Day: A History of Mental Health Services in Singapore, 1841–1993* (Singapore: Singapore University Press, 2001). In addition to these books there are also numerous journal articles, some of which will be referred to subsequently.
2. Sir A. Halliday, *A General View of the Present State of Lunatics, and Lunatic Asylums, in Great Britain and Ireland, and in Some Other Kingdoms* (London: Thomas and George Underwood, 1828), pp.79–80.
3. F.W. Hickling and R.C. Gibson, 'Philosophy and Epistemology of Caribbean Psychiatry', in F.W. Hickling and E. Sorel (eds.), *Images of Psychiatry: The Caribbean* (Kingston, Jamaica: Stephenson's Litho Press, 2005), pp.75–93 (quote, pp.75–6).
4. Hickling and Gibson, 'Philosophy and Epistemology', p.91.
5. F.W. Hickling and R.C. Gibson, 'The History of Caribbean Psychiatry', in F.W. Hickling and E. Sorel (eds.), *Images of Psychiatry*, pp.15–41.
6. I have previously discussed the custody/cure dichotomy in L. Smith, *'Cure, Comfort and Safe Custody': Public Lunatic Asylums in Early Nineteenth-Century England* (London: Leicester University Press, 1999).
7. R.C. Keller, *Colonial Madness: Psychiatry in French North Africa* (Chicago and London: University of Chicago Press, 2007), pp.5–11, 49–79; R. Keller,

'Madness and Colonization: Psychiatry in the British and French Empires, 1800–1962', *Journal of Social History* 35 (Winter 2001), 295–322.

8. B. Porter, *The Absent-Minded Imperialists: Empire, Society and Culture in Britain* (Oxford: Oxford University Press, 2004), pp.45–6, 228–29.

9. W.A. Green, *British Slave Emancipation: The Sugar Colonies and the Great Experiment* (Oxford: Clarendon Press, 1976, reprinted 1981), pp.84–94.

10. For discussion of moral management, see Chapters 1, 4, 5, and 8.

11. See Chapters 1, 6 and 8.

12. See Chapter 3.

1   Caribbean Institutions in Context

1. This was also true of Scotland and Ireland, where the asylum systems developed somewhat differently.

2. J. Andrews, A. Briggs, R. Porter, P. Tucker and K. Waddington, *The History of Bethlem* (London and New York: Routledge, 1997); C. Stevenson, *Medicine and Magnificence: British Hospital and Asylum Architecture, 1660–1815* (New Haven and London: Yale University Press, 2000).

3. J. Woodward, *To Do the Sick No Harm: A Study of the British Voluntary Hospital System to 1875* (London and Boston: Routledge and Kegan Paul, 1974).

4. L. Smith, *Lunatic Hospitals in Georgian England, 1750–1830* (London: Routledge, 2007).

5. W.L. Parry-Jones, *The Trade in Lunacy: A Study of Private Madhouses in England in the Eighteenth and Nineteenth Centuries* (London: Routledge and Kegan Paul, 1972).

6. R. Porter, *Mind Forg'd Manacles: A History of Madness in England From the Restoration to the Regency* (London: Athlone, 1987), pp.155–56; W. Battie, *A Treatise on Madness* (London: Whiston and White, 1758), pp.68–9.

7. M. Foucault, *History of Madness*, ed. J. Khalfa (London and New York: Routledge, 2006 – translation of 1972 edition), pp.44–77.

8. Foucault, *History of Madness*, pp.384–400, 503–8.

9. Porter, *Mind Forg'd Manacles*, pp.5–9, 110–11.

10. P. Bartlett and D. Wright (eds.), *Outside the Walls of the Asylum: the History of Care in the Community, 1750–2000* (London and New Brunswick: Athlone, 1999).

11. Porter, *Mind Forg'd Manacles*; A. Ingram, *Patterns of Madness in the Eighteenth Century, A Reader* (Liverpool: Liverpool University Press, 1998); A. Scull, *The Most Solitary of Afflictions: Madness and Society in Britain, 1700–1900* (New Haven and London: Yale University Press, 1993), pp.46–110.

12. Porter, *Mind Forg'd Manacles*, pp.184–87; Smith, *Lunatic Hospitals in Georgian England*, pp.141–49.

13. Battie, *A Treatise on Madness*, pp.68–9; J. Monro, *Remarks on Dr Battie's Treatise on Madness* (London: J. Clarke, 1758). Porter, *Mind Forg'd Manacles*, pp.207–8.

14. Porter, *Mind Forg'd Manacles*, pp.206–28; Smith, *Lunatic Hospitals in Georgian England*, pp.150–56; A. Digby, *Madness, Morality and Medicine: A Study of the York Retreat, 1796–1914* (Cambridge: Cambridge University Press, 1985).

15. Porter, *Mind Forg'd Manacles*, pp.6–7, 225; Foucault, *History of Madness*, pp.471–76, 482–91; A. Scull, 'Psychiatry and Social Control in the Nineteenth and Twentieth Centuries', *History of Psychiatry* 2 (1991), 149–69.

16. K. Jones, *A History of the Mental Health Services* (London: Routledge and Kegan Paul, 1972), pp.54–61; Smith, *'Cure, Comfort and Safe Custody'*, pp.20–6; British Parliamentary Papers (henceforth BPP) 1807, Vol. II, Report of Select Committee on the State of Criminal and Pauper Lunatics; 48 Geo. III, Cap.96, An Act for the Better Care and Maintenance of Lunatics, Being Paupers or Criminals in England, 1808.

17. These included Nottinghamshire and Bedfordshire (1812); Norfolk (1814); Lancashire (1816); Staffordshire and West Riding of Yorkshire (1818); and, Cornwall (1820).

18. Jones, *A History of the Mental Health Services*, pp.64–88, 101–8; Scull, *The Most Solitary of Afflictions*, pp.115–28.

19. BPP 1814–15, Vol. IV, Select Committee on the State of Madhouses; BPP 1816, Vol. VI, Select Committee on the State of Madhouses; BPP 1826–27, Vol. VI, Select Committee on Pauper Lunatics in the County of Middlesex and on Lunatic Asylums.

20. Digby, *Madness, Morality and Medicine*, pp.42, 62–3; Smith, *Lunatic Hospitals in Georgian England*, p.156.

21. R. Hunter and I. Macalpine, *Three Hundred Years of Psychiatry 1535–1860* (London: Oxford University Press, 1963), pp.648–52.

22. Smith, *'Cure, Comfort and Safe Custody'*, pp.227–31.

23. Ibid., pp.231–33; H. Martineau, 'The Hanwell Lunatic Asylum', *Tait's Edinburgh Magazine* (1834), pp.305–10. Ellis was subsequently rewarded with a knighthood – H. Warner Ellis, *'Our Doctor,' or Memorials of Sir William Charles Ellis* (London: Seeler, Jackson and Halliday, 1868).

24. L. Smith, 'The County Asylum in the Mixed Economy of Care', in J. Melling and B. Forsythe (eds.), *Insanity, Institutions and Society: A Social History of Madness in Contemporary Perspective* (London and New York: Routledge, 1999), pp.33–47.

25. R.G. Hill, *Total Abolition of Personal Restraint in the Treatment of the Insane: A Lecture on the Management of Lunatic Asylums and the Treatment of the Insane* (London: Simpkin, Marshall and Co., 1839).

26. Smith, *'Cure, Comfort and Safe Custody'*, pp.259–71; J. Conolly, *The Treatment of the Insane Without Mechanical Restraints* (London: Smith, Elder and Co., 1856).

27. Later Earl of Shaftesbury, the eminent social reformer.

28. The Metropolitan Commissioners had been formed under legislation in 1828 to licence and inspect private lunatic asylums in the London area – see Jones, *A History of the Mental Health Services*, pp.108–11; Scull, *The Most Solitary of Afflictions*, pp.129–32.

29. Scull, *The Most Solitary of Afflictions*, pp.155–64; *Report of the Metropolitan Commissioners in Lunacy to the Lord Chancellor* (London: Bradbury and Evans, 1844).

30. 8 & 9 Vic., Cap. 100, An Act for the Regulation of the Care and Treatment of Lunatics, 1845; 8 & 9 Vic., Cap. 126, An Act to Amend the Laws for the Provision and Regulation of Lunatic Asylums for Counties and Boroughs, and for the Maintenance and Care of Pauper Lunatics, in England, 1845;

Jones, *History of Mental Health Services*, pp.132–49; Scull, *Most Solitary of Afflictions*, pp.164–67.

31. Structured religious observance arguably formed a fourth element.
32. D.J. Mellett, *The Prerogative of Asylumdom: Social, Cultural and Administrative Aspects of the Institutional Treatment of the Insane in Nineteenth-Century Britain* (New York: Garland, 1982), pp.23–5.
33. Scull, *The Most Solitary of Afflictions*, pp.269–84, 365–70; P. McCandless, ' "Build! Build!" The Controversy Over the Care of the Chronically Insane in England, 1855–70', *Bulletin of the History of Medicine* 53 (1979), 553–74; Mellett, *The Prerogative of Asylumdom*, pp.28–41; J. Melling and B. Forsythe, *The Politics of Madness: The State, Insanity and Society in England, 1845–1914* (London and New York: Routledge, 1914), pp.46–73.
34. Scull, *The Most Solitary of Afflictions*, pp.334–70; E. Hare, 'Was Insanity on the Increase?', *British Journal of Psychiatry* 142 (1983), 439–55.
35. The population of England and Wales rose from 17,927,609 in 1851 to 25,974,439 in 1881, and to 32,527,843 in 1901.
36. J. Walton, 'Lunacy in the Industrial Revolution: A Study of Asylum Admissions in Lancashire, 1848–50', *Journal of Social History* 13 (1979), 1–22.
37. Scull, *The Most Solitary of Afflictions*, pp.344–70.
38. Ibid., pp.316–17, 328–31. The Metropolitan Imbecile Asylums, at Leavesden and Caterham, were opened in 1870, each with capacity to accommodate 2,000 patients.
39. S. Cherry, *Mental Health Care in Modern England: The Norfolk Lunatic Asylum/St Andrew's Hospital, 1810–1998* (Woodbridge and Rochester, NY: Boydell, 2003), pp.104–6, 135–36.
40. Scull, *The Most Solitary of Afflictions*, Chapter 6; Melling and Forsythe, *The Politics of Madness*, pp.54–73, 178–90.
41. The idea of a specific 'colonial psychiatry' was most recently advanced in S. Swartz, 'The Regulation of Colonial Lunatic Asylums and the Origins of Colonial Psychiatry, 1860–1864', *History of Psychology* 13, no.2 (2010), 160–77.
42. See Introduction, Endnote 1.
43. R. Porter, 'Introduction', in R. Porter and D. Wright (eds.), *The Confinement of the Insane: International Perspectives, 1800–1965* (Cambridge: Cambridge University Press, 2003), pp.1–19; M. Vaughan, 'Introduction', in Mahone and Vaughan (eds.), *Psychiatry and Empire*, pp.1–16.
44. A. Scull, *Social Order: Mental Disorder: Anglo-American Psychiatry in Historical Perspective* (London: Routledge, 1989); Scull, 'Psychiatry and Social Control in the Nineteenth and Twentieth Centuries'.
45. Mills, *Madness, Cannabis and Colonialism*, pp.36–40, 79, 107–25, 175–78; Bell, *Mental and Social Disorder in Sub-Saharan Africa*, p.21; Sadowsky, *Imperial Bedlam*, pp.4, 22–5, 29.
46. M. Vaughan, *Curing their Ills: Colonial Power and African Illness* (Cambridge: Polity Press, 1991), pp.101, 120; Mahone and Vaughan, *Psychiatry and Empire*, pp.1–10; Swartz, 'The Regulation of British Colonial Lunatic Asylums', p.172. James Mills has acknowledged that there was no 'great incarceration' in India, with the numbers of asylum patients remaining low in comparison to the population – *Madness, Cannabis and Colonialism*, p.13.

47. Sadowsky, *Imperial Bedlam*, pp.22–4, 29; McCulloch, *Colonial Psychiatry and the 'African Mind'*, pp.13–14, 20–1; M. Vaughan, 'Idioms of Madness: Zomba Lunatic Asylum, Nyasaland in the Colonial Period', *Journal of Southern African Studies* 3, no.2 (April 1993), 218–38; D. Wright, J. Moran and S. Douglas, 'The Confinement of the Insane in Victorian Canada: The Hamilton and Toronto Asylums', in Porter and Wright (eds.), *The Confinement of the Insane: International Perspectives*, pp.100–28; Yeong, *Till the Break of Day*, pp.8–9; Jackson, *Surfacing Up*, pp.38, 40, 43; Moran, *Committed to the State Asylum*, pp.4, 16–19, 25, 49–50; Parle, *States of Mind*, pp.32, 89–91; C. Coleborne, 'Making "Mad" Populations in Settler Colonies: The Work of Law and Medicine in the Creation of the Colonial Asylum', in D. Kirby and C. Coleborne (eds.), *Law, History, Colonialism: The Reach of Empire* (Manchester: Manchester University Press, 2001), pp.106–22; M. Jones, *The Hospital System and Health Care: Sri Lanka, 1815–1960* (New Delhi: Orient Blackswan, 2009), pp.181–82.

48. McCulloch, *Colonial Psychiatry*, pp.13, 20–1, 30; J. Leckie, 'Unsettled Minds: Gender and Madness in Fiji', in Mahone and Vaughan (eds.), *Psychiatry and Empire*, pp.99–123; Vaughan, *Curing Their Ills*, pp.121–24.

49. Sadowsky, *Imperial Bedlam*, pp.10, 25, 29–33; McCulloch, *Colonial Psychiatry*, pp.12–15, 21–4; Vaughan, 'Idioms of Madness', pp.223–24; Vaughan, *Curing Their Ills*, pp.121–24; Ernst, *Mad Tales From the Raj*, pp.41–50, 69–70, 77–82; Parle, *States of Mind*, pp.91–94; Yeong, *Till the Break of Day*, pp.9–15; S. Swartz, 'The Great Asylum Laundry: Space, Classification, and Imperialism in Cape Town', in L. Topp, J. Moran, J. Andrews (eds.), *Madness, Architecture and the Built Environment: Psychiatric Spaces in Historical Context* (London: Routledge, 2007), pp.193–213; Swartz, 'The Regulation of British Colonial Lunatic Asylums', pp.160–61, 166–70; Monk, *Attending Madness*, pp.27–35, 41–8; H.J. Deacon, 'Madness, Race and Moral Treatment: Robben Island Lunatic Asylum, Cape Colony, 1846–90', *History of Psychiatry* VII (1996), 287–97; Jackson, *Surfacing Up*, pp.44, 54, 60; Jones, *The Hospital System and Health Care*, pp.182–7. An official survey of facilities and conditions in lunatic asylums throughout the empire, conducted in 1863, revealed how widespread were the problems – British National Archives (henceforth TNA), CO 854/7 (1864), fos.447–67, 'Colonial Hospitals and Lunatic Asylums'.

50. H. Deacon, 'Insanity, Institutions and Society: The Case of the Robben Island Lunatic Asylum, 1846–1910', in Porter and Wright (eds.), *The Confinement of the Insane*, pp.20–53; C. Coleborne, 'Passage to the Asylum: The Role of the Police in Committals of the Insane in Victoria, Australia, 1848–1900', in Porter and Wright (eds.), pp.129–48; Leckie, 'Unsettled Minds', pp.101–2; W. Ernst, 'Out of Sight and Out of Mind: Insanity in Early Nineteenth-Century British India', in Melling and Forsythe (eds.), *Insanity, Institutions and Society*, pp.245–67; Jackson, *Surfacing Up*, pp.12, 36–7, 62; Bell, *Mental and Social Disorder*, pp.16, 182–5; Mills, *Madness, Cannabis and Colonialism*, pp.67–80.

51. A. Kirk-Greene, *Britain's Imperial Administrators, 1858–1966* (Basingstoke: Macmillan, 2000).

52. W. Ernst, 'Colonial Policies, Racial Politics and the Development of Psychiatric Institutions in Early Nineteenth-Century British India', in W. Ernst

and B. Harris (eds.), *Race, Science and Medicine, 1700–1960* (London and New York: Routledge, 1999), pp.80–100; J. Clarke, 'Asylums in Alien Places: The Treatment of the European Insane in British India', in W. Bynum, R. Porter and M. Shepherd (eds.), *The Anatomy of Madness: Essays in the History of Psychiatry*, Vol. III, *The Asylum and Its Psychiatry* (London and New York: Routledge, 1988), pp.48–70; Ernst, 'Madness and Colonial Spaces – British India, c.1800–1947', in Topp, Moran and Andrews (eds.), *Madness, Architecture and the Built Environment*, pp.215–38; W. Ernst, 'Institutions, People and Power: Lunatic Asylums in Bengal, c.1800–1900', in B. Pati and M. Harrison (eds.), *The Social History of Health and Medicine in Colonial India* (London: Routledge, 2009), pp.129–50.

53. S. Garton, *Medicine and Madness: A Social History of Insanity in New South Wales* (Kensington, New South Wales: New South Wales University Press, 1988), p.18.

54. Swartz, 'The Great Asylum Laundry', pp.195–97.

55. See below, Chapter 2.

56. Ernst, 'Out of Sight and Out of Mind', pp.247–49; Mills, *Madness, Cannabis and Colonialism*, pp.12–13.

57. Deacon, 'Insanity, Institutions and Society', pp.20–3. Cape Colony later became part of South Africa.

58. Garton, *Medicine and Madness*, pp.18, 23; Coleborne, *Madness in the Family*, pp.23–5; Coleborne, 'Making "Mad" Populations', p.110; Coleborne, 'Passage to the Asylum', pp.130–32.

59. Moran, *Committed to the State Asylum*, pp.19–28, 49–64; Wright et al., 'The Confinement of the Insane in Victorian Canada', pp.106–8.

60. M. Baker, 'Insanity and Politics: The Establishment of a Lunatic Asylum in St John's, Newfoundland, 1836–1855', *Newfoundland Quarterly* 77 (1981), 27–31.

61. Yeong, *Till the Break of Day*, pp.9–11.

62. Bell, *Mental and Social Disorder*, pp.44–8.

63. Jones, *The Hospital System and Health Care*, pp.179–83; J. Conolly, *The Construction and Government of Lunatic Asylums and Hospitals for the Insane* (London: John Churchill, 1847; re-printed London: Dawson, 1968), p.183.

64. Coleborne, 'Making "Mad" Populations', p.110; Coleborne, *Madness in the Family*, p.26.

65. See below, Chapters 2, 5.

66. Yeong, *Till the Break of Day*, pp.13, 16; Bell, *Mental and Social Disorder*, p.55.

67. Leckie, 'Unsettled Minds', pp.99, 102.

68. McCulloch, *Colonial Psychiatry*, p.12.

69. J.H. Mills, *Cannabis Britannica: Empire, Trade and Prohibition* (Oxford: Oxford University Press, 2003), pp.183–84.

70. Sadowsky, *Imperial Bedlam*, pp.10, 28–30.

71. McCulloch, *Colonial Psychiatry*, pp.14–15; Jackson, *Surfacing Up*, pp.23–4, 41–9.

72. Vaughan, *Curing Their Ills*, pp.102, 121; McCulloch, *Colonial Psychiatry*, pp.20–1.

73. Deacon, 'Insanity, Institutions and Society', pp.20, 48; S. Marks, 'The Microphysics of Power: Mental Nursing in South Africa in the First Half

of the Twentieth Century', in Mahone and Vaughan (eds.), *Psychiatry and Empire*, pp.67–98; Parle, *States of Mind*, pp.83–6.

74. Keller, *Colonial Madness: Psychiatry in French North Africa*, pp.23–9, 35–8, 41–4.

75. This is often referred to as the 'Stress Vulnerability Model' of the development of psychosis, following J. Zubin and B. Spring, 'Vulnerability – A New View of Schizophrenia', *Journal of Abnormal Psychology* 86, no.2 (1977), 103–26.

76. F. Fanon, *The Wretched of the Earth* (Harmondsworth: Penguin, 1985 edition); Mahone and Vaughan, *Psychiatry and Empire*, p.1.

77. McCulloch, *Colonial Psychiatry and the 'African Mind'*, pp.52–60; Vaughan, *Curing Their Ills*, pp.110–20; S. Mahone, 'East African Psychiatry and the Practical Problems of Empire', in Mahone and Vaughan (eds.), *Psychiatry and Empire*, pp.41–66.

78. Vaughan, 'Idioms of Madness', pp.230–33; Leckie, 'Unsettled Minds', pp.102–3; Jackson, *Surfacing Up*, pp.68–97; Mills, *Madness, Cannabis and Colonialism*, pp.67–96, 130–42; Jones, *The Hospital System and Health Care*, pp.184–86, 190–1.

79. Sadowsky, *Imperial Bedlam*, pp.3, 65–77, 89–92.

80. Coleborne, *Madness in the Family*.

81. Mills, *Madness, Cannabis and Colonialism*, pp.130–47; Yeong, *Till the Break of Day*, pp.154–58; Moran, *Committed to the State Asylum*, pp.115–29; Jones, *The Hospital System and Health Care*, pp.185–86; Coleborne, *Madness in the Family*, pp.59–61, 113–21.

82. A. McCarthy and C. Coleborne (eds.), *Migration, Ethnicity and Mental Health: International Perspectives, 1840–2010* (New York and London: Routledge, 2012); C. Cox and H. Marland (eds.), *Migration, Health and Ethnicity in the Modern World* (London: Palgrave Macmillan, 2013). This issue will be discussed further in Chapter 6, in relation to Trinidad and former British Guiana.

83. Jackson, *Surfacing Up*, pp.64–9, 76–97.

84. Wright et al., 'The Confinement of the Insane in Victorian Canada', pp.120–22; A. McCarthy, 'Ethnicity, Migration and the Lunatic Asylum in Early Twentieth-Century Auckland, New Zealand', *Social History of Medicine* 21, no.1 (2008), 47–65; Coleborne, 'Making "Mad" Populations in Settler Colonies', pp.112–15; E. Malcolm, 'Mental Health and Migration: The Case of the Irish, 1850s–1950s', in McCarthy and Coleborne (eds.), *Migration, Ethnicity and Mental Health*, pp.15–38.

85. Sadowsky, *Imperial Bedlam*, pp.65–9, 89–92; Vaughan, *Curing Their Ills*, pp.106–7; Yeong, *Till the Break of Day*, p.40.

86. These issues are explored in depth by McCulloch, *Colonial Psychiatry*, especially pp.1–7, 46–63, 107–19; see also Sadowsky, *Imperial Bedlam*, pp.98–110; Bell, *Mental and Social Disorder*, pp.1–7; Vaughan, 'Idioms of Madness', pp.226–33; Vaughan, *Curing Their Ills*, pp.107–20; Deacon, 'Madness, Race and Moral Treatment', pp.287, 293–95; S. Swartz, 'Changing Diagnoses in Valkenburg Asylum, Cape Colony, 1891–1920: A Longitudinal View', *History of Psychiatry* 6 (1995), 431–51; Mahone, 'East African Psychiatry and the Practical Problems of Empire', pp.43–54; Parle, *States of Mind*, pp.15–16, 43–4, 293–94.

87. Ernst, *Mad Tales From the Raj*, pp.157–66; Coleborne, 'Making "Mad" Populations', pp.114–17; Yeong, *Till the Break of Day*, pp.40–5, 112–13; L. Barry and C. Coleborne, 'Insanity and Ethnicity in New Zealand: Maori Encounters With the Auckland Mental Hospital, 1860–1900', *History of Psychiatry* 22, no.3 (2011), 285–301; P. Martyr, ' "Behaving Wildly": Diagnoses of Lunacy Among Indigenous Persons in Western Australia, 1870–1914', *Social History of Medicine* 24, no.2 (2011), 316–33. For the West Indies, see Chapter 6.

88. Jones, *The Hospital System and Health Care*, pp.187–90; Coleborne, *Madness in the Family*, pp.38–40; Barry and Coleborne, 'Insanity and Ethnicity in New Zealand', pp.286–87; Leckie, 'Unsettled Minds', pp.100–3. For the West Indies, see Chapter 7.

89. H. Deacon, 'Racial Categories and Psychiatry in Africa: The Asylum on Robben Island in the Nineteenth Century', in Ernst and Harris (eds.), *Race, Science and Medicine, 1700–1960*, pp.101–22; S. Marks, ' "Every Facility that Modern Science and Enlightened Humanity Have Devised": Race and Progress in a Colonial Hospital, Valkenberg Mental Asylum, Cape Colony, 1894–1910', in Melling and Forsythe (eds.), *Insanity, Institutions and Society*, pp.268–91; Marks, 'The Microphysics of Power', pp.70–2; Deacon, 'Insanity, Institutions and Society', pp.21–3, 45, 48; Parle, *States of Mind*, pp.97–9, 286–88, 292–93; S. Swartz, 'The Black Insane in the Cape, 1891–1920', *Journal of Southern African Studies* 21, no.3 (September 1995), 399–415.

90. McCulloch, *Colonial Psychiatry*, pp.12–23; Jackson, *Surfacing Up*, pp.43–61.

91. Ernst, 'Colonial Policies, Racial Politics and the Development of Psychiatric Institutions in Early Nineteenth-Century British India', pp.80–91; Ernst, *Mad Tales From the Raj*, chapters 1–2; Ernst, 'Madness and Colonial Spaces', pp.216–23; Mills, *Madness, Cannabis and Colonialism*, pp.2–13.

92. Deacon, 'Madness, Race and Moral Treatment', pp.288–91, 294; Deacon, 'Racial Categories and Psychiatry in Africa', pp.110–18; Marks, ' "Every Facility that Modern Science and Enlightened Humanity Have Devised" ', pp.272–74.

93. Bell, *Mental and Social Disorder*, p.49; Leckie, 'Unsettled Minds', p.116; Yeong, *Till the Break of Day*, pp.103–13; Jackson, *Surfacing Up*, pp.49–51; Mills, *Madness, Cannabis and Colonialism*, pp.39–40, 118–25; Jones, *The Hospital System and Health Care*, pp.192–93; Moran, *Committed to the State Asylum*, pp.92–5; Swartz, 'The Black Insane in the Cape, 1891–1920', pp.411–12.

94. Coleborne, *Madness in the Family*, pp.32–5.

95. Jones, *The Hospital System and Health Care*, pp.188–90 – Plaxton later moved to be medical superintendent of the Jamaica Lunatic Asylum in 1887 – see below, Chapter 4.

96. Yeong, *Till the Break of Day*, pp.17–19, 40–5, 79, 89, 104–6.

97. Marks, 'Every Facility', pp.272–81.

98. Parle, *States of Mind*, pp.43–5, 85–6, 96, 105–8.

99. Porter, 'Introduction', pp.13–18; Vaughan, 'Introduction', pp.5–6; Marks, 'The Microphysics of Power', pp.72–85; Marks, 'Every Facility', pp.275–78; Yeong, *Till the Break of Day*, pp.15–19, 66–77, 88–99; Moran, *Committed to the State Asylum*, pp.30–2, 40–1, 89–91; Mills, *Madness, Cannabis*

and Colonialism, pp.154–61; Monk, *Attending Madness*, pp.83–92, 147–51, 203–10.

100. Swartz, 'The Regulation of British Colonial Lunatic Asylums and the Origins of Colonial Psychiatry', pp.165–69.
101. Bell, *Mental and Social Disorder*, pp.15–17, 44–55; McCulloch, *Colonial Psychiatry*, pp.12–24; Sadowsky, *Imperial Bedlam*, pp.10, 25, 31–7; Vaughan, *Curing Their Ills*, pp.120–24; Jackson, *Surfacing Up*, pp.44–60.
102. For example – H. Thomas, *The Slave Trade: The History of the Atlantic Slave Trade 1440–1870* (London: Picador, 1997); B.W. Higman, *Slave Populations of the British Caribbean, 1807–1834* (Baltimore: Johns Hopkins University Press, 1984); T. Burnard, *Mastery, Tyranny and Desire: Thomas Thistlewood and His Slaves in the Anglo-Jamaican World* (University of North Carolina Press, 2004); J. Walvin, *Britain's Slave Empire* (Stroud: Tempus, 2007); V. Brown, *The Reaper's Garden: Death and Power in the World of Atlantic Slavery* (Cambridge and London: Harvard University Press, 2008).
103. H.McD. Beckles, *Britain's Black Debt: Reparations for Caribbean Slavery and Native Genocide* (Mona, Jamaica: University of West Indies Press, 2013).
104. See Introduction.
105. Slavery was also present in places such as the Cape Colony and Mauritius, though never on a scale comparable to that in the West Indies.
106. Green, *British Slave Emancipation*, pp.35–64, 170, 298–306; D. Hall, 'The Flight From the Estates Reconsidered: The British West Indies, 1838–1842', in H. Beckles and V. Shepherd (eds.), *Caribbean Freedom: Economy and Society From Emancipation to the Present* (London: James Currey, 1993), pp.55–63; J. Millette, 'The Wage Problem in Trinidad and Tobago, 1838–1938', in B. Brereton and K.A. Yelvington (eds.), *The Colonial Caribbean in Transition: Essays on Post-Emancipation Social and Cultural History* (Jamaica: University of West Indies Press, 1999), pp.55–76.
107. Green, *British Slave Emancipation*, pp.229–35, 249–51; G. Heuman, 'The British West Indies', in A. Porter (ed.), *The Oxford History of the British Empire*, Vol. III, *The Nineteenth Century* (Oxford: and New York: Oxford University Press, 1999), pp.470–93; B. Brereton, *A History of Modern Trinidad, 1783–1962* (Port of Spain, London: Heinemann, 1981), pp.82–3; B.L. Moore, *Race, Power and Social Segmentation in Colonial Society: Guyana After Slavery 1838–1891* (Montreux, Switzerland: Gordon and Breach, 1987), pp.33–4.
108. Green, *British Slave Emancipation*, pp.253–87; Heuman, 'The British West Indies', pp.484–85; Brereton, *A History of Modern Trinidad*, pp.96–106. The Jamaican planters also imported several thousand people from India in order to address their perceived labour problems, though never on the same scale as in Trinidad and British Guiana.
109. Heuman, 'The British West Indies', pp.488–90; H. Beckles, *Great House Rules: Landless Emancipation and Workers' Protest in Barbados 1838–1938* (Oxford: James Currey, 2004), pp.108–13, 158–61; Brereton, *A History of Modern Trinidad*, pp.87, 106–7.
110. Brereton, *A History of Modern Trinidad*, pp.91–5, 199–201; Millette, 'The Wage Problem in Trinidad and Tobago', pp.64–7.
111. W. Marshall, ' "We Be Wise to Many Things": Blacks' Hopes and Expectations of Emancipation', in Beckles and Shepherd (eds.), *Caribbean Freedom*, pp.12–20; S. Wilmot, 'Emancipation in Action: Workers and Wage Conflict

in Jamaica 1838–1848', in Beckles and Shepherd (eds.), *Caribbean Freedom*, pp.48–54; O.N. Bolland, 'Systems of Domination After Slavery: The Control of Land and Labour in the British West Indies After 1838', in Beckles and Shepherd (eds.), *Caribbean Freedom*, pp.107–23; Hall, 'The Flight From the Estates Reconsidered', pp.55–61; Heuman, 'The British West Indies', pp.480–82; G. Heuman, 'From Slavery to Freedom: Blacks in the Nineteenth-Century British West Indies', in P.D. Morgan and S. Hawkins (eds.), *Black Experience and the Empire* (Oxford: Oxford University Press, 2004), pp.141–65; D. Wood, *Trinidad in Transition: The Years After Slavery* (Oxford: Oxford University Press, 1968), pp.49–53; Brereton, *A History of Modern Trinidad*, pp.77–81; Moore, *Race, Power and Social Segmentation*, pp.37–8; Beckles, *Great House Rules*, pp.34–6, 44–51.

112. Beckles, *Great House Rules*, pp.61, 95–113, 137, 158–63; Millette, 'The Wage Problem in Trinidad and Tobago', pp.62–4; B.L. Moore, *Cultural Power, Resistance and Pluralism: Colonial Guyana 1838–1900* (Mona, Jamaica: University of West Indies Press: 1995), pp.86–9; Green, *British Slave Emancipation*, pp.307–14; B. Brereton, *Race Relations in Colonial Trinidad* (Cambridge: Cambridge University Press, 1979), pp.116–22; B.L. Moore and M.A. Johnson, *Neither Led Nor Driven: Contesting British Cultural Imperialism in Jamaica, 1865–1920* (Mona, Jamaica: University of West Indies Press, 2004), pp.10–11.

113. Green, *British Slave Emancipation*, pp.261–63, 313–14; Beckles, *Great House Rules*, pp.77–9, 108; M.J. Newton, *The Children of Africa in the Colonies: Free People of Color in Barbados in the Age of Emancipation* (Baton Rouge: Louisiana State University Press, 2008), pp.225–38; W. Marshall, 'Peasant Development in the West Indies Since 1838', in Beckles and Shepherd (eds.), *Caribbean Freedom*, pp.99–106; Wood, *Trinidad in Transition*, pp.65–6; Brereton, *A History of Modern Trinidad*, pp.96–7; Brereton, *Race Relations in Colonial Trinidad*, pp.110–12. Admissions to lunatic asylums in Trinidad and British Guiana reflected patterns of migration within the Caribbean – see Chapter 6.

114. Beckles, *Great House Rules*, pp.176–78; H. Johnson, 'The Black Experience in the British Caribbean in the Twentieth Century', in Morgan and Hawkins (eds.), *Black Experience and the Empire*, pp.317–46.

115. Green, *British Slave Emancipation*, pp.261–87; Heuman, 'The British West Indies', pp.484–85; Millette, 'The Wage Problem in Trinidad and Tobago', pp.61–3; B.L. Moore, 'Leisure and Society in Postemancipation Guyana', in Brereton and Yelvington (eds.), *The Colonial Caribbean in Transition*, pp.108–25; Moore, *Race, Power and Social Segmentation*, pp.42–7, 161–62; Moore, *Cultural Power, Resistance and Pluralism*, pp.7–8; Brereton, *A History of Modern Trinidad*, pp.98–113; Brereton, *Race Relations in Colonial Trinidad*, pp.176–92; Moore and Johnson, *Neither Led Nor Driven*, pp.245–54; L. Roopnarine, *Indo-Caribbean Indenture: Resistance and Accommodation, 1839–1920* (Kingston, Jamaica: University of West Indies Press, 2007); L. Roopnarine, 'The Indian Sea Voyage Between India and the Caribbean During the Second Half of the Nineteenth Century', *The Journal of Caribbean History* 44, no.1 (2010), 48–74.

116. K.O. Laurence, 'The Development of Medical Services in Trinidad and British Guiana, 1841–1873', in Beckles and Shepherd (eds.), *Caribbean*

Freedom, pp.269–73; J. De Barros, 'Dispensers, Obeah and Quackery: Medical Rivalries in Post-Slavery British Guiana', *Social History of Medicine* 25, no.2 (August 2007), 243–61.

117. Some of these plantation hospitals or 'hothouses' offered crude facilities, though others were more acceptable – J. Sturge and T. Harvey, *The West Indies in 1837: Being the Journal of a Visit to Antigua, Montserrat, Dominica, St. Lucia, Barbados, and Jamaica, Undertaken for the Purpose of Ascertaining the Actual Condition of the Negro Population of Those Islands* (London: Hamilton, Adams & Co., 1838), pp.173–85, 206, 214–29, 269, 296–338.

118. Green, *British Slave Emancipation*, pp.309–11; N.J. Wilkins, 'Doctors and Ex-Slaves in Jamaica 1834–1850', *Health, Disease and Medicine in Jamaica*, in *Jamaican Historical Review* XVII (1991), 19–30; De Barros, 'Dispensers, Obeah and Quackery', pp.244–45.

119. K. Brathwaite, *The Development of Creole Society in Jamaica 1770–1820* (Oxford: Oxford University Press, 1971; Miami: Ian Randle, 2005), pp.284–90; Wilkins, 'Doctors and Ex-Slaves', pp.23; 29–30.

120. Laurence, 'The Development of Medical Services in Trinidad and British Guiana', p.269; Green, *Britain's Slave Colonies*, p.310; R. Schomburgk, *The History of Barbados* (London: Longman, Brown, Green and Longmans, 1847), p.128; TNA, CO 854/7, 'Colonial Hospitals and Lunatic Asylums'.

121. Green, *Britain's Slave Colonies*, pp.309–10; D. Challenger, 'A Benign Place of Healing? The Contagious Diseases Hospital and Medical Discipline in Post-Slavery Barbados', in J. De Barros, S. Palmer and D. Wright (eds.), *Health and Medicine in the Circum-Caribbean, 1800–1968* (London: Routledge, 2009), pp.98–120.

122. M. Jones, *Public Health in Jamaica 1850–1940: Neglect, Philanthropy and Development* (Kingston, Jamaica: University of West Indies Press, 2013); Beckles, *Great House Rules*, pp.61–2; Green, *British Slave Emancipation*, pp.311–14; P.D. Curtin, *Death by Migration: Europe's Encounter With the Tropical World in the Nineteenth Century* (Cambridge: Cambridge University Press, 1989), pp.69–72, 130–31; J. De Barros and S. Stilwell, 'Introduction: Public Health and the Imperial Project', in De Barros and Stitwell (eds.), *Colonialism and Health in the Tropics – Caribbean Quarterly* 49, no.4 (December 2003), 1–11; J. De Barros, 'Sanitation and Civilization in Georgetown, British Guiana', in De Barros and Stillwell (eds.), pp.65–86; Wood, *Trinidad in Transition*, pp.27–8; Brereton, *Race Relations in Colonial Trinidad*, pp.116–18.

123. J. De Barros, '"Improving the Standards of Motherhood": Infant Welfare in Post-Slavery British Guiana', in De Barros, Palmer and Wright (eds.), *Health and Medicine in the Circum-Caribbean*, pp.165–94; De Barros, '"A Laudable Experiment": Infant Welfare Work and Medical Intermediaries in Early Twentieth-Century Barbados', in R. Johnson and A. Khalid (eds.), *Public Health in the British Empire: Intermediaries, Subordinates, and the Practice of Public Health, 1850–1960* (New York and London: Routledge, 2012), pp.100–17.

124. Jones, *Public Health in Jamaica*, pp.33–63; Brereton, *Race Relations in Colonial Trinidad*, p.119. For Barbados, see the 'Half-Yearly Reports of the Poor Law Inspector', in TNA, CO 31, from 1881 onwards.

125. Moore and Johnson, *Neither Led Nor Driven*, pp.205–44; Brereton, *A History of Modern Trinidad*, pp.122–26; J. De Barros, ' "Working Cutlass and Shovel": Labour and Redemption at the Onderneeming School in British Guiana', in G. Heuman and D.V. Trotman (eds.), *Contesting Freedom: Control and Resistance in the Post-Emancipation Caribbean* (Oxford: Macmillan, 2005), pp.39–64.

126. D. Paton, *No Bond But the Law: Punishment, Race and Gender in Jamaican State Formation* (Durham and London: Duke University Press, 2004); D.V. Trotman, *Crime in Trinidad: Conflict and Control in a Plantation Society, 1838–1900* (Knoxville: University of Tennessee Press, 1986); S. Boa, 'Discipline, Reform or Punish? Attitudes Towards Juvenile Crimes and Misdemeanours in the Post-Emancipation Caribbean, 1838–88', in Heuman and Trotman (eds.), *Contesting Freedom*, pp.65–86.

127. Green, *British Slave Emancipation*, pp.65–95; Kirk-Greene, *Britain's Imperial Administrators*, pp.36–45; Porter, *The Absent-Minded Imperialists*, pp.45–6; Beckles, *Great House Rules*, pp.33, 70–2, 114–17.

128. Marshall, ' "We Be Wise to Many Things" ', pp.15–19; Hall, 'The Flight From the Estates Reconsidered', pp.55–7.

129. Beckles, *Great House Rules*, pp.45–56, 139; Heuman, 'From Slavery to Freedom', pp.158–62; G. Heuman, *'The Killing Time': The Morant Bay Rebellion in Jamaica* (London: Macmillan, 1994), pp.38–41; G. Heuman, ' "Is this What You Call Free": Riots and Resistance in the Anglophone Caribbean', in Heuman and Trotman (eds.), *Contesting Freedom*, pp.104–17; M. Craton, 'Continuity Not Change: the Incidence of Unrest Among Ex-Slaves in the British West Indies, 1838–1876', in Beckles and Shepherd (eds.), *Caribbean Freedom*, pp.192–206; D. Trotman, 'Capping the Volcano: Riots and their Suppression in Post-Emancipation Trinidad', in Heuman and Trotman (eds.), *Contesting Freedom*, pp.118–41; H. Carter, *Labour Pains: Resistance and Protest in Barbados 1838–1904* (Kingston, Jamaica: Ian Randle, 2012), pp.89–94, 105–30.

130. Heuman, *'The Killing Time'*; C. Hall, *Civilising Subjects: Metropole and Colony in the English Imagination* (Cambridge: Polity Press, 2002), pp.23–5, 243–60; J. Evans, *Edward Eyre, Race and Colonial Governance* (Dunedin, New Zealand: University of Otago Press, 2005), pp.113–43; Craton, 'Continuity Not Change', pp.196–98; C. Bolt, *Victorian Attitudes to Race* (London: Routledge Kegan Paul, 1971), pp.75–108. The circumstances of the rebellion are summarised from these sources.

131. Evans, *Edward Eyre*, pp.136–37; Heuman, *'The Killing Time'*, pp.177–80. The assembly, with all its faults, had at least comprised some incipient democratic elements, including Black and Coloured members.

132. H. Carter, 'The Bridgetown Riot of 1872', in A. Thompson (ed.), *In the Shadow of the Plantation: Caribbean History and Legacy* (Kingston, Jamaica: Ian Randle, 2002), pp.334–48.

133. The summary of events is distilled from – Beckles, *Great House Rules*, pp.141–56; G. Belle, 'The Abortive Revolution of 1876 in Barbados', in Beckles and Shepherd (eds.), *Caribbean Freedom*, pp.181–91; J. Pope-Hennessy, *Verandah: Some Episodes in the Crown Colonies, 1867–1889* (London: George Allen and Unwin, 1964), pp.163–81; Carter, *Labour Pains*, pp.132–63; P. Howell and D. Lambert, 'Sir John Pope-Hennessy and

Imperial Government: Humanitarianism and the Translation of Slavery in the Imperial Network', in D. Lambert and A. Lester (eds.), *Colonial Lives Across the British Empire: Imperial Careering in the Long Nineteenth Century* (Cambridge: Cambridge University Press, 2006), pp.228–56 – whilst *en route* to Hong Kong in 1877 Pope-Hennessy wrote that 'in all my experience I was never in a community where there was such deliberate oppression of the masses as in the community of Barbadoes', p.243.

134. Moore and Johnson, *Neither Led Nor Driven*, pp.1–5, 311–13; Bolt, *Victorian Attitudes to Race*, pp.75–105; D.A. Lorimer, *Colour, Class and the Victorians: English Attitudes to the Negro in the Nineteenth Century* (Leicester: Leicester University Press, 1978), pp.11–14, 122–28, 131–61;Hall, *Civilising Subjects*, pp.23–5, 209–16, 255–60; Paton, *No Bond But the Law*, pp.132–33, 141–54; Porter, *The Absent-Minded Imperialists*, pp.98–100, 129, 180–86.

135. Green, *British Slave Emancipation*, pp.12–23, 295–96, 319–22; G. Heuman, *Between Black and White: Race, Politics and the Free Coloureds in Jamaica, 1792–1865* (Westport, CT: Greenwood Press, 1981), pp.72–9; Beckles, *Great House Rules*, pp.63–9; Brereton, *Race Relations in Colonial Trinidad*, pp.193–212.

136. Heuman, 'The Killing Time', pp.63–6, 104, 146–50; Newton, *The Children of Africa in the Colonies*, pp.178–81, 238–54; M. Newton, 'Race for Power: People of Colour and the Politics of Liberation in Barbados, 1816–c1850', in Heuman and Trotman (eds.), *Contesting Freedom*, pp.20–38.

137. Brereton, *A History of Modern Trinidad*, pp.106–13; Brereton, *Race Relations in Colonial Trinidad*, pp.176–92; Moore, 'Leisure and Society in Postemancipation Guyana', pp.121–23; R.J. Moore, 'Colonial Images of Blacks and Indians in Nineteenth-Century Guyana', in Brereton and Yelvington (eds.), *The Colonial Caribbean in Transition*, pp.126–58.

138. For meticulous expositions of the nature and dynamics of culture in the nineteenth-century British West Indies, see the works of Brian Moore – *Cultural Power, Resistance and Pluralism: Colonial Guyana 1838–1900*; Moore and Johnson, *Neither Led Nor Driven*.

139. Moore, *Cultural Power Resistance and Pluralism*, pp.295–300; Moore and Johnson, *Neither Led Nor Driven*, pp.140–257; Brereton, *Race Relations in Colonial Trinidad*, pp.94–108, 156–66, 176–92.

140. Moore, *Cultural Power Resistance and Pluralism*, pp.137–54, 299–300; Moore and Johnson, *Neither Led Nor Driven*, pp.51–89, 96–135, 167–205; B.L. Moore and M.A. Johnson, 'Married But Not Parsoned: Attitudes to Conjugality in Jamaica, 1865–1920', in Heuman and Trotman (eds.), *Contesting Freedom*, pp.197–214.

141. For critical analysis of the concepts of 'Creole' and 'Creolisation', see the essays in V. Shepherd and G.L. Richards, *Questioning Creole: Creolisation Discourses in Caribbean Culture* (Oxford: James Currey, 2002).

142. Moore, *Cultural Power, Resistance and Pluralism*, pp.295–307; Moore and Johnson, *Neither Led Nor Driven*, pp.272–74, 314–25; Brereton, *Race Relations in Colonial Trinidad*, pp.171–74.

143. Brereton, *Race Relations in Colonial Trinidad*, pp.166–74, 199–212; Brereton and Yelvington, *The Colonial Caribbean in Transition*, 'Introduction', pp.8–16; Beckles, *Great House Rules*, pp.158–78; Heuman, 'The British West Indies', pp.487–91.

2 The Early Lunatic Asylums

1. R.B. Sheridan, *Doctors and Slaves: A Medical and Demographic History of Slavery in the British West Indies, 1680–1834* (Cambridge, London and New York: Cambridge University Press, 1985).
2. Halliday, *A General View of the Present State of Lunatics, and Lunatic Asylums,* p.80.
3. M.H. Beaubrun et al., 'The West Indies', in J.G. Howells (ed.), *World History of Psychiatry* (New York: Brunner/Mazel, 1975), pp.507–27.
4. B. De Las Casas, *Historia de Las Indias*, Vol. II, cited in Beaubrun et al., 'The West Indies', p.508; Hickling and Sorel, *Images of Psychiatry: The Caribbean,* 'Introduction', p.4.
5. Beckles, *Britain's Black Debt*, pp.37–75, 91–5.
6. Sheridan, *Doctors and Slaves*, pp.112–13.
7. House of Commons Sessional Papers of the Eighteenth Century, Vol. 73, SC on the Slave Trade, Minutes of Evidence, p.85, 6 May 1790.
8. Sheridan, *Doctors and Slaves*, p.112.
9. House of Commons Sessional Papers, Vol. 72, SC on the Slave Trade, Minutes of Evidence, pp.581, 591, 8 March 1790.
10. Sheridan, *Doctors and Slaves*, pp.111, 117.
11. House of Commons Sessional Papers, Vol. 72, pp.561–63, 6 March 1790.
12. Ibid., pp.575–76. William Wilberforce cited Wilson's evidence in graphically describing how 'melancholy' and 'insanity' brought about many deaths, when describing the iniquities of the slave trade in 1792 – *The Parliamentary History of England From the Earliest Period to the Year 1803*, Vol. XXIX (London: Hansard, 1817), 2 April 1792, p.1069.
13. House of Commons Sessional Papers, Vol. 72, p.569, 6 March 1790.
14. Ibid., p.569.
15. Ibid., pp.567–68.
16. Ibid., p.568.
17. Sheridan, *Doctors and Slaves*, p.134.
18. C. Caines, *Letters on the Cultivation of the Otaheite Cane … and Also a Speech on the Slave Trade, the Most Important Feature in West Indian Cultivation* (London: Messrs Robinson, 1801), pp.262–65.
19. Sheridan, *Doctors and Slaves*, pp.32–5.
20. 'A Professional Planter' (D. Collins), *Practical Rules for the Management and Medical Treatment of Negro Slaves, in the Sugar Colonies* (London: J. Barfield, 1811), pp.293–95.
21. J. Thomson, *A Treatise on the Diseases of Negroes, as They Occur in the Island of Jamaica With Observations on the Country Remedies* (Jamaica: Alex Aikman, 1820), pp.46–7.
22. Thomson, *A Treatise on the Diseases of Negroes*, p.130.
23. Thomas Dancer, *The Medical Assistant, or Jamaica Practice of Physic: Designed Chiefly for the Use of Families and Plantations* (Kingston, Jamaica: Alex Aikman, 1801), pp.198–99, 206–7.
24. Burnard, *Mastery, Tyranny and Desire: Thomas Thistlewood and His Slaves in the Anglo-Jamaican World.*
25. Burnard, *Mastery, Tyranny and Desire*, pp.218–21, 263.
26. Ibid., pp.178, 263–4.

27. Brown, *The Reaper's Garden*, pp.132–33.
28. Cited in Sheridan, *Doctors and Slaves*, p.133.
29. *Royal Gazette* (of Jamaica), 13 December 1817.
30. *Royal Gazette*, 15 February 1817.
31. Ibid., 26 July 1817.
32. Beaubrun et al., 'The West Indies', p.508. For descriptions of estate hospitals in Jamaica just prior to emancipation, and their use as places of punishment and confinement, see Sturge and Harvey, *The West Indies in 1837*, pp.173–75, 184–85, 206, 214–16, 220, 224, 229, 243, 269–70, 296–98, 316–17, 329–30, 335–38.
33. See Chapter 1.
34. BPP 1826, Vol. XXVI, Second Report of the Commissioner of Inquiry into the Administration of Civil and Criminal Justice in the West Indies, p.22.
35. BPP 1826/7, Vol. XXIV, Third Report of the Commissioner of Inquiry into the Administration of Civil and Criminal Justice in the West Indies, p.106.
36. BPP 1826/7, Vol. XXIII, Report of His Majesty's Commissioners of Legal Inquiry on the Colony of Trinidad, p.136.
37. BPP 1830–31, Vol. XII, Gaols, West Indies: Copies of Correspondence Relative to the State of Gaols in the West Indies and the British Colonies in South America, p.91.
38. See below for early asylum provision in Jamaica and Barbados.
39. TNA, CO 28/118, fos.134–42, 7 December 1836, Murray McGregor to Glenelg; BPP 1837, Vol. LIII, Papers Presented to Parliament, in Explanation of the Measures Adopted by Her Majesty's Government, For Giving Effect to the Act for the Abolition of Slavery Throughout the British Colonies, Part IV, pp.347–51.
40. BPP 1841, Vol. III, *Papers Relative to the West Indies*, 1841. Part II. Jamaica, pp.24, 64.
41. CO 28/175, fos.46–7, 4 November 1851, Letter from Hector Gavin M.D.
42. BPP 1837/8, Vol. XL, Report of Captain J.W. Pringle on Prisons in the West Indies, Barbados; Montserrat, Antigua; Tortola, pp.12, 20–1, 24, 26.
43. CO 7/137, 20 August 1869, Nicholson to Baynes.
44. CO 7/122, 23 January 1864, Admissions, Cures, Deaths, 1842–54, at Rat Island; Antigua Archives, Minutes of Legislative Council, 1845–47, pp.543–44, Report by Dr Nicholson, 1 February 1847.
45. BPP 1847/8, Vol. XLVI, Reports to Secretary of State on Past and Present State of Her Majesty's Colonial Possessions, 1847, pp.70–1; CO 28/175, fos.49–51, 7 November 1851.
46. BPP 1850, Vol. XXXVI, Reports to Secretary of State, 1849 (Part 1), p.24; BPP 1851, Vol. XXXIV, Reports to Secretary of State, 1850, p.65.
47. BPP 1852, Vol. XXXI, Reports to Secretary of State, 1851, p.103.
48. CO 321/3, fos.93–5, 11 March 1874, Rennie to Rawson.
49. BPP 1852, Vol. XXXI, Reports to Secretary of State, 1851, p.142; BPP 1860, Vol. XLIV, Reports to Secretary of State, 1858, pp.106, 108.
50. BPP 1849, Vol. XXXIV, Reports to Secretary of State, 1848, pp.204–5.
51. BPP 1852–53, Vol. LXII, Reports to Secretary of State, 1852, p.90; BPP 1854–55, Vol. XXXVI, Reports to Secretary of State, 1853, pp.129, 132, 135, 140–41.

52. BPP 1856, Vol. XLII, Reports to Secretary of State, 1854, pp.111, 116–17; BPP 1857, Vol. X, Reports to Secretary of State, 1855, pp.161–62.
53. BPP 1889, Vol. LIV, Papers Relating to Her Majesty's Colonial Possessions, 67, St Lucia, p.17.
54. BPP 1847/8, Vol. XLVI, Reports to Secretary of State, 1847, pp.169, 176 (quote).
55. BPP 1849, Vol. XXXIV, Reports to Secretary of State, 1848, p.286.
56. CO 295/169, fos.34–8, 26 August 1849.
57. Library of the Society of Friends, M.S. Vol. S.22, 'John Candler: West Indies Journal 1849, 1850', fo.48, 1st Month 6th; 'John Candler's Visit to Trinidad', *Caribbean Studies* 4, no.3 (October 1964), 66–71; J.A. Borome, 'John Candler's Visit to America, 1850', *The Bulletin of Friends Historical Association* (Philadelphia) 48, no.1 (Spring 1959), 21–62; Digby, *Madness, Morality and Medicine: A Study of the York Retreat*, p.112.
58. BPP 1852–53, Vol. LXVII, Copies of Extracts of Despatches Relative to the Condition of the Sugar Growing Colonies, p.143.
59. *Port of Spain Gazette*, 4 March 1857.
60. *Port of Spain Gazette*, 19 February 1859.
61. *Port of Spain Gazette*, 4 February 1863, 4 March 1865.
62. BPP 1875, Vol. LI, Papers Relating to Her Majesty's Colonial Possessions, 1875, Part III, p.91.
63. BPP 1878, Vol. LVI, Papers Relating to Her Majesty's Colonial Possessions, 1876 and 1877, p.69.
64. BPP 1881, Vol. LXIV, Papers Relating to Her Majesty's Colonial Possessions, 1879, pp.172, 175; CO 321/20 (1879–80), fos.193, 209, 234, 441.
65. CO 321/54, fos.136–51, 18 August 1882, 'Transfer of Lunatics From St Vincent & Tobago to Grenada Asylum'.
66. Brathwaite, *The Development of Creole Society in Jamaica 1770–1820*, pp.284–86; Wilkins, 'Doctors and Ex-Slaves in Jamaica 1834–1850', p.23; Sheridan, *Doctors and Slaves*, p.269.
67. Brathwaite, *The Development of Creole Society*, pp.286–87.
68. L.Q. Bowerbank, *A Circular Letter to the Individual Members of the Legislative Council and of the House of Assembly of Jamaica, Relative to the Public Hospital and Lunatic Asylum of Kingston* (Kingston, Jamaica, 1858), p.43. Bowerbank cites returns from 1842–43, showing that one patient had been confined in the lunatic asylum for 28 years.
69. TNA, CO 140/104, Votes of the Assembly of Jamaica, 9 December 1818, pp.114–15, 11 December 1818, p.137.
70. CO 140/104, 8 December 1819, p.125.
71. E.N. Bancroft, *A Letter to the Hon. Hector Mitchel, Chairman of the Committee of Public Accounts, Representing the Total Unfitness of the Present Asylum for Lunatics, and the Urgent Necessity for Building a New Lunatic Asylum, in a Proper Situation* (Kingston, Jamaica: Jordon, Osborn & Co., 1840), p.4; 'The Jamaica Lunatic Asylum', *Journal of Mental Science* VI (1859), 157–67. An additional block was added during the 1840s – CO 137/361, 17 June 1861, pp.18–19, evidence of Dr James Scott; J. Macfadyen, 'On Medical Topography as Connected With the Choice of a Site for a Lunatic Asylum in a Tropical Country', *Edinburgh Medical and Surgical Journal* LXXI (1849), 114–25.

72. Bowerbank, *A Circular Letter to the Individual Members of the Legislative Council*, p.41.
73. Bowerbank, *A Circular Letter to the Individual Members of the Legislative Council*, pp.38, 42; Bancroft, *A Letter to the Hon. Hector Mitchel*, p.4.
74. Bowerbank, *A Circular Letter to the Individual Members of the Legislative Council*, p.42.
75. Bowerbank, *A Circular Letter to the Individual Members of the Legislative Council*, pp.38–40; Bancroft, *A Letter to the Hon. Hector Mitchel*, p.10.
76. Sturge and Harvey, *The West Indies in 1837*, p.280. Interestingly, they noted that 'the most violent cases were Europeans'.
77. J. Scott, *A Reply to a Letter by Lewis Quier Bowerbank, M.D. Edinburgh, to the Commissioners of the Public Hospital and Lunatic Asylum of Kingston, Jamaica, Relative to the Present State and Management of Those Institutions* (Kingston and Spanish Town, Jamaica: Jordon and Osborn, 1858), pp.13–14. Scott became House Surgeon to the Public Hospital and Lunatic Asylum, and was deeply implicated in the scandals of 1858–61 – see Chapter 3.
78. J. Candler, *Extracts From the Journal of John Candler, Whilst Travelling in Jamaica, Part 1* (London: Harvey and Darton, 1840), pp.10–11. After returning to England, Candler became Superintendent of the York Retreat in 1841 – Borome, 'John Candler's Visit to America, 1850', p.22.
79. Bancroft, *A Letter to the Hon. Hector Mitchel*.
80. Ibid., pp.5–6.
81. CO 137/274, fos.390–94, 21 August 1843, Elgin to Stanley; C.M. Morissey, 'The Road to Bellevue: Conditions and Treatment of the Mentally Ill in Jamaica, 1776–1861', *Jamaica Journal* 22, no.3 (August–October 1989), 2–10.
82. Morissey, 'The Road to Bellevue', pp.5–6. It was another 14 years before the asylum opened – see Chapter 3.
83. Scott, *A Reply to a Letter*, p.14; Bowerbank, *A Circular Letter to the Individual Members of the Legislative Council*, p.43; CO 137/359, Public Hospital and Lunatic Asylum Commission, 14 May 1861, pp.73–4, evidence of Lewis Bowerbank, 23 May 1861, pp.320–21, evidence of David Ryan.
84. See Chapter 3.
85. Barbados Archives, St Michael's Vestry, Minutes, 25 March 1823. St Michael was the island's most populous parish, incorporating Bridgetown.
86. St Michael's Vestry, Minutes, 25 March 1823, 25 March 1824, 25 March 1827, 25 March 1828.
87. St Michael's Vestry, Minutes, 3 August 1829; T. Rolph, *A Brief Account, Made During a Visit to the West Indies, and a Tour Through the Unites States of America, in Parts of the Years, 1832–33* (Dundas, Upper Canada: Heyworth, Hackstaff, 1836), pp.42–3. Rolph observed three patients when visiting in 1832, two White and one 'Mulatto'.
88. St Michael's Vestry, Minutes, 2 November 1831, 13 February 1832, 25 March 1841, 13 March 1846.
89. CO 28/104, fos.27–8, 21 March 1829, Lyon to FitzRoy Somerset.
90. St Michael's Vestry, Minutes, 3 August 1829.
91. TNA, CO 28/118, fo.134, 7 December 1836, Murray McGregor to Glenelg.
92. *The Public Acts in Force Passed by the Legislature of Barbados, in the First, Second, Third, and Fourth Years of the Reign of Her Majesty, Queen*

Victoria, 1837–41 (Bridgetown, 1842), pp.224–29, 3 Victoria, Cap. XXVIII, 4 June 1840, 'An Act for the Better Care and Maintenance of Lunatics'.

93. St Michael's Vestry, Minutes, 27 October, 19 December 1842.

94. CO 28/144, fos.89–96, 2 November 1842, Grey to Stanley. The 'Slave Compensation Fund' had been established to compensate former slave owners for loss of their 'property'.

95. CO 33/54, Blue Book, 1844, fos.29, 56–7, 63; CO 33/55 (1845), fos.31, 65; CO 33/56 (1846), fo.27; R. Schomburgk, *The History of Barbados* (London: Longman, Brown, Green and Longmans, 1847), pp.129–30.

96. Schomburgk, *The History of Barbados*, p.130; CO 28/160 (1844), fos.216–21; CO 33/54 (1844), fos.56–7, 63.

97. Barbados Archives, Register of Patients, 1846–88, nos.1–20; CO 28/33, fo.202, extract from *The Globe*, 6 April 1893.

98. Library of the Society of Friends, 'John Candler: West Indies Journal 1849, 1850', fos.11–12, 12th Month 3rd; J.A. Borome (ed.), 'John Candler's Visit to Barbados, 1849', *Journal of the Barbados Museum and Historical Society* 28, no.4 (August 1961), 128–36; Borome, 'John Candler's Visit to America, 1850', pp.22–3.

99. CO 101/104, fo.165, Barbados Lunatic Asylum, Fifth Annual Report, Year Ending 31st March 1851; BPP 1852, Vol. XXXI, Reports to Secretary of State, 1851, p.66.

100. CO 28/196, fos.233–63, 16 May 1863, Walker to Newcastle.

101. See Chapter 5.

102. Both were established as colonies in the first half of the seventeenth century.

103. Antigua National Archives, Minutes of Legislative Council, 1839–40, pp.397–99, 28 May 1840; F. Lanaghan, *Antigua and the Antiguans: A Full Account of the Colony and Its Inhabitants From the Time of the Caribs to the Present Day* (London: Saunders and Otley, 1844), Vol. 1, p.256.

104. Antigua National Archives, The Laws of the Island of Antigua, pp.453–54, 5th Vic. 1, no.905, 1841, 'An Act for Establishing a Lunatic Asylum in This Island'.

105. TNA, CO 7/71, 15 January 1842, McPhail to Stanley; CO 7/74, 20 January 1843, Fitzroy to Stanley; CO 7/75, 1 June 1843, Fitzroy to Stanley.

106. Antigua National Archives, Minutes of Legislative Council, 1845–47, pp.543–44, 1 February 1847; CO 7/121, 4 November 1863, Hill to Newcastle, Attachments, IV; BPP 1843, Vol. XXXIII, Papers Relative to the Earthquake in the West Indies, pp.3–4, 9.

107. Antigua Legislative Council, 1845–47, pp.542–43, 12 January 1847.

108. Legislative Council, 1845–47, p.545, 3 February 1847.

109. 'John Candler: West Indies Journal 1849, 1850', fo.96, 3rd Month 11th; 'John Candler's Visit to Antigua', *Caribbean Studies* 5, no.3 (October 1965), 51–7.

110. Legislative Council, 1849–51, pp.259–60, 4 April 1850.

111. Legislative Council, 1855–59, pp.95–6, 17 April 1856, p.183, 4 August 1856.

112. Legislative Council, 1855–59, pp.249–50, 12 February 1857, p.270, 5 March 1857.

113. CO 7/122, 23 January 1864, 'Tabulated Statistical Statements Shewing the Working of the Lunatic Asylum at Rat Island From 1st February 1842 to 31st December 1863', 'General Remarks'.
114. CO 7/139, 11 March 1870, Pine to Granville, 9 July 1870, Pine to Granville. See Chapter 5 for subsequent developments in Antigua and the Leeward Islands.
115. BPP 1847, Vol. XXXVII, Reports to Secretary of State, 1846, p.44; BPP 1850, Vol. XXXVI, Reports to Secretary of State, 1849, Part 1, p.47.
116. 'John Candler's Visit to St Kitts', *Caribbean Studies* 5, no.4 (January 1966), 35–8.
117. BPP 1867, Vol. XLVIII, Reports to Secretary of State, 1865, pp.90–1.
118. BPP 1846, Vol. XXVII, British Colonies (West Indies and Mauritius) – Papers Relating to the Labouring Population of the British Colonies, p.276; BPP 1847, Vol. XXXVII, Reports to Secretary of State, 1846, p.43.
119. BPP 1850, Vol. XXXVI, Reports to Secretary of State, 1849, Part 1, p.54; BPP 1851, Reports to Secretary of State, 1850, p.110.
120. BPP 1867–68, Vol. XLVIII, Reports to Secretary of State, 1866, pp.97–8, 100. Eight lunatics were recorded in the institution, alongside 42 paupers. The lunatic asylum comprised ten confined cells and two small exercise yards – BPP 1871, Vol. XLVII, Reports to Secretary of State, 1869, p.90.
121. BPP 1860, Vol. XLIV, Reports to Secretary of State, 1858, pp.106–8.
122. TNA, CO 885/3/4, 'Colonial Hospitals and Lunatic Asylums', pp.2, 4, 5. It provided for no more than six patients.
123. CO 28/175, fos.49–51, 7 November 1851; CO 101/103, fos.169–76, 21 November 1850, Colebrook to Grey; CO 101/104, fos.158–65, 15 May 1851, Colebrook to Grey; CO 101/109, fo.282, 25 April 1855, Neale to Colebrook; BPP 1851, Vol. XXXIV, Reports to Secretary of State, 1850, p.65. The Rules were sent to the Commissioners in Lunacy in London for perusal.
124. CO 101/109, fos.271–84, 14 June 1855, Colebrooke to Russell.
125. CO 885/3/4, 'Colonial Hospitals and Lunatic Asylums', p.25; CO 101/119 (1863), Lunatic Asylum Returns, fos.201–7.
126. CO 101/129 (1869), Lunatic Asylum Returns, fos.251–59.
127. BPP 1857, Vol. X, Reports to Secretary of State, 1855, pp.161–62.
128. CO 885/3/4, 'Colonial Hospitals and Lunatic Asylums', p.26.
129. G.W. Des Voeux, *My Colonial Service in British Guiana, St Lucia, Trinidad, Fiji, Australia, Newfoundland and Hong Kong With Interludes* (London: John Murray, 1903), pp.127, 187–88, 247. A new asylum opened in 1874.
130. BPP 1847, Vol. XXXVII, Reports to Secretary of State, 1846, p.90; L. Gramaglia, 'Dr Robert Grieve (1839–1906), An "Apostle of Science"' – 'Introduction' to R. Grieve *The Asylum Journal* (Guyana: The Caribbean Press, 2010), Vol. 1, pp.xiii–xxxii.
131. 'John Candler: West Indies Journal 1849, 1850', fos.25–6, 12th Month 19th; 'John Candler's Visit to British Guiana', *Caribbean Studies* 4, no.2 (July 1964), 52–61. Candler's condemnatory report subsequently appeared in the *Anti-Slavery Reporter*, to the consternation of Governor Barkly – BPP 1852, Vol. XXXI, Reports to Secretary of State, 1851, pp.144–45.
132. BPP 1861, Vol. XL, Reports to Secretary of State, 1859, p.43; CO 137/345, 21 July 1859, Scott to Trench, fo.559.
133. CO 885/3/4, 'Colonial Hospitals and Lunatic Asylums', pp.4, 5, 24 (quote).

134. BPP 1871, Vol. XLVII, Reports to Secretary of State, 1869, pp.47–9. See Chapter 5 for subsequent development of the Berbice asylum.
135. BPP 1847, Vol. XXXVII, Reports to Secretary of State, 1846, p.112; BPP 1847/8, Vol. XLVI, Reports to Secretary of State, 1847, pp.165, 167.
136. BPP 1847/8, Vol. XLVI, Reports to Secretary of State, 1847, pp.167–69; BPP 1856, Vol. XLII, Reports to Secretary of State, 1854, p.35. The jail was moved to the site after its 'total destruction' in 1853.
137. BPP 1857, Vol. X, Reports to Secretary of State, 1855, p.73; BPP 1857/8, Vol. XL, Reports to Secretary of State, 1856, p.20.
138. BPP 1867, Vol. XLVIII, Reports to Secretary of State, 1865, p.14.
139. BPP 1868–69, Vol. XLIII, Reports to Secretary of State, 1867, pp.26–7.
140. BPP 1882, Vol. XLIV, Papers Relating to Her Majesty's Colonial Possessions, 1881, p.64.
141. CO 295/141 (1843), fos.112–14.
142. BPP 1847–48, Vol. XXII, Seventh Report From the Select Committee on Sugar and Coffee Planting, p.277; BPP 1852–53, Vol. LXVII, Copies of Extracts of Despatches Relative to the Condition of the Sugar Growing Colonies, pp.143, 226.
143. CO 295/183, fos.131–32, 7 June 1853, Treasury to Newcastle.
144. *Port of Spain Gazette*, 19 February 1859.
145. *Port of Spain Gazette*, 11 February 1860.
146. *Port of Spain Gazette*, 4 February 1863; 3 February 1864; 4 March 1865; BPP 1865, Vol. XXXVII, Reports to Secretary of State, 1863, pp.28–9.
147. Trinidad Archives, Proceedings of the Legislative Council (1877): Lunatic Asylum, Annual Report of the Medical Superintendent for 1876, pp.1–2. For subsequent development of the Trinidad Asylum, see Chapter 4.
148. Beckles, *Britain's Black Debt*, pp.143–59.
149. CO 7/139, 11 March 1870, Pine to Granville.

3 Scandal in Jamaica – The Kingston Lunatic Asylum

1. TNA, CO 137/360, Public Hospital and Lunatic Asylum Commission, 28 May 1861, evidence of Elizabeth Scott, pp.89–90.
2. A contextualised summary of events is available in M. Jones, 'The Most Cruel and Revolting Crimes: The Treatment of the Mentally Ill in Mid-Nineteenth Century Jamaica', *The Journal of Caribbean History* 42, no.2 (2008), 290–309.
3. For the scandal's wider significance, see S. Swartz, 'The Regulation of British Colonial Lunatic Asylums and the Origins of Colonial Psychiatry, 1860–1864', *History of Psychology* 13, no.2 (2010), 160–77.
4. R.M. Martin, *The British Colonies: Their History, Extent, Condition, and Resources*, Vol. 4, *Africa and the West Indies* (London, 1853), pp.76–7, cited in D. Hall, *Free Jamaica 1838–1865: An Economic History* (Aylesbury: Ginn and Company, 1981 edition – first published 1959), pp.9–10.
5. Hall, *Free Jamaica*, pp.2–8; P. Curtin, *Two Jamaicas: The Role of Ideas in a Tropical Colony 1830–1865* (Cambridge, MA: Harvard University Press, 1955), pp.179–91.

6. G. Heuman, *Between Black and White: Race, Politics, and the Free Coloreds in Jamaica, 1792–1865* (Westwood, CT: Greenwood Press, 1981), chapters 9–12.
7. Curtin, *Two Jamaicas*, pp.190–91; Heuman, *Between Black and White*, pp.163–66, 174–7; *New York Times*, 17 October 1861.
8. Bancroft, *A Letter to the Hon. Hector Mitchel.*
9. Bancroft, *A Letter to the Hon. Hector Mitchel*, pp.4–5.
10. Ibid., pp.5–7.
11. Ibid., pp.3, 10–11.
12. CO 137/273, fos.149–50, 13 January 1843, Robbins to Stanley.Bancroft died in 1843.
13. CO 137/273, fo.144, 11 February 1843, Elgin to Stanley; fos.146–48, 29 March 1843, Stanley to Elgin; fos.151–52, 12 December 1842, Robbins to Ossett; fos.152–56, 16 December 1842, Robbins to Stanley.
14. CO 137/274, fos.390–94, 21 August 1843, Elgin to Stanley.
15. CO 137/359, Public Hospital and Lunatic Asylum Commission, 15 May 1861, evidence of Lewis Bowerbank, p.88.
16. CO 137/359, 23 May 1861, evidence of David Ryan, pp.320–21.
17. Scott, *A Reply to a Letter by Lewis Quier Bowerbank*, pp.21–2.
18. CO 137/359, 15 May 1861, evidence of Bowerbank, p.74, 23 May 1861, evidence of David Ryan, p.321; CO 137/361, Public Hospital and Lunatic Asylum Commission, 19 June 1861, evidence of James Scott, p.196; Bowerbank, *A Circular Letter to the Individual Members of the Legislative Council and of the House of Assembly*, Appendix, pp.4–6.
19. Bowerbank, *A Circular Letter to the Individual Members of the Legislative Council and of the House of Assembly*, Appendix, pp.3–4.
20. Ibid., Appendix, pp.16–17.
21. Ibid., Appendix, p.5.
22. Ibid., Appendix, p.7.
23. Library of the Society of Friends, M.S. Vol. S.22, 'John Candler: West Indies Journal 1849, 1850', fo.165. Candler was relatively impressed with the state of the hospital, finding it in 'good order'.
24. BPP 1854, Vol. XLIII: Cholera (Jamaica). Copy of the Report Made by Dr Milroy to the Colonial Office, on the Cholera Epidemic in Jamaica, 1850–51, p.46.
25. M. Jones, *Public Health in Jamaica, 1850–1940*, pp.10, 13; C.D. Fryar, 'The Moral Politics of Cholera in Postemancipation Jamaica', *Slavery and Abolition* 34, no.4 (December 2013), 598–618.
26. BPP 1854, Vol. XLIII, Cholera (Jamaica), p.45; CO 137/359, Public Hospital and Lunatic Asylum Commission, p.40, 15 May 1861, evidence of Bowerbank.
27. BPP 1854, Vol. XLIII, Cholera (Jamaica), pp.45–6. For Milroy, see Mark Harrison, 'Milroy, Gavin (1805–1886)', *Oxford Dictionary of National Biography*, Oxford University Press, 2004; online edition, September 2010 [http://www.oxforddnb.com/view/article/18797, accessed 18 November 2013].
28. Bowerbank, *A Circular Letter to the Individual Members of the Legislative Council and of the House of Assembly*, Appendix, pp.9–10.
29. Bowerbank, *A Circular Letter*, Appendix, pp.12–14.

30. CO 137/346, 23 September 1859, Trench to Darling, fos.291, 308; CO 137/359, Public Hospital and Lunatic Asylum Commission, 15 May 1861, evidence of Lewis Bowerbank, pp.54–5; CO 137/366, 'Report on the Management of the Public Hospital', 20 November 1861, p.12; W.J. Gardner, *A History of Jamaica From the Discovery by Christopher Columbus to the Present Time* (London: Elliot Stock, 1873), p.469.

31. CO 137/338, fos.202–12, 9 August 1858, Darling to Bulwer Lytton.

32. Jones, 'The Most Cruel and Revolting Crimes', pp.299–300; Scott, *A Reply to a Letter by Lewis Quier Bowerbank*, p.2.

33. CO 137/338, 9 August 1858, Darling to Bulwer Lytton, fos.204–6. I have been unable to locate a copy of Bowerbank's original pamphlet; parts were reproduced elsewhere.

34. Scott, *A Reply to a Letter by Lewis Quier Bowerbank*.

35. Ibid., pp.13–15. Scott claimed that Bowerbank had rehashed evidence given by Dr Edward Bancroft in 1836 – pp.18–22.

36. L.Q. Bowerbank, *A Third Letter to the Commissioners of the Public Hospital and Lunatic Asylum of Kingston Jamaica, Relative to Dr Scott's Reply* (Kingston, Jamaica: Ford and Gall, 1858) – copy in CO 137/338, fos.255–74.

37. Bowerbank, *A Third Letter*, p.8.

38. Ibid., p.19.

39. CO 137/338, 9 August 1858, Darling to Bulwer Lytton, fos.207–9.

40. CO 137/338, fos.205–6.

41. CO 137/338, fos.219–21, 7 August 1858, Bowerbank to Darling; CO 137/339, fo.9, 24 August 1858, Bowerbank to Austen; fos.10–11, 30 August 1858, Austen to Bowerbank; fos.48–50, 10 September 1858, Bowerbank to Austin; fos.51–62, 16 September 1858, Austen to Bowerbank; fos.64–78, 21 September 1858, Bowerbank to Austin; fos.79–88, 27 September 1858, Austen to Bowerbank; CO 137/343, fos.360–61, 20 December 1858, Austin to Bowerbank; fos.362–65, 21 December 1858, Bowerbank to Austin; fo. 367, 27 December 1858, Austin to Bowerbank; fos.357–59, 10 January 1859, Bowerbank to Bulwer Lytton.

42. CO 137/339, fos.3–6, 9 September 1858, Darling to Bulwer Lytton; fos.42–4, 25 September 1858, Darling to Bulwer Lytton.

43. CO 137/339, fos.6, 44–6; for Henry Taylor, see Green, *British Slave Emancipation*, pp.84–6.

44. Bowerbank, *A Circular Letter to the Individual Members of the Legislative Council and the House of Assembly*.

45. CO 137/343, fos.192–95, 2 December 1858, Bowerbank to Austen; fos.198–99, 4 December 1858, Bowerbank to Austen; fos.220–21, 19 December 1858, Bowerbank to Austen.

46. CO 137/343, fos.360–61, 20 December 1858, Austen to Bowerbank; fos.362–65, 21 December 1858, Bowerbank to Austen; fo.367, 27 December 1858, Austen to Bowerbank; fos.357–59, 10 January 1859, Bowerbank to Bulwer Lytton; fos.348–56, 26 January 1859, Darling to Bulwer Lytton.

47. *The Lancet*, 23 October 1858, p.431.

48. *The Lancet*, 25 December 1858, pp.659–60.

49. *The Lancet*, 26 March 1859, p.325. Bulwer Lytton had a personal interest in mental health issues, having recently committed his allegedly insane wife to a private lunatic asylum – Scull, *The Most Solitary of Afflictions*, p.252.

50. *The Lancet*, 3 September 1859, p.243; D.J. Mellett, 'Bureaucracy and Mental Illness: The Commissioners in Lunacy 1845–90', *Medical History* 25 (1981), 221–50. The commissioners had no formal role regarding colonial asylums.

51. TNA, MH 50/10, Commissioners in Lunacy, p.182, 19 April 1859.

52. MH 50/10, p.192, 11 May 1859; p.195, 12 May 1859; p.209, 2 June 1859; p.214, 8 June 1859.

53. CO 137/345, fos.183–84, 19 July 1859, draft letter Newcastle to Darling; MH 50/10, p.258, 27 July 1859; *The Lancet*, 3 September 1859, p.243; *The Times*, 8 September 1859.

54. *The Lancet*, 6 August 1859, p.152.

55. *The Times*, 30 August 1859.

56. *The Times*, 1 September, 8 September 1859.

57. *The Lancet*, 3 September, 16 December 1859, 4 August 1860.

58. *Medical Times and Gazette*, 10 September 1859.

59. *Asylum Journal of Mental Science* VI (1859), pp.157–67, 'The Jamaica Lunatic Asylum'. The 'house journal' for British asylum doctors provided a comprehensive description of the Kingston asylum and its difficulties, and analysed the political forces in Jamaica and England that impeded effective remedies.

60. CO 137/345, fos.254–61, 9 July 1859, Darling to Bulwer Lytton; fos.262–63, appended note by Newcastle.

61. CO 137/346, fos.274–84, 10 October 1859, Darling to Newcastle.

62. Jones, 'The Most Cruel and Revolting Crimes', pp.295–98; CO 137/366, 'Report on the Management of the Public Hospital', p.6.

63. CO 137/346, fos.290–315, 23 September 1859, Trench to Darling.

64. Ibid., fo.291.

65. Ibid., fos.292–97, 314.

66. Ibid., fos.301–2.

67. Ibid., fos.307–8.

68. Ibid., fos.309–12.

69. Ibid., fos.308–9.

70. Shaftesbury was a prestigious figure, having overseen a series of important social reforms – G.B. Finlayson, *The Seventh Earl of Shaftesbury, 1801–1885* (London: Eyre Methuen, 1981).

71. CO 137/346, fos.324–26, 21 October 1859, Darling to Newcastle, fos.400–2, 10 November 1859, Darling to Newcastle.

72. CO 137/348, fos.8–17, 10 January 1860, Darling to Newcastle.

73. *Asylum Journal of Mental Science* VI (1859), p.165.

74. BPP 1860, Vol. XLV, Jamaica (Newspaper Correspondence), 18 October 1859, 'Extract From a Despatch From His Grace the Duke of Newcastle to Governor Darling'.

75. CO 137/349, fos.404–5, 31 May 1860, Darling to Newcastle; fo.410, 28 April 1860, Hutchings to Scott; fos.412–5, 15 May 1860, Scott to Hutchings. (Scott now referred to himself as 'Principal medical officer of Public Hospital and Lunatic Asylum'); fos.416–6, 11 October 1859, Scott to Trench; fo.419, 9 December 1859, Scott to Altman; fos.426–27, 25 May 1860, Hutchings to Austin; CO 137/352, fos.182–83, 20 August 1860, Commissioners in Lunacy to Colonial Office. For Elizabeth Green's baby, see note 130.

76. CO 137/350, fos.412–14, 26 June 1860, Hall to Trench; fo.415, 26 June 1860, Judith Ryan to Trench; fos.416–21, 28 June 1860, Evidence Taken

by D.P. Trench, Inspector and Director; fos.422–26, 29 June 1860, Hall to Trench.

77. CO 137/350, fos.410–11, 29 June 1860, Trench to Austin.
78. CO 137/350, fos.427–28, 4 July 1860, Austin to Trench.
79. *Official Documents on the Case of Ann Pratt, the Reputed Authoress of a Certain Pamphlet, Entitled "Seven Months in the Kingston Lunatic Asylum, and What I Saw There"* (Kingston and Spanish Town, Jamaica: Jordon and Osborn, 1860), pp.4–5. Copy in CO 137/350, fos.442–68.
80. *Official Documents on the Case of Ann Pratt*, pp.5–14.
81. A. Pratt, *Seven Months in the Kingston Lunatic Asylum, and What I Saw There* (Kingston, Jamaica: George Henderson, Savage & Co. 1860). Copy in CO 137/350, fos.429–41.
82. Jones, 'The Most Cruel and Revolting Crimes', pp.299–300. The pamphlet contained a dedication to Bowerbank. Ann stayed for a few days at his house after her discharge from the asylum, and he attended when Trench interviewed her – *Official Documents on the Case of Ann Pratt*, pp.7–8, 14.
83. Pratt, *Seven Months in the Kingston Lunatic Asylum*, pp.8–21.
84. CO 137/350, Austin to Trench, fo.403.
85. *Official Documents on the Case of Ann Pratt*, p.23.
86. *Official Documents on the Case of Ann Pratt*, pp.19, 49–54; Jones, 'The Most Cruel and Revolting Crimes', pp.298–99.
87. *Official Documents on the Case of Ann Pratt*, p.25.
88. *Official Documents on the Case of Ann Pratt*, p.22; CO 137/350, fo.396, 27 July 1860, Austin to Pratt; fo.397, 31 July 1860, Pratt to Austin; fos.398–99, 6 August 1860, Macneil to Austin; fo.400, 8 August 1860, Austin to Macneil; fos.402–4, 11 August 1860, Austin to Trench; fo.405, 26 July 1860, Austin to Kemble.
89. CO 137/350, fos.406–9, 8 August 1860, Kemble to Austin.
90. *Abuses in the Jamaica Lunatic Asylum. Magisterial Investigation into a Charge of Manslaughter, of Mrs Matilda Carey* (Kingston, Jamaica: Geo. Henderson, Savage & Co., 1860) – Copy in CO 137/363, fos.166–84.
91. *Official Documents on the Case of Ann Pratt*, pp.22–3; CO 137/355, fos.222–23, 1 September 1860, Austin to Trench.
92. R. Rouse, *New Lights on Dark Deeds: Being Jottings From the Diary of Richard Rouse, Late Warden of the Lunatic Asylum of Kingston. Edited by His Son* (Kingston, Jamaica: Gall and Myers, 1860) – Copy in National Library of Jamaica, Kingston. Rouse died shortly after his dismissal, in January 1859 (p.5).
93. Rouse, *New Lights on Dark Deeds*, pp.11, 13; CO 137/359, Public Hospital and Lunatic Asylum Commission, 15 May 1861, Evidence of Lewis Bowerbank, pp.9–12. Rouse bequeathed his manuscripts to Bowerbank, who took them with him on his voyage to England.
94. Rouse, *New Lights on Dark Deeds*, p.6. The 'friend' was doubtless Bowerbank, to whom the pamphlet was dedicated.
95. Rouse, *New Lights on Dark Deeds*, pp.11–19.
96. Ibid., pp.9, 14, 19–32.
97. CO 137/355, fos.224–25, 3 September 1860, Trench to Austin; fos.226–27, 1 September 1860, 'Order From the Inspector and Director to the

Matron of the Lunatic Asylum'; fos.228–33, 3 September 1860, Scott to Trench.

98. CO 137/355, fo.234, 3 September 1860, Austin to Trench.
99. CO 137/355, fos.235–38, 18 September 1860, Trench to Austin.
100. CO 137/355, fos.240–44, 7 September 1860, Scott to Trench.
101. CO 137/355, fos.245–49, 14 September 1860, 'Examination of Mary Clarke late a Nurse in the Female Lunatic Asylum'.
102. CO 137/355, fos.235–38, 18 September 1860, Trench to Austin; fo.250, 22 September 1860, Austin to Trench; fos.251–52, 1 October 1860, Trench to Austin; fo.253, 26 September 1860, Keech to Trench; fos.254–61, 27 September 1860, Scott to Trench.
103. CO 137/350, fos.469–70, draft letters from Colonial Office dated 19 and 22 October 1860; CO 137/352, fo.187, 16 November 1860, Commissioners in Lunacy to Colonial Office; MH 50/11, p.63, 10 October 1860.
104. CO 137/353, fos.385–86, 8 April 1861, Darling to Newcastle; Jones, 'The Most Cruel and Revolting Crimes', pp.296–97.
105. CO 137/353, fos.388–89, 16 May 1861, Newcastle to Darling (draft). Dr Keech was now house surgeon.
106. CO 137/355, fos.208–20, 20 June 1861, Darling to Newcastle. Interestingly, Darling gave 17th June as the date of the Enquiry's commencement, though it actually started on 14th May; 17th June was when Scott and others began their response to the allegations.
107. CO 137/359, Public Hospital and Lunatic Asylum Commission, fo.1; Jones, 'The Most Cruel and Revolting Crimes', pp.295, 306.
108. CO 137/359–363 – the evidence is contained within five bound folio volumes, in addition to the final report.
109. CO 137/359, 15–16 May 1861, pp.3–147.
110. CO 137/366, fos.245–51, 'Report on the Management of the Public Hospital', 20 November 1861, pp.1–2.
111. CO 137/361, 17–18 June 1861, pp.1–123, 19–20 June, pp.194–214; CO 137/362, 20 August 1861, pp.1–267.
112. CO 137/361, 19 June 1861, pp.125–93; CO 137/362, 21 August 1861, pp.285–88.
113. CO 137/362, 22 August 1861, pp.303–17.
114. CO 137/360, 29 May 1861, pp.125–59.
115. CO 137/363, pp.235–38, 'Narrative of Mary Bell'.
116. CO 137/360, 28 May 1861, pp.88–94.
117. CO 137/360, 3 June 1861, pp.227–38, 4 June, pp.242–50.
118. CO 137/363, pp.251–98, 'Henrietta Dawson's Statement'.
119. CO 137/366, fos.245–51, 'Report on the Management of the Public Hospital', 20 November 1861.
120. 'Report on the Management of the Public Hospital', pp.2–4.
121. Ibid., pp.6–8, 12. Dr Keech had regularly consorted in the hospital with women of 'loose character', whilst Dr Scott was financially indebted to David and Judith Ryan.
122. Ibid., pp.8–9.
123. Ibid., pp.9, 13.
124. Ibid., pp.2, 6, 12.
125. Ibid., pp.3–4, 10–11.
126. CO 137/363, Henrietta Dawson's Statement, p.260.

127. CO 137/366, p.10.
128. CO 137/363, pp.257, 279.
129. CO 137/360, 28 May 1861, p.89.
130. CO 137/360, 29 May 1861, pp.125–29 – he claimed that Letitia West, a quiet patent, died from her injuries. Elizabeth Green, whose baby died in the asylum, allegedly became pregnant by a patient nick-named 'Omnibus' whilst locked up on the male side – CO 137/360, 5 June 1861, p.211, evidence of Ann Brown.
131. CO 137/366, p.10.
132. Ibid.
133. CO 137/363, p.256.
134. CO 137/363, 'Narrative of Mary Bell', p.237.
135. CO 137/361, 19 June 1861, p.195, 22 August 1861, pp.307–9.
136. CO 137/366, pp.11–12; Jones, 'The Most Cruel and Revolting Crimes', pp.291–92.
137. CO 137/363, p.262.
138. CO 137/360, 28 May 1861, pp.89–90. See opening quote at beginning of chapter.
139. CO 137/360, 29 May 1861, pp.133–34.
140. Pratt, *Seven Months in the Kingston Lunatic Asylum*, pp.9–16, 21; Jones, 'The Most Cruel and Revolting Crimes', pp.291, 296.
141. CO 137/366, pp.10–12.
142. CO 137/360, 5 June 1861, pp.210.
143. Pratt, *Seven Months in the Kingston Lunatic Asylum*, p.13; *Official Documents on the Case of Ann Pratt*, p.13.
144. Pratt, *Seven Months in the Kingston Lunatic Asylum*, p.9; *Official Documents on the Case of Ann Pratt*, p.10.
145. CO 137/360, 5 June 1861, p.213.
146. CO 137/359, 22 May 1861, evidence of Alexander Naar de Graffe, pp.284–85 – De Graffe was an external contractor; he paid Steele for his work in money, tobacco, or alcohol; CO 137/361, 19 June 1861, evidence of James Scott, p.197 – Scott described Steele as 'very useful at the Institution, and did a vast quantity of work for the Institution'.
147. Rouse, *New Lights on Dark Deeds*, pp.20, 23, 25.
148. CO 137/360, 27 May 1861, evidence of Charlotte Campbell, p.59.
149. CO 137/366, p.10.
150. CO 137/361, 19 June 1861, p.202.
151. CO 137/359, 22 May 1861, p.264.
152. CO 137/359, 21 May 1861, p.260.
153. CO 137/360, 28 May 1861, p.91.
154. CO 137/360, 1 June 1861, p.197.
155. CO 137/363, 'Henrietta Dawson's Statement', pp.267–68.
156. CO 137/360, 27 May 1861, evidence of Charlotte Campbell, p.62. Charlotte had been Dr Keech's mistress, and resided in hospital staff quarters.
157. CO 137/363, pp.268–69.
158. CO 137/366, 'Report on the Management of the Public Hospital', p.11.
159. Green, *British Slave Emancipation*, pp.12–22, 295–7, 319–21; Heuman, *Between Black and White*, pp.72–9.

160. For the social characteristics of White Jamaican Creoles, see Brathwaite, *The Development of Creole Society in Jamaica*, chapters 9, 10, 13; Moore and Johnson, *Neither Led Nor Driven*, pp.140–48.
161. Rouse, *New Lights on Dark Deeds*, p.35.
162. Heuman, *Between Black and White*, pp.108, 161. Darling had served in several colonies. He had been Lieutenant Governor of St Lucia and of the Cape of Good Hope, and Governor of Newfoundland before becoming Governor of Jamaica in 1857. He also had previous experience in Jamaican politics, being briefly a member of the House of Assembly in 1838.
163. Heuman, *Between Black and White*, pp.163–66, 172–8. Jordon and Osborn both had involvement in the commission appointed to oversee the affairs of the hospital.
164. CO 137/343, 26 January 1859, Darling to Bulwer Lytton, fos.124–25, 134–44.
165. Ibid., fo.138.
166. Ibid., fos.139–40.
167. CO 137/343, fo.316, 25 January 1859, Leake to Austin.
168. CO 137/355, 20 June 1861, Darling to Newcastle, fos.210–12; Jones, 'The Most Cruel and Revolting Crimes', p.302.
169. CO 137/346, 23 September 1859, Trench to Darling, fo.291.
170. Bowerbank, *A Circular Letter to the Individual Members of the Legislative Council and of the House of Assembly*, p.10. The nature of the allegations is not clear.
171. CO 137/361, 22 August 1861, evidence of Judith Ryan, pp.315–16.
172. *The Times*, 30 August 1859; MH 50/10, p.182, 19 April 1859, p.188, 4 May 1859, p.192, 11 May 1859.
173. *Official Documents on the Case of Ann Pratt*, p.7.
174. CO 137/363, 20 August 1861, fo.178.
175. *Official Documents on the Case of Ann Pratt*, pp.32–3.
176. CO 137/363, fo.178.
177. Rouse, *New Lights on Dark Deeds*, p.28. 'F.L.A.' stood for Female Lunatic Asylum; the male patients' uniform trousers proclaimed 'M.L.A.'
178. *Official Documents on the Case of Ann Pratt*, p.12.
179. CO 137/360, 4 June 1861, evidence of Margaret Jane Ravenshill, p.257.
180. Pratt, *Seven Months in the Kingston Lunatic Asylum*, p.9.
181. CO 137/363, 'Henrietta Dawson's Statement', pp.285–88.
182. Rouse, *New Lights on Dark Deeds*, p.35.
183. Pratt, *Seven Months in the Kingston Lunatic Asylum*, p.12; *Official Documents on the Case of Ann Pratt*, pp.11–12.
184. Rouse, *New Lights on Dark Deeds*, p.27.
185. CO 137/355, 14 September 1860, Examination of Mary Clarke, fo.246.
186. Rouse, *New Lights on Dark Deeds*, pp.15–16.
187. CO 137/360, 4 June 1861, evidence of Aaron Da Costa, fos.277–81; CO 137/361, 19 June 1861, evidence of Dr James Scott, pp.198–200; CO 137/366, 'Report on the Management of the Public Hospital', p.10. There were unsubstantiated insinuations that Da Costa had been sodomised.
188. CO 137/360, 5 June 1861, evidence of Ann Brown, pp.208–9. Dr Scott later sought to discredit Ann Brown as a witness, suggesting that she was 'a strong woman, and useful in the Asylum, but sometimes the matron was

obliged to restrain her lest she should ill-treat any of the inmates' – CO 137/366, 20 February 1862, Scott to Trench, fo.444.

189. CO 137/366, 'Report on the Management of the Public Hospital', p.12.

190. CO 137/364, fos.417–18, 11 January 1862, Austin to Trench.

191. CO 137/366, fos.252–370, 18 January 1862, 'Dr Scott's defence statement to the Governor'; fos.371–524, 20 February 1862, Scott to Trench.

192. CO 137/366, fos.379–80.

193. CO 137/364, 28 February 1862, Darling to Newcastle, fos.407–10.

194. CO 137/365, fos.165–66, 20 March 1862, Darling to Newcastle; fo.216, 24 March 1862, Darling to Newcastle. Darling's recall from Jamaica proved permanent. It stemmed largely from Colonial Office disquiet at his continuing entanglement in party political controversy and rancorous arguments with Coloured politicians; Darling was appointed Governor of Victoria, Australia – Heuman, *Between Black and White*, pp.174–8. His replacement was the even more controversial Edward Eyre.

195. CO 137/365, fos.218–23, Scott to Newcastle.

196. CO 137/365, fos.243–62, 14 August 1862, draft despatch Henry Taylor to Edward Eyre (initialled by Newcastle); CO 137/368, fos.18–20, 9 October 1862, Eyre to Newcastle.

197. CO 137/366, p.12.

198. CO 137/364, fo.419, 28 January 1862, Austin to Trench.

199. CO 137/365, fos.247–50. Cited in Jones, 'The Most Cruel and Revolting Crimes', p.290.

200. CO 137/368, fos.18–20, 9 October 1862, Eyre to Newcastle.

201. CO 137/368, 3 December 1862, Eyre to Newcastle, fo.549; Jones, 'The Most Cruel and Revolting Crimes', pp.300, 302.

202. CO 137/387, fos.293–310, 25 January 1865, Eyre to Cardwell.

203. *Jamaica Gleaner*, 3 June 1892.

204. Heuman, *The Killing Time*, pp.12, 104, 146–50, 157; TNA, PRO 30/48/44, Cardwell Scrap Book, fos.1–9, 'Jamaica' by Lewis Bowerbank. Bowerbank acknowledges (fo.3) that Gordon had supported him in his campaign for reform of the hospital and asylum, until Dr Scott's dismissal.

205. Jones, *Public Health in Jamaica*, pp.75–80.

206. *Jamaica Gleaner*, 3 June 1892.

207. See Morissey, 'The Road to Belle Vue', pp.3–5. The early development of the asylum will be discussed more fully in Chapter 4.

208. CO 137/343, 26 January 1859, Darling to Bulwer Lytton, fos.126–29. The Duke of Newcastle accepted arguments for appointing an experienced medical officer from England, following advice from the Commissioners in Lunacy – CO 137/345, fos.171–82, 10 June 1859, Darling to Bulwer Lytton, fos.183–84, 19 July 1859, (draft letter) Newcastle to Darling.

209. CO 137/346, 10 October 1859, Darling to Newcastle, fos.278–80.

210. CO 137/348, 10 January 1860, Darling to Newcastle, fos.12–13.

211. CO 137/353, fos.47–9, 23 January 1861, Darling to Newcastle.

212. CO 137/367, fos.272–73, 6 August 1862, Eyre to Newcastle.

213. CO 137/365, fos.239–42, 26 July 1862, Memorandum from Taylor to Newcastle.

214. CO 137/365, fos.234–38, 17 September 1862, draft letters from Newcastle to Commissioners in Lunacy and College of Physicians. For consideration of

the empire-wide survey and its consequences, see Swartz, 'The Regulation of British Colonial Lunatic Asylums and the Origins of Colonial Psychiatry, 1860–1864'.

215. CO 137/365, fos.234–36; CO 885/3/4, 14 January 1864, 'Colonial Hospitals and Lunatic Asylums'; another copy in CO 854/7, fos.447–67; Jones, 'The Most Cruel and Revolting Crimes', pp.302–3.

216. CO 885/3/4, 'Colonial Hospitals and Lunatic Asylums', p.4.

217. CO 885/3/4, pp.2, 4–5, 24–6.

218. CO 885/3/4, pp.13–22; Swartz, 'The Regulation of Colonial Lunatic Asylums', pp.168–69; Jones, 'The Most Cruel and Revolting Crimes', p.303.

219. Swartz, 'The Regulation of Colonial Lunatic Asylums', pp.170–72.

220. BPP 1814–15, Vol. IV, Select Committee on the State of Madhouses; 1816, Vol. VI, Select Committee on the State of Madhouses; 1826–27, Vol. VI, Select Committee on Pauper Lunatics in the County of Middlesex and on Lunatic Asylums; *Report of the Metropolitan Commissioners in Lunacy to the Lord Chancellor* (1844); Scull, *The Most Solitary of Afflictions*, pp.115–74; K. Jones, *A History of the Mental Health Services*, pp.64–149.

221. Deacon, 'Madness, Race and Moral Treatment: Robben Island Lunatic Asylum, Cape Colony, 1846–90', pp.290–91.

222. Monk, *Attending Madness*, pp.41–9.

223. See Chapter 2.

224. CO 137/365, fos.239–42, 26 July 1862, Memorandum from Henry Taylor.

225. Morissey, 'The Road to Belle Vue', p.6; Hall, *Free Jamaica*, pp.81–104.

226. I am grateful to Peter Barham for emphasising this point in a conference presentation in Cardiff in November 2011. There were other patient testimonies, including that of Rosa Henry reproduced by Richard Rouse – *New Lights on Dark Deeds*, pp.23–4.

4 Reform – The Jamaica Lunatic Asylum

1. Bancroft, *A Letter to the Honourable Hector Mitchel*, p.3.

2. *The Builder*, 1 June 1844, p.282; Morissey, 'The Road to Belle Vue', pp.3–5; Conolly, *The Construction and Government of Lunatic Asylums*, Appendix, pp.182–83.

3. BPP 1849, Vol. XXXIV, Reports to Secretary of State on Past and Present State of Her Majesty's Colonial Possessions, 1848, p.89; Morissey, 'The Road to Belle Vue', p.5.

4. J. Macfadyen, 'On Medical Topography as Connected With the Choice of Site for a Lunatic Asylum in a Tropical Country', *Edinburgh Medical and Surgical Journal* LXXI (1849), 114–25.

5. CO 137/343, 26 January 1859, Darling to Bulwer Lytton, fos.126–28; CO 137/346, fos.316–18, 24 June 1859, Leahy to Trench.

6. *The Times*, 30 August 1859; *Journal of Mental Science* VI (1859), 163–64.

7. TNA: CO 137/353, fos.47–9, 23 January 1861, Darling to Newcastle; CO 137/367, 6 August 1862, Eyre to Newcastle, fo.272.

8. CO 137/382, fos.78–82, 10 May 1864, Eyre to Cardwell, fos.83–93, 12 December 1863, Lake to Trench.

9. CO 137/345, 10 June 1859, Darling to Bulwer Lytton, fos.179–80.

10. MH 50/10, Commissioners in Lunacy, 27 July 1859; CO 137/364, fos.278–91, 22 February 1862, Darling to Newcastle; CO 137/372, fos.225–28, 22 May 1863, Eyre to Newcastle. It was agreed also to recruit a 'Warden' and 'Matron' from England, at salaries of £100 and £84 respectively.

11. Lincolnshire Archives, HOSP/ST JOHN'S 1/1/3, Minute Book of Visitors, 12 October 1854, 29 July 1864.

12. *Tenth Annual Report of the Lincolnshire County Lunatic Asylum, at Bracebridge, Near Lincoln* (Lincoln: W. And B. Brooke,1863), pp.3–4, 11.

13. *Fifth Annual Report of the Lincolnshire County Lunatic Asylum* (Lincoln: W. And B. Brooke, 1858), p.12.

14. CO 137/364 (1862), fos.278–79; CO 137/372 (1863), fos.313–15. Allen was receiving just £70 per annum at the Lincolnshire asylum.

15. CO 137/375, fos.187–89, 23 October 1863, Eyre to Newcastle, fo.215, 2 November 1863, Eyre to Newcastle. Freshney and Miss Brown subsequently married – CO 137/481, fos.312–15, 27 June 1876, Grey to Carnarvon.

16. CO 137/382, 2 April 1864, Allen to Eyre, fos.306–9. Edward Eyre took over as governor from Charles Darling during 1863. He became the focus of bitter controversy for his uncompromising role in the Morant Bay rebellion of 1865 (see Chapter 1).

17. CO 137/382, 2 April 1864, fos.306–456.

18. Ibid., fos.310–73, 428–46.

19. Fo.313.

20. Fos.320–21.

21. Fos.317–18.

22. Fos.308, 324, 330, 334–35, 340, 356, 369, 438.

23. Fos.328, 332–33.

24. Fos.327–28.

25. Fo.371.

26. Fos.333–34.

27. E.g. fos.332, 338–44, 364–70.

28. Fo.319.

29. Fo.328.

30. Fo.330.

31. Fos.310, 339–44.

32. Fos.364–65.

33. Fo.378.

34. Allen's disparaging and racist attitudes to the patients will be considered further in Chapter 6.

35. Fos.439–44.

36. Fos.446–50.

37. Fos.450–55.

38. Fo.453.

39. Fos.376–77.

40. Fos.372–73.

41. Fos.347–48.

42. Fo.337.

43. Fos.348–52.

44. Fos.435–36.
45. Fo.436.
46. Fo.439 (quotes); CO 137/388, fos.143–45, 1 December 1864, Report of Chaplain, Wyndham F. Serres.
47. CO 137/382, fos.467–75, 4 May 1864, Lake to Trench.
48. CO 137/382, 28 May 1864, Trench to Austin, fos.509–10. Daniel Trench had retained his role of Inspector and Director of both the Kingston public hospital and the new lunatic asylum.
49. CO 137/382, 28 May 1864, Eyre to Cardwell, fos.303–4.
50. Ibid., fos.300–4, 28 May 1864, Eyre to Cardwell. See also note by Henry Taylor, 7 July 1864, fos.304–5, expressing satisfaction regarding Allen's appointment and the approval he had earned.
51. CO 137/383, 20 June 1864, Eyre to Cardwell, fos.393–95. Cardwell sent the plans and estimates to the Commissioners in Lunacy for their opinion – fo.427, 27 July 1864.
52. CO 137/388, fos.68–73, 2 February 1865, Eyre to Cardwell.
53. CO 137/388, fos.75–136, 21 October 1864, Allen to Austin.
54. Ibid., fos.85–90.
55. Fos.133–36.
56. Fo.92.
57. Fos.104–6.
58. Fos.92–3.
59. Fos.93–4.
60. Fos.97–102.
61. Fos.102–3.
62. CO 137/388, fos.68–73, 2 February 1865, Eyre to Cardwell.
63. CO 137/388, fos.68–72, 111–12, 125–36.
64. CO 137/389, fos.5–7, 10 March 1865, 'Extract From the Visitors Book, New Lunatic Asylum'.
65. *Jamaica Gleaner* (henceforth *JG*), 8 June 1867.
66. See Chapter 1.
67. CO 137/388, fos.375–76, 16 March 1865, Eyre to Cardwell, fos.378–83, 23 February 1865, Pennell to Austin; BPP 1867, Vol. XLVIII, Reports to Secretary of State on the Past and Present State of Her Majesty's Colonial Possessions, 1865, p.6.
68. BPP 1871, Vol. XLVII, Reports to the Secretary of State, 1869, p.15; CO 137/445, fos.439–40, 9 February 1869, Commissioners in Lunacy to Sandford; fo.442, 28 February 1869, draft letter from Colonial Office to Grant. Sir John Grant had replaced Edward Eyre after the Morant Bay debacle. Grant was clearly much less in sympathy with Allen. By late 1869 relations between the two men had become extremely strained – *Jamaica Gleaner*, 3 November 1869.
69. BPP 1871, Vol. XLVII, p.15; BPP 1871, Vol. XLII, Reports to the Secretary of State, 1870, p.19.
70. *Jamaica: Annual Reports of the Superintending Medical Officers* (Kingston: General Printing Office, 1898–1907), Annual Report on the Lunatic Asylum, Year Ended 31 March 1907, p.15. [Copies in Library of London School of Hygiene and Tropical Medicine.]

71. BPP 1875, Vol. LI, Papers Relating to Her Majesty's Colonial Possessions, 1875, Part I, p.32, Part III, p.45; BPP 1877, Vol. LIX, Papers Relating to Her Majesty's Colonial Possessions, 1875–77, Part II, p.14; BPP 1881, Vol. LXIV, Papers Relating to Her Majesty's Colonial Possessions, 1879, p.146; BPP 1882, Vol. XLIV, Papers Relating to Her Majesty's Colonial Possessions, 1881, p.27; CO 140/83, Report of the Medical Superintendent and Director of the Jamaica Lunatic Asylum for the Year 1878/79, p.25; *JG*, 13 January 1877.

72. CO 137/484, fos.396–433, Report of Medical Superintendent of the Jamaica Lunatic Asylum for 1875/76.

73. CO 140/81, Report of Medical Superintendent for 1879/80, pp.196–98; *JG*, 20 June 1880.

74. CO 137/484, Allen to Colonial Secretary, 31 May 1876, fos.589–90; CO 140/86, Annual Report of the Jamaica Lunatic Asylum, year ended 30th September 1883, p.167; BPP 1873, Vol. L, Report on Leprosy and Yaws in the West Indies, Addressed to Her Majesty's Secretary of State for the Colonies, by Gavin Milroy, M.D., p.50.

75. CO 140/83, Report of Medical Superintendent for 1878/79, p.25.

76. Ibid., pp.24–5.

77. CO 137/477, fos.190–99, 30 July 1874, Inspection of Lunatic Asylums in the West Indies; CO 137/478, fos.327–35, 4 May 1874, Proposed Appointment of Dr Allen as Inspector for Lunatic Asylums in the West Indies; CO 31/65, Proceedings of Barbados House of Assembly, 1874–75, Appendix B, pp.3–4; CO 321/1, fos.285–91, 5 August 1874, Proposed Appointment of Dr Allen as Medical Inspector of Lunatic Asylums in West Indies. For the appointment of Inspectors elsewhere, see Coleborne, *Madness in the Family*, pp.33–5; Swartz, 'The Great Asylum Laundry', p.200.

78. CO 137/481, fos.346–59, 3 June 1876, Allen to Colonial Secretary; CO 31/67, Proceedings of Barbados House of Assembly, Appendix BBB, p.95, 18 October 1876, Circular from Earl of Carnarvon.

79. CO 137/481, fos.343–45, 28 June 1876, Grey to Carnarvon; CO 137/482, fos.563–64, 3 October 1876 – a critical senior civil servant suggested that it was 'now pretty well known what sort of man he is'.

80. CO 295/280, fos.187–89, 26 February 1878, 'Dr Allen's Appt as Travelling Inspector of Lunatic Asylums in the West Indies'.

81. CO 140/81, Report of Medical Superintendent for 1879/80, p.197.

82. CO 140/86, Annual Report of the Jamaica Lunatic Asylum, Year Ended, 30 September 1883, p.167.

83. CO 140/81, Report of the Medical Superintendent for the Year 1879/80, pp.194–95.

84. CO 140/84, Report of the Medical Superintendent for the Year 1880/81, p.97.

85. Ibid., p.99.

86. CO 140/86, Report of the Medical Superintendent for the Year 1881/82, p.171.

87. CO 140/86, Annual Report, 30 September 1883, pp.165–66.

88. CO 140/89, Annual Report, 30 September 1884, pp.163–64; Annual Report, 30 September 1885, pp.64–5; CO 140/94, Annual Report, 30 September

1886, p.46. Departures to Panama led the asylum band to break up, contributing to a decline in entertainment provision for the patients.

89. CO 140/89, Annual Report, 30 September 1885, pp.66–8.

90. *JG*, e.g. 12 August, 17 November 1880, 13 September 1883.

91. *JG*, 13 June, 13 July 1885, 5 March, 22 September 1886, 19 January 1887.

92. CO 137/527, fos.292–305, 31 August 1886, Retirement of Dr Allen; *JG*, 23 October 1886; C140/94, Annual Report, 30 September, 1887, p.205.

93. CO 137/527, fos.292–93, 303–5; CO 137/528, fos.495–506, 24 November 1886, Appt of Med Super Lunatic Asylum; CO 137/533, fos.461–73, 4 January 1887, Appointment of Med Super, Lunatic Asylum.

94. M. Jones, *The Hospital System and Health Care: Sri Lanka, 1815–1960*, pp.188–90.

95. CO 140/94, Annual Report, 30 September 1887, pp.205–8.

96. BPP 1887, Vol. LVII, Papers Relating to Her Majesty's Colonial Possessions, 1885 and 1886, p.34.

97. *Jamaica: Annual Reports*, Annual Report on the Lunatic Asylum, Year Ended 31 March 1898, p.8.

98. BPP 1889, Vol. LIV, Papers Relating to Her Majesty's Colonial Possessions, 56, Jamaica, p.16; CO 140/99, Annual Report, 30 September 1888, pp.98–100.

99. CO 140/99, Annual Report, 30 September 1889, p.162.

100. Ibid., p.161.

101. *Jamaica: Annual Reports of the Superintending Medical Officers*, Annual Report on the Lunatic Asylum, Year Ended 31 March 1907, p.5.

102. *JG*, 1 August 1890.

103. *Daily Gleaner* (henceforth *DG*), 27 October 1891.

104. *DG*, 29 July, 9 November 1892; 29 September 1893.

105. *DG*, 29 July, 9 November 1892.

106. *DG*, 16 July, 9 November 1892.

107. *DG*, 20 March 1893.

108. *DG*, 29 September 1893, 2 May 1894. Crane had previously been surgeon-general in Trinidad, where he was a strong supporter of reforms – see Chapter 5.

109. *DG*, 17 September 1894.

110. *DG*, 16 July 1895, 29 May 1896.

111. *DG*, 31 October 1895, 29 May 1896.

112. Annual Report, 31 March 1898, pp.1–2; *DG*, 18 September 1897, 16 December 1898; BPP 1899, Vol. LXI, *Annual Series of Colonial Reports*, 261, Jamaica, p.13.

113. *DG*, 22 November 1899.

114. *DG*, 20 December 1901, 12 September 1902.

115. *DG*, 5 January, 10 July, 7 October, 3 December 1903, 21 January 1904.

116. *DG*, 28 March 1904. Williams had previously acted as medical superintendent in Plaxton's absence. He had been a medical officer in Jamaica since 1893, and was generally well regarded.

117. *DG*, 7 June, 9 August, 30 September 1904.

118. *Jamaica: Annual Reports*, Report on the Lunatic Asylum, Year Ended 31 March 1905, pp.36–7.

119. *Jamaica: Annual Reports*, Lunatic Asylum, 31 March 1906, pp.1–2.

120. *Jamaica: Annual Reports*, Lunatic Asylum, 31 March 1907, pp.1–2 – the 'foreign press' gave 'lurid accounts' of 'unspeakable enormities' committed by escaped lunatics.
121. *DG*, 15 September 1908, 19 August 1910.
122. *DG*, 6 September 1912.
123. *DG*, 22 September 1911.
124. *DG*, 6 September 1912.
125. *DG*, 27 August 1914.
126. *JG*, 18 August 1913.
127. Heuman, *'The Killing Time'*.
128. C140/86, Annual Report, 30 September 1883, p.167; C140/89, Annual Report, 30 September 1885, pp.65–67.
129. Scull, *The Most Solitary of Afflictions*, pp.277–93, 303–33; Mahone and Vaughan, *Psychiatry and Empire*, pp.5–6; Marks, ' "Every Facility that Modern Science and Enlightened Humanity Have Devised" ', pp.275–277; Yeong, *Till the Break of Day*, pp.19, 77, 90; Parle, *States of Mind*, pp.97–100, 111–13.

5 Colonial Asylums in Transition

1. CO 885/3/4, 'Colonial Hospitals and Lunatic Asylums', pp.1, 4–5, 25–7.
2. BPP 1871, Vol. XLVII, Reports to Secretary of State on the Past and Present State of Her Majesty's Colonial Possessions, 1869, pp.47–9.
3. CO 111/394, 25 July 1872, 'Report on Lunatic Asylum. Gavin Milroy to Under Secretary of State for the Colonies.' Colonial Office officials were unconvinced of the need for a visit to Jamaica, pointing out that the British Guiana government had previously been 'furnished by H.M. Govt with a mass of documentary instruction which if diligently read would afford a complete knowledge of the principles at least of Asylum management.' – note, 12 August 1872.
4. BPP 1875, Vol. LI, Papers Relating to Her Majesty's Colonial Possessions, Part III, pp.67, 74.
5. L. Gramaglia, 'Colonial Psychiatry in British Guiana: Dr Robert Grieve', in K. White (ed.), *Configuring Madness: Representation, Context and Meaning* (Oxford: Inter-Disciplinary Press, 2009), pp.191–206; L. Gramaglia, 'Dr Robert Grieve (1839–1906), An "Apostle of Science" ' – 'Introduction', in R. Grieve (eds.), *The Asylum Journal* (Guyana: The Caribbean Press, 2010), Vol. 1, pp.xiii–xxxii.
6. BPP 1875, Vol. LI, Papers Relating to Her Majesty's Colonial Possessions, Part III, p.74.
7. *The Asylum Journal for 1881. Conducted by the Medical Superintendent of the Public Lunatic Asylum for British Guiana*, April 1881, p.18; BPP 1888, Vol. LXXII, Papers Relating to Her Majesty's Colonial Possessions, 1884 and 1885, 39, British Guiana, p.42; BPP 1893–94, Vol. LIX, Annual Series of Colonial Reports, 55, British Guiana, 1891, p.34.
8. For example, *The Asylum Journal* (henceforth *TAJ*), 2 January 1882, p.86; 15 April 1882, p.20; 15 June 1882, p.38; 15 August 1882, p.54; 15 May 1883, p.29; 15 June 1883, p.38; 15 October 1884, p.58; 15 June 1885, p.104.

[Copies of *The Asylum Journal* are available in the British Library and the Wellcome Library.]

9. *TAJ*, 1 August 1881, p.46; 15 April 1882, p.20; 15 June 1883, p.38; 15 May 1885, pp.95–6.
10. BPP 1871, Vol. XLVII, Reports to Secretary of State, 1869, pp.47–9.
11. *TAJ*, 1 September 1881, pp.50–1.
12. BPP 1875, Vol. LI, Papers Relating to Her Majesty's Colonial Possessions, Part III, p.77; *TAJ*, 1 July 1881, p.34.
13. *TAJ*, 1 November 1881, pp.65–6.
14. In late 1885 Grieve left to become the colony's acting surgeon-general, as well as Medical Officer to the Immigration Service – *TAJ*, 16 November 1885, p.137.
15. *TAJ*, 1 January 1882, p.86; 15 May 1882, p.30.
16. *TAJ*, 1 July 1881–15 September 1884.
17. *TAJ*, 1 March 1881, p.2.
18. *TAJ*, 16 January 1882, p.91; 15 January 1883, p.91; 15 June 1883, p.38; 15 January 1884, p.91; 15 January 1885, p.71; 15 January 1886, p.149.
19. *TAJ*, 1 March 1881, p.2.
20. *TAJ*, 1 April 1881, pp.7–8. This aspect will be considered more fully in Chapter 7.
21. *TAJ*, 1 April 1881, pp.8–9.
22. Gramaglia, 'Dr Robert Grieve, An "Apostle of Science"'. The journal detailed aspects of the asylum's operation, and also included articles on general mental health issues as well as wider public health matters.
23. *TAJ*, 1 March 1881, pp.4–5. The asylum housed almost twice as many male patients as females, reflected in the numbers employed.
24. *TAJ*, 1 July 1881, p.39.
25. *TAJ*, 15 April 1882, pp.23–4.
26. *TAJ*, 15 June 1882, p.39. There are problems with these figures, as they probably included people only able to work a few hours per week.
27. *TAJ*, 15 March 1883, pp.14–15.
28. *TAJ*, 1 September 1881, p.55; 1 November 1881, p.70; 16 June 1883, p.39; 15 July 1883, pp.46–7; 15 January 1884, pp.91–2; 15 August 1884, p.50; 15 December 1885, p.144.
29. *TAJ*, 15 December 1883, p.63; 15 January 1886, p.150.
30. *TAJ*, 15 March 1883, p.15; 15 May 1883, p.31; 15 May 1885, p.96; 15 July 1885, p.112.
31. *TAJ*, 15 May 1882, p.31; 15 February 1883, p.6; 15 January 1884, p.92; 15 July 1884, p.47.
32. *TAJ*, 15 January 1885, pp.69, 71.
33. *TAJ*, 1 March 1881, p.4; 1 April 1881, p.14; 1 May 1881, pp.21–2; 1 September 1881, pp.53–4; 15 March 1882, p.14; 15 April 1883, pp.21–2; 15 May 1883, p.29; 15 January 1884, p.91.
34. *TAJ*, 1 April 1881, p.13; 1 May 1881, p.21; 1 August 1881, p.45; 1 September 1881, p.54; 1 October 1881, p.62; 15 July 1882, p.46; 15 September 1883, p.62 – visitors even came from distant Georgetown; 15 November 1884, pp.61–2.
35. BPP 1888, Vol. LXXII, Papers Relating to Her Majesty's Colonial Possessions, 39, British Guiana, Reports on the Blue Books for 1884 and 1885, p.29; *TAJ*,

15 January 1886, p.147; Gramaglia, 'Dr Robert Grieve', pp.xix–xx. Grieve's 'acting' appointment was soon made permanent. He continued to take an interest in asylum affairs, retaining overall responsibility for publication of its annual reports.

36. *TAJ*, 15 August 1884, p.49.
37. *British Guiana: Reports of the Surgeon-General* (Georgetown, Demerara: C.K. Jardine, Printer to the Government), 1895–96, pp.42, 45; 1896–97, p.38; 1899–1900, pp.39, 43–4. [Copies in Library of LSHTM].
38. *Reports of the Surgeon-General* – 1895–96, pp.42, 45; 1899–1900, p.43.
39. *Reports of the Surgeon-General* – 1895–96, pp.42–4; 1896–97, pp.38–9; 1898–99, pp.43–4; 1899–1900, pp.39–40.
40. See Chapter 2.
41. *Port of Spain Gazette*, 19 February 1859. Murray also acted as medical officer to the Royal Gaol.
42. *Port of Spain Gazette*, 11 February 1860.
43. *Port of Spain Gazette*, 4 February 1863.
44. *Port of Spain Gazette*, 3 February 1864; 4 March 1865; BPP 1865, Vol. XXXVII, Reports to Secretary of State, 1863, pp.28–9.
45. BPP 1871, Vol. XLVII, Reports to Secretary of State, 1869, p.66; BPP 1871, Vol. XLII, Reports to Secretary of State, 1870, p.68; BPP 1873, Vol. XLVIII, Papers Relating to Her Majesty's Colonial Possessions, 1873, p.89.
46. CO 295/76, fo.465, Annual Report for 1875; Trinidad Archives, Proceedings of the Legislative Council (1877): Lunatic Asylum, Annual Report of the Medical Superintendent for 1876, pp.1–2.
47. CO 295/76, fos.461–62, 25 March 1876, Irving to Carnarvon, and note on fo.466.
48. CO 295/78, fos.295–96, note 27 April 1877.
49. CO 295/78, fos.144–45, 9 March 1877, Des Voeux to Carnarvon; fos.191–94, 4 April 1877, Des Voeux to Carnarvon; fos.295–96, 27 April 1877. Murray was granted a pension of £500 per annum, equivalent to his salary as medical attendant to the asylum and jail. Des Voeux succeeded Governor Irving until his return in early 1878.
50. *Port of Spain Gazette*, 19 February 1859; CO 295/78, fos.196–201, 4 April 1877, Crane to Colonial Secretary – the resident superintendent received a substantial £350 per annum plus board.
51. CO 295/78, fos.191–94, 4 April 1877, Des Voeux to Carnarvon; fos.196–98, 4 April, 1877, Crane to Colonial Secretary; fos.200–1, 14 February 1877, Martin to Crane. Martin had come to Trinidad in 1876, as part of a Colonial Office scheme.
52. CO 295/78, fos.527–28, 25 August 1877, Des Voeux to Carnarvon; fos.529–30, 25 July 1877, Crane to Colonial Secretary.
53. CO 295/81, fos.113–14, 25 May 1878, Irving to Hicks Beach.
54. BPP 1877, Vol. LIX, Papers Relating to Her Majesty's Colonial Possessions, 1876, p.46; BPP 1878–79, Vol. L, Papers Relating to Her Majesty's Colonial Possessions, 1877, Part 1, pp. 27,31; BPP 1882, Vol. XLIV, Papers Relating to Her Majesty's Colonial Possessions, 1881, p.103.
55. CO 295/87, fos.106–8, 8 July 1880, Irving to Kimberley; CO 295/88, fos.9–10, 13 September 1880, Martin to Crane; CO 298/36, Trinidad Legislative Council (1881), no.15, Report of Medical Superintendent for 1880, pp.1–2.

56. CO 295/91, fos.426–29, 1 November 1881, Freeling to Kimberley; fos.431–36, 29 October 1881, Crane to Freeling. Over recent months Martin had taken on extra duties. In addition to medical superintendence of the asylum and prison, he was Health Officer for Shipping, Medical Inspector for Immigrants, and Medical Attendant of the Police Force. He had temporarily taken on the role of Medical Officer for the troops at St James's barracks during a fever outbreak. His premature death was 'clearly attributable to his zeal and devotion to his duties'.

57. CO 295/91, 1 November 1881, Freeling to Kimberley, fos.427–28; fos.707–10, 28 November 1881, Freeling to Kimberley; fos.711–17, 28 November 1881, Crane to Freeling.

58. London Metropolitan Archives, Metropolitan Asylum District: Board Minutes, Vol. IX, 1875/76, p.105; Vol. X, 1876/77, Medical Superintendent's Report, 1 October 1876, p.549; Vol. XV, 1881/82, Medical Superintendent's Report, 31 December 1880, p.88; Vol. XVI, 1882/83, Medical Superintendent's Report, 31 December 1881, p.48; Vol. XV, 1883/84, Medical Superintendent's Report, 31 December 1882, p.390.

59. London Metropolitan Archives, MAB/0251, Committee Minutes, 16 June 1882; Board Minutes, Vol. XV, 1883/84, p.390.

60. CO 298/39, Report of Medical Superintendent for 1882, pp.1–3.

61. CO 298/39, Report for 1882, p.2; Beaubrun et al, 'The West Indies', p.511.

62. Report for 1882, pp.2–3.

63. *Journal of Mental Science*, XXX (1884), 141–44.

64. *Trinidad: Reports of the Surgeon-General on the Medical Service and Medical Institutions of the Country*, 1887–1914 [Copies in library of LSHTM]; Reports of Medical Superintendent, 1882–1914 [Copies in TNA, CO 298, Trinidad Legislative Council].

65. CO 298/36, Report of Medical Superintendent for 1880, p.1; CO 298/40, Report of Medical Superintendent for 1883, pp.1–2; CO 298/41, Report of Medical Superintendent for 1884, pp.1–4; *Reports of the Surgeon-General* – 1887, pp.12–13; 1889, p.86; 1890, pp.89–90; 1891, p.48; 1892, pp.44–45; 1893, p.43; 1894, p.48; 1897, pp.63–4; 1900, p.87.

66. CO 298/41, Report of the Medical Superintendent for 1884, p.4. Seccombe also protested directly to the Trinidad governor and Colonial Office via Crane, the colonial surgeon – CO 295/310, fos.298–301, 13 March 1886, Seccombe to Crane.

67. *Report of Surgeon-General*, 1894, p.48.

68. *Reports of Surgeon-General* – 1893, p.43; 1894, p.48; 1897, p.63; 1899, p.57.

69. *Reports of Surgeon-General* – 1891, p.48; 1892, p.45; 1897, p.63.

70. CO 298/40, Report of Medical Superintendent for 1883, p.2; CO 298/41, Report of Medical Superintendent for 1884, pp.1–4; CO 298/42, Report of Medical Superintendent for 1885, pp.13, 16–18; CO 298/43, Report of Medical Superintendent for 1886, pp.2, 9; *Reports of Surgeon-General* – 1887, pp.12–15; 1888, p.21; 1889, p.86; 1890, p.90. See also CO 295/94, fos.398–400, 2 August 1882, Neale to Kimberley; CO 295/95, fos.57–60, 22 November 1882, Freeling to Kimberley.

71. *Report of Surgeon-General*, 1889, p.86.

72. CO 298/40, Report of Medical Superintendent for 1883, p.3.

73. CO 298/40, Report of Medical Superintendent for 1883, pp.2–3; CO 298/41, Report of Medical Superintendent for 1884, pp.1, 4; CO 298/42, Report of Medical Superintendent for 1885, pp.17–18; CO 298/43, Report of Medical Superintendent for 1886, p.9. See also CO 295/303, 19 August 1884, Neale to Derby, fo.67; CO 295/310, fos.296–97, 13 March 1886, Crane to Robinson; fos.298–301, 13 March 1886, Seccombe to Crane.
74. *Report of Surgeon-General*, 1887, p.14.
75. *Reports of Surgeon-General* – 1891, p.47; 1892, p.44; 1895, p.56; 1900, pp.98–101.
76. E.g. CO 295/303, 19 August 1884, Neale to Derby, fos.68–9.
77. *Report of Surgeon-General*, 1891, p.48.
78. CO 295/355, fos.331–33, 18 July 1894, Lovell to Napier Broome.
79. *Report of Surgeon-General*, 1894, p.48.
80. CO 298/40, Report of Medical Superintendent for 1883, p.3; *Reports of Surgeon-General* – 1887, p.14; 1889, p.86; 1892, p.45; 1893, p.43; BPP 1897, Vol. LIX, Annual Series of Colonial Reports, no.201, 1896, p.18. The asylum built at St Ann's continues to function as Trinidad's main psychiatric in-patient facility.
81. CO 295/87, fos.106–8, 8 July 1880, Irving to Kimberley; fos.109–10, 28 February 1880, Crane to Irving; CO 295/89, fos.550–60, 29 March 1881, Freeling to Kimberley; CO 295/90, fos.627–67, 27 August 1881, Freeling to Kimberley; CO 295/94, fos.422–33, 5 August 1882, Plans for New Lunatic Asylum; CO 295/98, fos.149–78, 4 September 1883, New Lunatic Asylum; CO 295/99, fos.276–94, 6 November 1883, New Lunatic Asylum, Plans and Estimates; CO 295/345, fos.312–24, 17 May 1893, New Lunatic Asylum – Plans.
82. CO 295/90, 27 August 1881, Freeling to Kimberley, fos.635–36; Kimberley endorsed the argument that the proposed expenditure was too lavish – fo.627–28.
83. CO 295/345, fo.324, 18 July 1893, draft letter, Ripon to Napier Broome; CO 295/386, fos.299–312, 17 June 1898, Knollys to Chamberlain.
84. CO 295/386, 17 June 1898, Knollys to Chamberlain, fo.304.
85. CO 295/386, 17 June 1898, Knollys to Chamberlain, fos.305–37; BPP 1900, Vol. LIV, Annual Series of Colonial Reports, no.303, 1899, p.19.
86. *Report of Surgeon-General*, 1900, p.87.
87. *Report of Surgeon-General*, 1900, pp.87, 93.
88. *Report of Surgeon-General*, 1913/14, p.52.
89. *Report of Surgeon-General*, 1903/4, p.55.
90. *Report of Surgeon-General*, 1904/5, p.52.
91. *Reports of Surgeon-General* – 1905/6, p.52; 1906/7, p.57; 1907/8, p.91; 1908/9, p.110; *Jamaica Daily Gleaner*, 8 November 1909 – citing the *Port of Spain Gazette* in referring to the 'scandal' arising from overcrowding at the asylum.
92. *Reports of Surgeon-General*, 1908/9, pp.52, 110; CO 295/450, fos.26–43, 4 March 1909. Knaggs to Crewe.
93. *Reports of Surgeon-General* – 1901/2, p.57; 1904/5, p.52; 1910/11, p.27; 1912/13, p.34; CO 295/450, fos.29–38.
94. *Reports of Surgeon-General* – 1910/11, p.27; 1911/12, p.28; 1912/13, p.34

95. See Chapter 2 for the origins and early development of the Barbados asylum.

96. CO 885/3/4, 'Colonial Hospitals and Lunatic Asylums', p.25.

97. CO 28/209 (1869), fos.158–72.

98. CO 28/208, fos.342–43, 23 July 1869, Rawson to Granville; CO 28/209, fos.127–30, 8 October 1869, Rawson to Granville, 15 October 1869, fos.158–61, Rawson to Granville; Barbados Archives, E.A. Stoute Scrapbook, Vol. 1, p.44, 'Glimpse of Old Barbados', 7 June 1970.

99. CO 31/62, Proceedings of the Barbados Legislative Council and Assembly, p.121, 11 August, 19 September 1871, p.133, 5, 14 October 1871, p.138, 21 November 1871.

100. CO 28/208, fos.342–43, 23 July 1869, Rawson to Granville; CO 28/209, fos. 127–30, 8 October 1869, Rawson to Granville; CO 28/218, fo.178, 4 March 1873, Rawson to Kimberley; CO 31/64, Proceedings of House of Assembly, 1873–74, p.13, 15 July 1873, p.34, 2 December 1873, Appendix F, pp.2–3, 28 June, 5 July 1873.

101. CO 31/65, Barbados Assembly, 1874–75, p.4, 7 July 1874, p.12, 28 July 1874, Appendix B, pp.3–4; BPP 1876, Vol. LIII, Papers Relating to the Late Disturbances in Barbados, p.33; CO 321/18, fos.308–9, 14 April 1877, Rawson to Carnarvon – an aggrieved Rawson, now retired, claimed Allen had exceeded his brief in Barbados.

102. Barbados Archives, PAM C 72, Report of the Commission on Poor Relief, 1875–77, Appendix, Report of Thomas Allen, M.D., Medical Superintendent and Director of the Jamaica Lunatic Asylum, to His Excellency the Governor of Barbados. See also CO 31/66, Barbados Assembly, 1875–76, Appendix E; CO 31/67, 1875–76, Appendix D.

103. Report of Thomas Allen, M.D., pp.63–6, XXV.

104. Ibid., pp.2–12, 21–3, 27–31, 38, 41, 46.

105. Ibid., pp.6–7, 11–27, 37–8, 42–4, 49–50, XXIIIXXIV.

106. CO 321/18, 14 April 1877, fo.309.

107. CO 321/1, fos.345–47, 13 October 1874, Rawson to Carnarvon.

108. CO 31/66, Barbados Assembly, 1875–76, Appendix C, p.1.

109. The Male and Female Casebooks and Medical Superintendent's Journals are held in the Barbados Archives – see Chapters 6–8.

110. CO 321/15, fos.180–82, 22 August 1877, Dundas to Carnarvon; BPP 1878–79, Vol. L, Papers Relating to Her Majesty's Colonial Possessions, 1877, Part I, p.47.

111. CO 31/69, Barbados Assembly, 1878–79, Appendix K, p.38. The case books show instances of people spending several months in jails attached to police stations before admission to the asylum.

112. CO 31/69, Appendix K, pp.39–40.

113. CO 31/69, Barbados Assembly, 1878/79, p.29, 25 March 1879, p.33, 1 April 1879.

114. CO 321/43, fos.636–51, 9 May 1881, Robinson to Kimberley; CO 31/71, Barbados Assembly, 1881–82, pp.11–12, 7 June 1881.

115. CO 321/43, fo.646, p.5, 19 April 1881.

116. CO 31/72, Barbados Assembly, 1882–83, p.67, 16 January 1883.

117. CO 31/73, Barbados Legislative Council, 1883–84, p.1, 24 July 1883.

118. CO 31/73, Barbados Assembly, 1883–84, 23 October 1883, Half-Yearly Report of the Poor Law Inspector, January–June 1883, pp.3–4; 22 April 1884, Half-Yearly Report (henceforth HYR), July–December 1883, p.4.
119. CO 31/74, Barbados Assembly, 1884–85, 6 October 1885, HYR, January–June 1885, pp.4–5. The case books and journals give evidence of numerous fights and disturbances.
120. CO 31/75, Barbados Assembly, 1886–87, 28 September 1886, HYR, January–June 1886, pp.6–8. Patients' racial and social class characteristics will be considered in Chapter 6.
121. CO 31/76, Barbados Assembly, 1887–88, 10 May 1887, HYR, July–December 1886, p.8.
122. CO 31/76, Barbados Assembly, 1887–88, 26 April 1887, Report from Commissioners to Governor, pp.5–6.
123. CO 31/76, Barbados Assembly, 1887–88, 4 October 1887, HYR, January–July 1887, p.29.
124. CO 31/77, Barbados Assembly, 1888–89, 9 October 1888, HYR, January–July 1888, pp.8–9, 9 April 1889, HYR, July–December 1888, p.17.
125. CO 31/79, Barbados Assembly, 1890–91, 13 October 1891, HYR, January–June 1891, pp.16–17.
126. CO 31/80, Barbados Assembly, 1892–93, 26 April 1892, HYR, July–December 1891, pp.18–22 (quote p.21).
127. CO 31/80, Barbados Assembly, 1892–93, 28 March 1893, HYR, July–December 1892, pp.20–23.
128. Beckles, *Great House Rules*, pp.60–9, 114–17; Carter, *Labour Pains*, pp.36–43.
129. Belle, 'The Abortive Revolution of 1876 in Barbados'; Beckles, *Great House Rules*, pp.135–57; Carter, *Labour Pains*, pp.132–63. The case books and superintendent's journal show that the Confederation Riots directly affected the asylum, accounting for several admissions as well as patients distressed by disturbances nearby. See Chapter 1 for the riots.
130. The Special Commission comprised the principal medical officer of the garrison, an officer from the Royal Engineers, a 'local medical gentleman of high standing', and a Bridgetown merchant – BPP 1878–79, Vol. L, Papers Relating to Her Majesty's Colonial Possessions, 1877, Part I, p.189.
131. CO 31/68, Barbados Assembly, 1877–78, Appendix OO, pp.5–20, 7 September 1878.
132. CO 31/69, Barbados Assembly, 1878–79, p.7, 14 January 1879.
133. CO 31/69, Appendix W, pp.23–4.
134. CO 31/71, Barbados Assembly, 1881–82, 26 April 1881, pp.4–7, 12 July 1881, pp.2–3.
135. CO 31/71, 21 March 1882, pp.1–2.
136. CO 31/73, Barbados Assembly, 1883–84, 28 August 1883, p.26.
137. CO 31/73, 5 February 1884, p.2.
138. CO 31/75, 1886–87, Barbados Legislative Council, 9 February 1886, p.2.
139. CO 31/75, 1886–87, 22 June 1886, pp.1–5.
140. CO 31/73, 6 July 1886, p.1.
141. CO 31/76, 1887–88, Barbados Assembly, p.38, 26 July 1887, 26 April 1887, pp.5–9, Barbados Legislative Council, p.31, 7 February 1888; Stoute Scrapbook, Vol. 1, p.44, 'Glimpse of Old Barbados', 14 June 1970.
142. CO 31/78, 1889–90, Barbados Assembly, p.63, 22 July 1890.

143. CO 28/233, fos.199–202, 7 April 1893, Hay to Ripon; BPP 1893–94, Vol. LIX, Annual Series of Colonial Reports, 83, Barbados, p.5; Barbados Archives, Medical Superintendent's Journal, 5 April 1893. Through its Jenkinsville location the asylum acquired the popular name 'Jenkins', which has persisted to the present.
144. Medical Superintendent's Journal, 10–13 July 1893; CO 31/81, 1893–94, Barbados Assembly, 10 April 1894, HYR, July–December 1893, p.17; Beaubrun et al, 'The West Indies', p.516.
145. CO 31/81, pp.18–19.
146. CO 31/87, 1897–98, Barbados Assembly, 7 June 1898, HYR, July–December 1897, pp.25–7.
147. CO 31/88, 1899–1900, Barbados Assembly, 15 May 1900, HYR, July–December 1899, pp.14–15, 9 October 1900, HYR, January–June 1900, pp.25–7; CO 31/90, 1901–2, Barbados Assembly, 22 April 1902, HYR, July–December 1901, p.27.
148. CO 28/268, fos.50–63, 26 July 1907, Report on Pellagra Epidemic.
149. CO 31/101, Barbados Assembly, 1914–15, 16 June 1914, HYR, July–December 1910, p.5; 19 January 1915, HYRs, July–December 1911, p.6, July–December 1912, p.5, January–June 1913, p.4, January–June 1914, p.4.
150. CO 31/89, 1900–1, Proceedings of Assembly, 14 May 1901, HYR, July–December 1900, pp.22–3; 19 November 1901, HYR, January–June 1901, p.28.
151. CO 31/90, 1901–2, Barbados Assembly, 22 April 1902, HYR, July–December 1901, p.28.
152. CO 28/256, fos.96–99, 10 January 1902, Hodgson to Chamberlain; fo.123, 10 January 1902, Hutson to Colonial Secretary; CO 28/257, fos.63–4, Hodgson to Chamberlain. Governor Hodgson favoured an outside appointment but could not persuade reluctant politicians to sanction the additional expenditure. Dr John Hutson, the Poor Law Inspector, supported recruitment of a local man who could receive training in England.
153. CO 31/91, 1902–3, Barbados Assembly, 21 April 1903, HYR, July–December 1902, p.50; 10 November 1903, HYR, January–June 1903, p.7.
154. CO 31/92, Barbados Assembly, 1904–5, 15 November 1904, HYR, January–June 1904, pp.7–8; 11 April 1905, HYR, July–December 1904, pp.9–10.
155. CO 31/95, Barbados Assembly, 1906–7, 17 December 1907, HYR, July–December 1906, p.7; 26 May 1908, HYR, January–June 1907.
156. CO 31/92, Barbados Assembly, 1903–4, 15 November 1904, HYR, January–June 1904, p.7; CO 31/93, Barbados Assembly, 1905–6, 17 April 1906, HYR, July–December 1905, p.7; CO 31/101, Barbados Assembly, 1914–15, 19 January 1915, HYRs, July–December 1911, p.7, January–June 1912, p.6, July–December 1912, p.6, January–June 1913, p.5, July–December 1913, p.6.
157. CO 31/92, Barbados Assembly, 1904–5, 11 April 1905, HYR, July–December 1904, p.10.
158. CO 31/93, Barbados Assembly, 1905–6, 12 September 1905, HYR, January–June 1905, p.47; 17 April 1906, HYR, July–December 1905, p.32.
159. CO 31/96, Barbados Assembly, 31 August 1909, HYR, July–December 1907, p.8; HYR, January–June 1908, p.49; CO 31/101, Barbados Assembly,

1914–15, 19 January 1915, HYRs, July–December 1911, p.7, January–June 1912, p.6, January–June 1914, p.5.
160. CO 885/3/4, 'Colonial Hospitals and Lunatics Asylums', pp.1–2, 4–5, 25–7.
161. CO 7/139, 11 March 1870, Pine to Granville.
162. CO 28/208, fos.342–6, 23 July 1869, Rawson to Granville; CO28/209, fos.127–30, 8 October 1869, Rawson to Granville; CO 7/139, 9 July 1870, Pine to Granville.
163. Antigua Archives, Legislative Council, 1845–47, 9 February 1847, p.544.
164. CO 7/137, 27 September 1869, Pine to Granville; 20 August 1869, Nicholson to Baynes.
165. CO 7/139, 11 March, 9 July 1870.
166. BPP 1878, Vol. LVI, Papers Relating to Her Majesty's Colonial Possessions, 1877, pp.75, 98; BPP 1886, Vol. XLV, Papers Relating to Her Majesty's Colonial Possessions, 1885, p.216. At the end of 1877, 14 of the 47 patients in the Antigua asylum were from other islands. There were 15 patients in the St Kitts asylum. The lunatic wards at 'The Chateau' in Montserrat were extensively repaired during 1885.
167. BPP 1884, Vol. XLVI, Report of the Royal Commission to Enquire into the Public Revenues, Expenditure, Debts, and Liabilities of the Islands of Jamaica, Grenada, St Vincent, Tobago, and St Lucia, and the Leeward Islands, Part III, The Leeward Islands, pp.15, 23, 81.
168. BPP 1892, Vol. LV, Annual Series of Colonial Reports, 22, Leeward Islands, 1890, pp.28–30.
169. BPP 1893–94, Vol. LV, Annual Series of Colonial Reports, 51, Leeward Islands, 1891, pp.38–9.
170. BPP 1893–94, Vol. LIX, Annual Series of Colonial Reports, 77, Leeward Islands, 1892, pp.50–1. No patients died at The Ridge during the year.
171. BPP 1894, Vol. LVI, Annual Series of Colonial Reports, 112, Leeward Islands, 1893, p.29.
172. BPP 1896, Vol. LVII, Annual Series of Colonial Reports, 167, Leeward Islands, 1894, p.18.
173. H.A. Tempany, *Antigua B.W.I.: A Handbook of General Information, 1911* (London: Waterlow & Sons, 1911), pp.38–9; Beaubrun et al, 'The West Indies', p.523.
174. CO 1069/411, Colonial Office Photographic Collection, nos.15, 25–7, 33, 37–8, 46, available in the series 'Caribbean Through a Lens', at http:// www .flickr.com/photos/nationalarchives/sets/72157630634941210/.
175. CO 1069/411, nos. 35–6.
176. CO 101/109, fo.271, 14 June 1855, Colebrooke to Russell; fos.273–77, 27 February 1855, Heale to Colebrooke.
177. CO 101/129 (1869), fos.251–53, Grenada Lunatic Asylum Return; fos.254–58, Plans of Asylum.
178. CO 28/208, fos.342–46, 23 July 1869, Rawson to Granville; CO 28/209, fos.127–30, 8 October 1869, Rawson to Granville.
179. C 101/130, fos.63–4, 24 March 1870, Mundy to Rawson.
180. *St George's Chronicle and Grenada Gazette*, 28 October 1871.
181. *St George's Chronicle and Grenada Gazette*, 30 December 1871, 30 March 1872.
182. *St George's Chronicle and Grenada Gazette*, 21 September 1872.

183. BPP 1873, Vol. XLVIII, *Papers Relating to Her Majesty's Colonial Possessions,* 1873, p.142.
184. CO 321/2, fos.51–3, Returns of Lunatic Asylum, 1873.
185. CO 321/21, fo.183, Report by R.W. Harley, 18 June 1877.
186. CO 321/16, fos.47–50, 5, 8 May 1877, Graham to Strahan.
187. CO 321/21, fos.174–76, 6 May 1878, Harley to Strahan, fos.177–79, Returns of Lunatic Asylum, 1877; CO 321/27, fos.531–33, Reports by R.W. Harley, 18 June 1877 to 19 January 1878; BPP 1878, Vol. LVI, Papers Relating to Her Majesty's Colonial Possessions, 1876 and 1877, pp.58, 62.
188. CO 321/21, fos.180–82, 7 February 1878, Massiah to Harley.
189. CO 321/28, fos.71–2, 8 May 1879, Strahan to Hicks Beech; fos.191–93, Lunatic Asylum Returns, 1878; fos.197–98, Massiah to Harley.
190. CO 321/20, fos.440–43, 8 October 1878, Dundas to Hicks Beach. Tobago's lunatics were still housed in the jail. St Vincent established a small asylum in 1873 at Fort Charlotte, intended as only temporary – CO 321/3, fos.93–5, 11 March 1874, Rennie to Rawson; BPP 1875, Vol. LI, Papers Relating to Her Majesty's Colonial Possessions, 1875, Part II, p.47; BPP 1877, Vol. LIX, Papers Relating to Her Majesty's Colonial Possessions, 1875–77, Part 1, p.56.
191. CO 321/27, fos.278–79, 28 November 1878, Hicks Beach to Dundas.
192. CO 321/20, fos.195–96, 31 March 1880, Strahan to Hicks Beach; fos.197–99, 9 July 1879, Harley to Strahan; fos.201–3, 7 July 1879, Chadwick to Mitchell; fo.205, 3 July 1879, Massiah to Mitchell.
193. CO 321/20, fos.210–12, 4 November 1879, Harley to Strahan; fos.213–14, 10 February 1880, Strahan to Harley; fos.215–17, 24 March 1880, Harley to Strahan; fo.218, 19 March 1880, Massiah to Maling – Massiah noted an increase of staff was necessary because 'the New Arrivals have been far more troublesome than any of our previous inmates'.
194. BPP 1899, Vol. LXI, Annual Series of Colonial Office Reports, 247, Grenada, 1897, p.20.
195. CO 321/54, fos.136–51, 18 August 1882, 'Transfer of Lunatics From St Vincent & Tobago to Grenada Asylum'; BPP 1884–85, Vol. LII, Papers Relating to Her Majesty's Colonial Possessions, 1883–84, pp.130, 138–39, 145–47.
196. BPP 1884–85, Vol. LII, Papers Relating to Her Majesty's Colonial Possessions, 1883–84, p.138.
197. See Chapter 2.
198. CO 321/2, fos.511–12, 12 November 1874, Rawson to Carnarvon; fos.513–16, 2 November 1874, Des Voeux to Rawson; CO 321/6, fos.674–80, 27 October 1875, Des Voeux to Freeling; CO 321/12, fo.279, 22 June 1876, Pope Hennessy to Carnarvon; fos.281–82, 30 April 1876, Des Voeux to Pope Hennessy; fos.283–84, 17 January 1876, Meagher to Des Voeux; CO 321/16, fos.319–22, 3 August 1877, Dundas to Carnarvon; fos.323–24, 16 June 1877, Des Voeux to Dundas; fos.328–30, St Lucia: Lunatic Asylum Returns, 1876; fos.331–33, 14 June 1877, Dennehy to Dix; CO 321/22, fos.129–31, 26 March 1878, Dix to Strahan; Barbados Archives, PAM C 72, Report of Thomas Allen, M.D., to the Governor of Barbados, pp.53–4; Des Voeux, Sir G.W., *My Colonial Service* (London: John Murray, 1903), pp.187, 247, 254. Des Voeux, who oversaw the new asylum's opening, considered it a great improvement on its predecessor.

199. BPP 1889, Vol. LIV, Papers Relating to Her Majesty's Colonial Possessions, 67, St Lucia, 1888, p.17. Fifteen lunatics were left in the jail.
200. BPP 1890–91, Vol. LV, Annual Series of Colonial Office Reports, 10, St Lucia, 1890, p.16; St Lucia Archives, Colonial Reports – Annual, 1891, p.22; 1893, p.20; 1903, p.25; 1907, p.17. Inevitably, some patients remained in the Soufriere house.
201. BPP 1892, Vol. LV, Annual Series of Colonial Office Reports, 23, Grenada, 1890, p.16; BPP 1893–94, Vol. LIX, Annual Series of Colonial Office Reports, 52, Grenada, 1891, p.10; BPP 1893–94, Vol. LIX, 82, Grenada, 1892, p.18. Tobago patients were now being sent to Trinidad.
202. BPP 1900, Vol. LIV, Colonial Reports – Annual, 280, Grenada, 1898, p.25
203. *Grenada: Medical Reports on the Charitable Institutions, &c* (St George: Government Printing Office), Report on the Lunatic and Poor Asylums, 1899, pp.12–13, 1900, p.6. [Copies in Library of LSHTM.]
204. *Medical Reports on the Charitable Institutions*, Reports on the Lunatic and Poor Asylums, 1901, 1902, p.20, 1903, p.17.
205. Reports on the Lunatic Asylum, 1904, pp.15–16, 1905, pp.55–6
206. Report on the Lunatic Asylum, 1906.
207. Reports on the Lunatic Asylum, 1908–10.
208. Reports on the Lunatic Asylum, 1911–14.
209. Report on the Lunatic Asylum, 1912; St Lucia Archives, Colonial Reports – Annual, 1909, p.17, 1910, p.18, 1912–13, p.15.
210. The asylum at Fort Matthew on Richmond Hill still functioned as Grenada's mental hospital until bombed during the U.S. invasion of 1983, killing several patients. A visit to the derelict building in 2009 revealed that many of its worst features had remained virtually unaltered for over 100 years. The St Lucia asylum later became 'Golden Hope' Mental Hospital, and continued operating until replaced by a new Chinese-built institution in 2009.
211. See Chapter 2.
212. BPP 1870, Vol. XLIX, Reports to the Secretary of State, 1868, p.21; BPP 1871, Vol. XLVII, Reports to the Secretary of State, 1869, p.28.
213. BPP 1875, Vol. LI, Papers Relating to Her Majesty's Colonial Possessions, 1875, Part II, p.23.
214. CO 321/20, fos.158–59, 15 March 1879, hand-written note by Colonial Office official.
215. CO 137/481, fos.343–60, 28 June 1876, Grey to Carnarvon (quotes from fos.351–52); BPP 1875, Vol. LI, Papers Relating to Her Majesty's Colonial Possessions, 1875, Part II, p.25; BPP 1877, Vol. LIX, Papers Relating to Her Majesty's Colonial Possessions, 1875–77, Part I, p.19.
216. CO 321/20, fos.158–59, 15 March 1879. Allen's visit cost about £50.
217. BPP 1878, Vol. LVI, Papers Relating to Her Majesty's Colonial Possessions, 1876 and 1877, pp.7, 14.
218. BPP 1882, Vol. XLIV, Papers Relating to Her Majesty's Colonial Possessions, 1881, p.64.
219. *British Honduras: Medical Reports*, 1896, p.1. [Copies in Library of LSHTM.]
220. *Medical Reports*, 1897, p.1.
221. *Medical Reports*, 1898, p.5, 1899, p.3; BPP 1900, Vol. LIV, Colonial Reports – Annual, 278, British Honduras, 1898, p.23. The paupers were removed to the New Town Barracks.

222. *Medical Reports*, 1906, p.5.
223. *Medical Reports*, 1907, p.3, 1908, p.3.
224. *Medical Reports*, 1912, p.3, 1913, p.4.
225. E. Pinder, *Letters on the Labouring Population of Barbados* (Barbados: National Cultural Foundation, Barbados Heritage Reprint Series, 1858, reprinted 1990), p.16.

6 Pathways to the Asylum

1. J.K. Walton, 'Lunacy in the Industrial Revolution: A Study of Asylum Admissions in Lancashire, 1848–50', *Journal of Social History* 13 (1979), 1–22; D. Wright, 'Getting Out of the Asylum: Understanding the Confinement of the Insane in the Nineteenth Century', *Social History of Medicine* X (April 1997), 137–55.
2. R. Adair, J. Melling and B. Forsythe, 'Migration, Family Structure and Pauper Lunacy in Victorian England: Admissions to the Devon County Pauper Lunatic Asylum, 1845–1900', *Continuity and Change* 12 (1997), 373–401; Melling and Forsythe, *The Politics of Madness: Insanity and Society in England, 1845–1914*.
3. Mills, *Madness, Cannabis and Colonialism*, pp.43–102, 130–42; Coleborne, *Madness in the Family*, pp.65–106; Jackson, *Surfacing Up*, pp.57–117.
4. Leckie, 'Unsettled Minds: Gender and Madness in Fiji', pp.102–12; McCarthy and Coleborne, *Migration, Ethnicity and Mental Health*; McCarthy, 'Ethnicity, Migration and the Lunatic Asylum in Early Twentieth-Century Auckland, New Zealand'; Barry and Coleborne, 'Insanity and Ethnicity in New Zealand: Maori Encounters With the Auckland Mental Hospital'; Martyr, ' "Behaving Wildly": Diagnoses of Lunacy Among Indigenous Persons in Western Australia'.
5. See also – M. Finnane, 'Asylums, Families and the State', *History Workshop Journal* 20 (Autumn 1995), 134–48; C. Coleborne, 'Families, Patients and Emotions: Asylums for the Insane in Colonial Australia and New Zealand, c1880–1910', *Social History of Medicine* 19 (2006), 425–42; C. Coleborne, ' "His Brain was Wrong, his Mind Astray": Families and the Language of Insanity in New South Wales, Queensland and New Zealand, 1880s–1910', *Journal of Family History* 31 (2006), 45–65.
6. Dr Thomas Allen initiated the keeping of case books in Jamaica and it continued for many years, but they have been lost within recent years. There have been reports of surviving case books in Trinidad, but to the author's knowledge these have not yet been located.
7. Barbados Archives: Barbados Lunatic Asylum, Case Book, Men 1875–1916; Case Book, Women, 1875–1916.
8. *TAJ*, 15 December 1882–15 December 1885.
9. CO 137/382 (1864), fos.379–425.
10. It was customary in the nineteenth century to separate the 'causes' of insanity into 'physical' and 'moral' – the latter comprising social, psychological or emotional factors.
11. In many instances 'causes' remained unidentified. As noted in Barbados in 1886, there was 'no clue to the history of the ailment' in nearly 60 per cent

of cases – CO 31/76, Barbados Assembly, 10 May 1887, Half-Yearly Report of the Poor Law Inspector (HYR), July–December 1886, p.8.

12. CO 7/122, 23 June 1864, A. Nicholson, M.D., 'Remarks'.
13. *TAJ*, 15 January 1884, pp.89–90.
14. CO 298/42, Trinidad Legislative Council, Annual Report of the Colonial and Criminal Lunatic Asylum, 31 March 1886, p.27. The gender disparities in the Trinidad and British Guiana asylums reflected the preponderance of males among immigrant indentured labourers.
15. CO 140/94, Jamaica Lunatic Asylum, Annual Report, year ended 30 September 1886, p.52.
16. The numbers and proportion of tradesmen admitted were higher in Jamaica, reflecting a more developed urban economy.
17. Calculated from Barbados Lunatic Asylum, Male Case Book, Female Case Book. After 1879 the data becomes irregular.
18. Admissions also included one 'lady' and one 'governess'.
19. CO 295/90, fos.629–40, 27 August 1881, Freeling to Kimberley (quote, fo.635).
20. Barbados National Archives, PAM C 72, 'Report of Thomas Allen to His Excellency the Governor of Barbados' (1876), p.59.
21. 'Report of Thomas Allen', p.55.
22. CO 140/84, Jamaica Lunatic Asylum, Report of Medical Superintendent for the Year 1880/81, p.97.
23. CO 140/86, Jamaica Lunatic Asylum, Annual Report, year ended 30 September 1883, p165. Allen also observed that the asylum's 'servants' were 'for the most part obliged to be drawn from the same ranks'.
24. CO 7/122, attachments with 23 January 1864, Hill to Newcastle.
25. Moore and Johnson, *Neither Led Nor Driven*, pp.7, 245; Roopnarine, *Indo-Caribbean Indenture*, p.6.
26. Beckles, *Great House Rules*, pp.63–9; Newton, *The Children of Africa in the Colonies*, pp.204–21.
27. For the social and economic state of Barbados see Beckles, *Great House Rules*, pp.158–75; Carter, *Labour Pains*, pp.40–83. Barbados did have a considerably larger White minority than other islands.
28. Moore, *Cultural Power, Resistance and Pluralism*, pp.7–8.
29. Roopnarine, *Indo-Caribbean Indenture*, pp.27, 91–6. Women accounted for about a quarter of immigrants to the West Indies from the Indian subcontinent.
30. Moore, *Cultural Power, Resistance and Pluralism*, pp.7–15; Brereton, *A History of Modern Trinidad, 1783–1962*, pp.96–113; L. Gramaglia, 'Migration and Mental Illness in the British West Indies 1838–1900: The Cases of Trinidad and British Guiana', in Cox and Marland (eds.), *Migration, Health and Ethnicity in the Modern World*, pp.61–82.
31. Moore, *Cultural Power*, pp.11–12; Newton, *The Children of Africa in the Colonies*, pp.234–45, 273–76; Beckles, *Great House Rules*, pp.77–9; Carter, *Labour Pains*, pp.117, 143, 186–93; Gramaglia, 'Migration and Mental Illness in the British West Indies', pp.64, 69–74.
32. Gramaglia, 'Migration and Mental Illness in the British West Indies', p.68.
33. The figures probably disguise many Trinidad-born people of Indian origin.

34. Brereton, *Race Relations in Colonial Trinidad*, pp.110–5, 121, 127–28; Newton, *The Children of Africa*, pp.235–37; Carter, *Labour Pains*, pp.191–93.
35. Coleborne and McCarthy, *Migration, Ethnicity and Mental Health*.
36. *TAJ*, 2 May 1881, pp.19–20; Gramaglia, 'Migration and Mental Illness in the British West Indies', pp.72–3.
37. *TAJ*, 15 January 1882, p.12; 15 May 1882, p.30.
38. *TAJ*, 15 November 1882, pp.73–7.
39. J.S. Donald, 'Notes on Lunacy in British Guiana', *Journal of Mental Science* XX (1876–77), 76–81.
40. McCulloch, *Colonial Psychiatry and 'The African Mind'*.
41. Cannabis was variously referred to in reports as 'ganja', 'ganga', 'ganje', or 'Indian hemp'.
42. Mills, *Madness, Cannabis and Colonialism*, pp.43–65; J.H. Mills, *Cannabis Britannica: Empire, Trade and Prohibition* (Oxford: Oxford University Press, 2003).
43. *TAJ*, 2 January 1882, pp.84–5; Gramaglia, 'Migration and Mental Illness in the British West Indies', pp.75–7.
44. *TAJ*, 16 July 1883, pp.44–5; 15 September 1883, pp.60–1.
45. CO 298/40, Report of Medical Superintendent for 1883, p.2.
46. CO 298/42, Report of Medical Superintendent for 1885, p.16; *Trinidad: Reports of Surgeon-General*, 31 March 1889, p.20; 31 March 1890, pp.85–6.
47. CO 28/232, fos.305–6, HYR, January–June 1892.
48. CO 31/74, Barbados Assembly, 23 March 1885, HYR, July–December 1884, p.6.
49. CO 31/87, Barbados Assembly, 23 October 1898, HYR, January–June 1898, p.19.
50. Trinidad Archives, Proceedings of the Legislative Council, 1877, Council Paper No. 1, 'Lunatic Asylum. Annual Report of the Medical Superintendent for 1876', p.1.
51. CO 298/40, Report of Medical Superintendent for 1883, p.2; CO 298/41, Report of Medical Superintendent for 1884, p.1; CO 298/42, Report of Medical Superintendent for 1885, p.16; *Trinidad: Reports of Surgeon-General*, 31 March 1892, p.47; 31 March 1898, p.63.
52. *Reports of Surgeon-General*, 31 March 1890, p.86.
53. *TAJ*, 2 January 1882, p.84; 15 July 1882.
54. CO 137/388, 21 October 1864, Allen to Austin, fo.80.
55. *DG*, 19 September 1902.
56. Bowerbank, *A Circular Letter to the Individual Members of the Legislative Council*, Appendix, p.4.
57. CO 137/349, fos.412–15, 15 May 1860, Scott to Hutchings.
58. CO 137/353, fos.332–33, 8 November 1860, Trench to Austin.
59. CO 137/382, 12 December 1864, Lake to Trench, fos.84–5.
60. CO 140/86, Annual Report, Year Ended 30 September 1882, p.161; CO 140/86, Annual Report, Year Ended 30 September 1883, p.164; CO 140/89, Annual Report, Year Ended 30 September 1885, p.63.
61. CO 140/84, Report of Medical Superintendent for 1880/81, pp.96–7.
62. *JG*, 19 January 1887.
63. *Jamaica: Annual Reports of the Superintending Medical Officers*, Year Ended 30 March 1907, p.2.

64. *TAJ*, 1 March 1881, p.9.
65. Donald, 'Notes on Lunacy in British Guiana', pp.78–9; Gramaglia, 'Migration and Mental Illness in the British West Indies', pp.69–70.
66. CO 298/36, Report of Medical Superintendent for 1880, p.1.
67. CO 298/40, Report of Medical Superintendent for 1883, p.1; CO 298/42, Report of Medical Superintendent for 1885, p.16; *Reports of Surgeon-General* – 31 March 1889, p.20; 31 March 1891, p.89; 31 March 1897, p.59; 31 March 1909, p.110.
68. P.L.V. Welch, 'From Laissez Faire to Disinterested Benevolence: The Social and Economic Context of Mental Health Care in Barbados, 1870–1920', *The Journal of Caribbean History* 32 (1998), 121–42.
69. CO 31/78, Barbados Assembly, 1 April 1890, HYR, July–December 1889, p.28.
70. CO 31/80, Barbados Assembly, 28 March 1893, HYR, July–December 1892, p.21; CO 31/84, Barbados Assembly, 17 November 1896, HYR, January–June 1896, p.25; CO 31/86, Barbados Assembly, 23 March 1897, HYR, July–December 1896, p.14; CO 31/88, Barbados Assembly, 9 October 1900, HYR, January–June 1900, p.25.
71. CO 31/89, Barbados Assembly, 19 November 1901, HYR, January–June 1901, p.28.
72. Barbados Archives – Medical Superintendent's Journal (henceforth MSJ), 3 August 1878; Case Book, Men (henceforth CBM), p.80, no.834, 3, 26 August 1878.
73. CBM, p.69, no.817, 13, 27 March 1878.
74. *DG*, 6 September 1912.
75. *Report of Surgeon-General*, 1910/11, p.27.
76. Parle, *States of Mind*, pp.172–76, 230–39; Mills, *Madness, Cannabis and Colonialism*, pp.76–9.
77. Many of the case studies relate to Barbados, the rest being from Jamaica and British Guiana. Surnames have been anonymised, unless published in newspapers.
78. P.L.V. Welch, 'Gendered Health Care: Legacies of Slavery in Health Care Provision in Barbados Over the Period 1870–1920', *Caribbean Quarterly* 49, no.4 (December 2003), 104–29; Welch, 'From Laissez Faire to Disinterested Benevolence', p.141.
79. CBM, p.41, no.757, 25 January 1877.
80. CBM, p.103, no.870, 26 June 1879.
81. Barbados National Archives, Case Book, Women (henceforth CBW), p.63, no.831, 13 July 1878.
82. CBW, p.64, no.840, 13 September 1878.
83. CBW, p.69, no.861, 4 April 1879. During her second admission, in 1873, her husband was also an inmate at the asylum and 'their cow too was brought in along with them'.
84. CO 137/382, 2 April 1864, Allen to Eyre, fos.401–25.
85. CO 137/382, fo.418.
86. CO 137/382, fos.417–18.
87. CBM, p.44, 24 December 1876.
88. For the Confederation Riots, see Chapter 1.
89. CBM, p.29, no.723, 9 May 1876.

90. CBM, p.47, no.765, 25 March 1877. The case book provides considerable detail regarding the circumstances and progress of Reverend G., as for Dr Alexander G. (endnote 92). Dr Charles Hutson, when medical superintendent, evidently devoted particular attention to the cases of middle class White patients.

91. This is only speculation; it is conceivable that more attempted suicides by Black or Coloured people were successful or went unnoticed by the authorities.

92. CBM, p.43, no.762, 17 February 1877.

93. CBW, p.3, no.677, 27 May 1875.

94. CBW, p.49, no.799, 26 November 1877.

95. CBM, p.65, no.804, 23 December 1877.

96. See Laurence, 'The Development of Medical Services in Trinidad and British Guiana', pp.272–73.

97. CBM, p.65, 26–31 December 1877.

98. CBW, p.6, no.685, 23 July 1875.

99. CBW, p.37, no.759, 15 February, 24 April, 10 July, 30 September 1877. The baby was born in the asylum three months later, but the family refused to have them home. The child remained for two months before placement with a paid foster mother, but died a few weeks later.

100. CBW, p.78, no.882, 25 July 1879.

101. CBM, p.58, no.787, 16 August 1877.

102. CBM, p.117, no.918, 10 October 1879.

103. CBW, p.81, no.888, 25 July 1879.

104. CBW, p.16, no.706, 25 January 1876.

105. CBW, p.32, no.746, 7 November 1876.

106. CBW, p.57, no.813, 9 February 1878.

107. CBM, p.64, no.802, 1 December 1877.

108. CBM, p.14, no.692, 15 November 1875.

109. Moore and Johnson, *Neither Led Nor Driven*, pp.79–86; P. Bryan, *The Jamaican People, 1880–1902* (London: Macmillan Caribbean, 1991), pp.41–5.

110. Moore and Johnson, *Neither Led Nor Driven*, p.79. The American folklorist Martha Beckwith, who met Bedward in 1920, stated that he was born in 1859, though this appears unlikely – M.W. Beckwith, *Black Roadways: A Study of Jamaican Folk Life* (Chapel Hill, North Carolina: University of North Carolina Press, 1929), p.167.

111. CO 137/566, fos.277–301, 28 May 1895, Blake to Ripon, 'Proceedings of one Bedward'. The suggestion that he had 'exhibited signs of an unsound mind' in 1883 appears in a printed summary, fos.281–85.

112. Beckwith, *Black Roadways*, pp.167–68.

113. DG, 4 October, 30 November 1893; Moore and Johnson, *Neither Led Nor Driven*, pp.79–82 – as they point out, no heed was paid to medical confidentiality in the detail provided.

114. Beckwith, *Black Roadways*, p.168.

115. CO 137/566, 28 May 1895, Blake to Ripon, fo.278.

116. DG, 30 April 1895.

117. DG, 2, 23 May 1895; CO 137/566, fos.278–80, 286–301; Moore and Johnson, *Neither Led Nor Driven*, p.83. 'Bedwardism' continued to thrive

after this episode. He was again arrested for incitement, tried, and acquitted on grounds of insanity in 1921, remaining in the lunatic asylum until his death in 1930 – Moore and Johnson, *Neither Led Nor Driven*, pp.83–6; Beckwith, *Black Roadways*, pp.168–71.

118. *DG*, 26 June 1899; Moore and Johnson, *Neither Led Nor Driven*, pp.71–5, 86–9.
119. *DG*, 3 May 1899.
120. *DG*, 26 June 1899.
121. *DG*, 12 September 1902; 14 September, 13 November 1906, 7 October 1919; Moore and Johnson, *Neither Led Nor Driven*, pp.72, 85–6.
122. CO 137/382, fo.394. Halfway Tree, now a district of Kingston, was the main town in St Andrew parish.
123. CO 137/382, fos.412–13.
124. CBW, p.10, no.697, 9 December 1875.
125. CO 137/388, 21 October 1864, Allen to Austin, fo.81.
126. D. Wright, 'The Certification of Insanity in Nineteenth-Century England and Wales', *History of Psychiatry* 9 (1998), 267–90.
127. Mills, *Madness, Cannabis and Colonialism*, p.12, 67–8; McCulloch, *Colonial Psychiatry and 'The African Mind'*, p.26.
128. Coleborne, 'Passage to the Asylum'.
129. *The Public Acts in Force Passed by the Legislature of Barbados, 1837–41*, 3 Victoria, Cap. XXVIII, 4 June 1840, 'An Act for the Better Care and Maintenance of Lunatics', pp.225–26; *The Public Acts Passed in the Fifth Year of the Reign of Her Majesty* (Bridgetown, 1842), pp.21–4, 6 Victoria, Cap. VII, 20 December 1842, 'An Act to Amend an Act of This Island, Entitled "An Act for the Better Care and Maintenance of Lunatics"'. The legislation was consolidated in 1853 – CO 28/196, 16 May 1863, Response to Circular on Hospitals and Lunatic Asylums, fos.237–38.
130. Beaubrun et al., 'The West Indies', in Howells (ed.), *World History of Psychiatry*, pp.509–10; F.W. Hickling and H.D. Maharajh, 'Mental Health Legislation', in Hickling and Sorel (eds.), *Images of Psychiatry: The Caribbean*, pp.43–74; BPP 1875, Vol. LI, Papers Relating to Her Majesty's Colonial Possessions, 1875, Part 1, p.11.
131. CO 137/359, Public Hospital and Lunatic Asylum Commission, 15 May 1861, evidence of Lewis Bowerbank, pp.45–6; CO 140/86, Annual Report of Jamaica Lunatic Asylum, Year Ended 30 September 1882, p.156.
132. BPP 1871, Vol. XLVII, Reports to Secretary of State on the Past and Present State of Her Majesty's Colonial Possessions, 1869, p.49; *TAJ*, 2 May 1881, p.18; 15 November 1882, p.74.
133. Beaubrun et al., 'The West Indies', pp.510, 515; CO 295/75, fos.242–49, 18 December 1875, Irving to Carnarvon; CO 295/76, fos.477–86, 7 April 1876, Irving to Carnarvon; CO 295/77, fos.624–30, 19 February 1876, Commissioners in Lunacy to Colonial Office; CO 295/78, fos.144–47, 9 March 1877, Des Voeux to Carnarvon; *Reports of Surgeon-General*, 31 March 1891, p.89.
134. CO 7/71, 15 January 1842, McPhail to Stanley; CO 7/74, 20 January 1843, Fitzroy to Stanley; CO 7/121, 4 November 1863, Hill to Newcastle, enclosure – 'Lunatic Asylum Antigua, Rules and Regulations'; Antigua National Archives, The Laws of the Island of Antigua, pp.729–34, no.205,

3 December 1863, 'An Act to make Provision for the keeping up and Management of the Lunatic Asylum'.

135. CO 101/19, 12 May 1863, Walker to Newcastle, fos.182, 198.

136. CBW, p.19, no.712, Sarah Elizabeth K., 28 February 1876; p.32, no.746, Mary Louisa Y., 7 November 1876; p.44, no.791, Mary Lee T., 14 September 1877; p.56, no.810, Henrietta H., 6 February 1878.

137. CO 31/62, Barbados Legislative Council, 21 November 1871, pp.138–39.

138. MSJ, 3, 9 May 1883.

139. CO 140/86, Annual Report of Jamaica Lunatic Asylum, Year Ended 30 September 1883, p.163; CO 140/89, Annual Report, Year Ended 30 September 1885, p.63.

140. Pratt, *Seven Months in the Kingston Lunatic Asylum*, pp.6–7.

141. *Official Documents on the Case of Ann Pratt*, p.29.

142. CO 137/388, 21 October 1864, Allen to Austin, Report for 1 October 1863 to 30 September 1864, fos.105–6.

143. Ibid., fo.107.

144. Fos.106–8.

145. Fo.109.

146. *JG*, 13 June 1872.

147. *JG*, 4 October 1883, 12 June 1886; *DG*, 17 September 1894, 17 August 1896.

148. *DG*, 16 September 1893.

149. *Jamaica: Annual Reports of the Superintending Medical Officers*, Annual Report on Lunatic Asylum, Year Ended 31 March 1899, p.1; BPP 1900, Vol. LIV, Colonial Reports – Annual, no.283, Jamaica, 1898–99, p.43; *DG*, 20 December 1901.

150. *DG*, 16 May 1914.

151. CO 298/39, Report of Medical Superintendent for 1882, p.2.

152. CO 298/40, Report of Medical Superintendent for 1883, p.1.

153. CO 298/41, Report of Medical Superintendent for 1884, p.3; CO 298/42, Report of Medical Superintendent for 1885, p.16.

154. *Trinidad: Report of Surgeon-General*, 1895, p.56.

155. Report of Thomas Allen to the Governor of Barbados, Appendix, X–XI, XIV–XV; Barbados National Archives, Pam X 27 – Colonel John Elliott, C.B., *Rules, Regulations and Standing Orders for the General Government of the Non-Commissioned Officers and Other Constables of the Barbados Police Force, 1882* (Barbados: 'West Indian' Office, 1882), p.83. According to these regulations female lunatics transferred to the asylum were to be accompanied by 'a female friend or relative, and if none such be found, the Public female scrubber'.

156. MSJ, 23 February 1883.

157. CO 31/80, Barbados Assembly, 28 March 1893, HYR, July–December 1892, p.21. Original italics.

158. CO 31/81, Barbados Assembly, 10 April 1894, HYR, July–December 1893, p.18.

159. CO 7/71, 15 January 1842, McPhail to Stanley, attachment 10 January 1842, Rule no. 4. The language was subsequently moderated, with the addition that they should be 'undressed with gentleness' – CO 7/43, 14 September 1842, Daniell to Fitzroy, Rule no. 4; Laws of the Island of Antigua, no.205, 3 December 1863, p.731.

160. Rouse, *New Lights on Dark Deeds*, p.28; Pratt, *Seven Months in the Kingston Lunatic Asylum*, pp.18–19; CO 137/360, Public Hospital and Lunatic Asylum Commission, 28 May 1861, p.88, evidence of Elizabeth Scott; CO 137/360, 5 June 1861, p.291, evidence of Mary Ann Yates; CO 137/363, 20 August 1861, fo.178, evidence of Henrietta Dawson.
161. CO 31/81, Barbados Assembly, 10 April 1894, HYR, July–December 1893, p.17; CO 298/42, Trinidad, Report of Medical Superintendent for 1885, p.16; CO 140/99, Annual Report of Jamaica Lunatic Asylum, Year Ended 30 September 1889, p.161; Yeong, *Till the Break of Day*, p.39; Mills, *Madness, Cannabis and Colonialism*, pp.12–13; Coleborne, *Madness in the Family*, p.36; Parle, *States of Mind*, pp.16–18.

7 The Patient Challenge

1. See Chapter 6.
2. Barbados National Archives, PAM C 72, Report of Thomas Allen to His Excellency the Governor of Barbados (1876), p.61.
3. J. Patterson Smith, 'The Liberals, Race and Political Reform in the British West Indies, 1866–1874', *The Journal of Negro History* 79 (Spring 1994), 131–46; Bolt, *Victorian Attitudes to Race*, pp.75–108, 209–16; Lorimer, *Colour, Class and the Victorians*, pp.131–61.
4. For Allen's attitudes on race see also Chapters 4 and 6.
5. *JG*, 13 June 1872.
6. CO 140/84, Jamaica Lunatic Asylum, Report of Medical Superintendent 1880/81, p.99.
7. CO 140/86, Report of Medical Superintendent 1881/82, p.171.
8. CO 137/344, fos.7–8, Annual Report of Public Hospital and Lunatic Asylum, 1858, Appendix, pp.17–18.
9. CO 137/382, 12 December 1863, Lake to Trench, fo.99. Admissions of 'Coolies' reflected the immigration of indentured servants from the Indian subcontinent. About 37,000 were brought to Jamaica, numbers that never approached those in Trinidad and Guiana – Moore and Johnson, *Neither Led Nor Driven*, pp.7, 245–57; Roopnarine, *Indo-Caribbean Indenture*, p.6.
10. CO 140/94, Report of Medical Superintendent 1885/86, p.45.
11. BPP 1895, Vol. LXIX, Annual Series of Colonial Reports, 14, Jamaica, 1893–94, p.21.
12. *DG*, 22 November 1899. Blumer was medical superintendent of the Utica State Hospital.
13. CO 31/75, Barbados Assembly, 22 June 1886, Report to Sir C.C. Lees, Governor, from Dr C.J. Manning, p.5.
14. CO 31/75, Barbados Assembly, 28 September 1886, HYR, January–June 1886, p.6. 'Semi-respectable' patients were presumably either poor Whites or middle class Coloured people.
15. CO 31/76, Barbados Assembly, 4 October 1887, HYR, January–June 1887, p.29.
16. CO 31/77, Barbados Assembly, 9 October 1888, HYR, January–June 1888, p.9.
17. *DG*, 22 November 1899.

18. Welch, 'Gendered Health Care', p.117; Welch, 'From Laissez Faire to Disinterested Benevolence', p.141.
19. Barbados Archives, CBM, p.12, no.684, 21 July 1875.
20. CBM, p.18, no.702, 9 January 1876.
21. CBW, p.5, no.682, 4 July 1875.
22. CBW, p.48, no.797, 22 November 1877.
23. Barbados Archives, Medical Superintendent's Journals, 1875–1893.
24. Report of Thomas Allen to the Governor of Barbados, pp.5, 28.
25. CBM, p.47, no.765, 25 March 1877–31 December 1878, Rev. N.H.G.; CBM, p.65, no.804, 23–31 December 1877, Dr Alex G.; CBW, p.53, no.806, 11 January 1878–23 May 1881, Elizabeth Susannah T.; MSJ, 16–23 December 1883, re Mr Henry H.; 19 November–2 December 1884, re Miss Catherine J.W.; 22 February–14 March 1890, re Mr Robert S.B.; 24 April 1890–21 October 1891, re Dr Julius S.; 8–28 February 1891, re Miss N.
26. CBM, p.39, no.756, 22 January 1877, 13 July, 27 October 1878.
27. MSJ, 14, 28 August 1891.
28. MSJ, 11 October 1883.
29. MSJ, 20 May 1889.
30. MSJ, 5 September 1889.
31. MSJ, 2, 5 August 1890.
32. MSJ, 11 August 1891.
33. Donald, 'Notes on Lunacy in British Guiana', pp.76–7; Gramaglia, 'Migration and Mental Illness in the British West Indies 1838–1900', pp.68–71.
34. Donald, 'Notes on Lunacy in British Guiana', pp.77–8.
35. Ibid., p.78.
36. Ibid., p.79. Portuguese people from Madeira formed the majority of the asylum's White patients.
37. *TAJ*, 15 May 1882, pp.28–9.
38. *TAJ*, 15 November 1882, pp.73–7.
39. *TAJ*, 15 November 1882, p.77.
40. *TAJ*, 15 February 1883, pp.1–4.
41. Sir Frederick Treves, *The Cradle of the Deep: An Account of a Voyage to the West Indies* (London: Smith, Elder & Co., 1912), p.15. Treves, the King's physician, was describing a scene in an 'open quadrangle covered with grass' at the Barbados Lunatic Asylum.
42. CO 885/3/4, 14 January 1864, 'Colonial Hospitals and Lunatic Asylums', Appendix, p.37, Note VI.
43. See below for illustrative examples.
44. See Chapter 6.
45. CO 137/388, 21 October 1864, Allen to Austin, fo.79.
46. CO 140/81, Report of Medical Superintendent 1879/80, p.194.
47. CO 140/84, Report of Medical Superintendent 1880/81, p.97.
48. *Jamaica: Annual Reports*, Report on the Lunatic Asylum, Year Ended 31st March 1906, p.2.
49. *Journal of Mental Science* XXX (1884), 144.
50. *Reports of the Surgeon-General*, 31 March 1896, p.56; CO 295/355, fos.329–30, 18 July 1894, Napier Bourne to Ripon; CO 295/366, fos.79–80, 3 October 1895, Knollys to Chamberlain. Specific wards for lunatics or

'imbeciles' were a relatively common feature of workhouses in late nineteenth century England – J. Reinarz and L. Schwarz (eds.), *Medicine and the Workhouse* (Rochester, New York and Woodbridge: 2013), pp.113–16, 151–57, 204–6.

51. *TAJ*, 16 April 1883, pp.20–1.
52. *TAJ*, 15 June 1885, pp.102–3. General Paralysis of the Insane (GPI) was a chronic, deteriorating condition, later discovered to be linked with syphilis.
53. CBW, p.18, no.708, 7 February, 19 March, 29 June 1876, 7 October 1878, 17 July 1879; MSJ, 22 March, 29 June, 15 November 1876.
54. For the circumstances leading to Alonzo W's admission, see Chapter 6.
55. CBM, p.79, no.833, 18, 23, 31 July, 17 October, 31 December 1878, 17 December 1879, 27 June 1880; MSJ, 30 July, 1 August, 31 October, 2 November 1878. Not all cases of epilepsy proved fatal, some showing improvement sufficient for them to be discharged – CBM, p.18, no.702, Charles G., admitted 9 January 1876; p.45, no.763, Joseph I., admitted 8 March 1877; CBW, p.45, no.792, Mary F., admitted 16 September 1877.
56. For circumstances of his admission, see Chapter 6.
57. CBM, p.47, no.765, 25, 31 March, 31 May, 30 June, 30 September, 31 October, 31 December 1877, 1 August, 31 December 1878. The case book provides more detail on this particular case than any other.
58. Bancroft, *A Letter to the Hon. Hector Mitchel*.
59. Rouse, *New Lights on Dark Deeds*, pp.10–11.
60. Ibid., pp.27–8.
61. CO 137/360, Public Hospital and Lunatic Asylum Commission, 31 May 1861, pp.162, 167.
62. CO 137/382, 12 December 1863, Lake to Trench, fos.88–90.
63. CO 137/382, 2 April 1864, Allen to Eyre, fos.306–51, 378–410. See also Chapter 4.
64. CO 137/382, fos.319–20, 328, 378, 380, 386, 388, 403, 405.
65. The list comprised 104 patients out of a total 169.
66. CO 137/382, fo.379.
67. Ibid., fo.380. JLC and WH had initially been in the old Kingston asylum and were subsequently transferred.
68. Fos.384–85.
69. Fos.399–400.
70. Fo.406.
71. Fos.408–9.
72. Fos.416–17.
73. CO 137/388, 21 October 1864, Allen to Austin, fo.80.
74. E.g. JG, 13 June 1872 – Allen observed that Jamaican patients were typified by 'ungovernable passion', and were 'vindictive and savage'.
75. CBW, p.33, no.750, 19, 20, 30 January, 25 February, 31 March, 30 April, 5 May, 2, 30 June, 30 September 1877, 18 April, 2 December 1879, 19 October 1880, 22 October, 19 November 1881, 23 March 1882; MSJ, 14, 19 October 1880.
76. CBM, p.40, no.755, 22, 25, 30 January, 31 March, 30 April, 31 May, 30 June, 1 September, 30 November 1877, 31 January, 17 October, 31 December 1878; MSJ, 9, 30 March, 1 April, 11, 22 May 1878.

77. Barbados Lunatic Asylum, Register of Patients, 1846–88, no.118, 23 December 1854, no.149, 20 July 1856. She was described as a 'Hawker of Cakes'.
78. CBW, p.85, no.893, 26 July, 1 August, 24 October 1879, 1, 15, 22 September, 9 October, 13 November 1880, 31 March, 28 August 1881, 29 May 1882; MSJ, 16 September–20 November 1880, 31 March 1881, 21 April, 29 May 1882, 16 February 1883.
79. MSJ, 3, 4 June 1883.
80. MSJ, 7 February, 8 July, 20, 30 August 1884. The Journals for 1885–87 are not available.
81. MSJ, 30 April, 7, 27 November, 17 December 1888.
82. MSJ, 25 April, 6 June–10 October 1889, 28 March, 3 September–29 November 1890, 2, 26 February, 23 April, 18 May, 3 September 1891.
83. CO 137/360, Public Hospital and Lunatic Asylum Commission, 28 May 1861, pp.89–90, 29 May 1861, pp.133–34.
84. CO 137/355, fos.224–25, 3 September 1860, Trench to Austin; fos.226–27, 1 September 1860, 'Order From the Inspector and Director to the Matron of the Lunatic Asylum'.
85. CO 137/355, fos.228–33, 3 September 1860, Scott to Trench.
86. CO 137/382, Allen to Eyre, 2 April 1864, fos.350–51.
87. CO 137/382, fo.414.
88. CO 137/382, fos.419–20.
89. CO 137/382, fo.437.
90. CO 137/388, 21 October 1864, Allen to Austin, fo.89.
91. CO 137/517, fos.307–19, 22 September 1884, 'Compassionate Allowance for Mrs Scotland'.
92. *DG*, 22 September 1911.
93. *Journal of Mental Science* XXX (1884), p.143; CO 298/39, Report of Medical Superintendent for 1882, pp.2–3.
94. For example – in August 1878 there was 'Terrible fighting on female side', and in October 1880 there were serious disturbances on the female Lower Ward – MSJ, 17 August 1878, 8–14 October 1880.
95. MSJ, 14, 20 June 1876; CBW, p.19, no.712, 28, 29 February, 21 June, 31 July 1876. Field was standing in for Hutson, the medical superintendent.
96. MSJ, 30 June, 14, 18 July 1876; CBM, p.27, no.718, 20 April, 28 June, 12, 15 July 1876.
97. MSJ, 15 July 1876.
98. MSJ, 29 September 1882, 1 January 1883; CBW, p.28, no.742, 28 October 1876, 22 June 1880. Arabella died of tuberculosis three months after the incident, aged 30.
99. MSJ, 18 January 1883.
100. MSJ, 22, 26 March 1883.
101. MSJ, 19 November 1880; CBM, p.39, no.756, 13 July, 31 December 1878.
102. MSJ, 7 September 1883.
103. CBF, p.59, no.814, 28 February 1878, 11 June, 14 July 1879, 1, 23 May 1881.
104. MSJ, 3 January 1883.
105. MSJ, 26 February 1889. Chandler was badly bitten by patients on at least two other occasions – MSJ, 5 May 1888, 8 June 1891.
106. MSJ, 11 February 1890. This was not the same M. as referred to in note 97.

107. MSJ, 30 May 1888.
108. MSJ, 12–28 October 1888.
109. MSJ, 27 May, 5 June–11 September 1889.
110. CO 31/78, Barbados Assembly, 1 April 1890, HYR, July–December 1889, p.29.
111. MSJ, 22 November 1890–18 April 1891.
112. MSJ, 29 October–14 December 1891.
113. MSJ, 13, 14 February, 5–21 April 1882. It was suspected that Betty W. (see above) started destroying rooms after hearing about Mary Anne C.
114. MSJ, 24–28 April, 2, 12 May, 19–29 June, 7, 14 July, 14 August 1882.
115. MSJ, 5 June, 3 July 1883.
116. There were doubtless many incidents in the intervening years, but the journals are unavailable.
117. MSJ, 21 March, 13 May, 21, 22 July, 27 November 1888.
118. MSJ, 14 July, 3 August 1893.
119. CO 101/30, fos.63–6, 24 March 1870, Mundy to Rawson, and 12 May 1870, draft response from Colonial Office.
120. CO 101/30, fos.67–8, 15 March 1870, Newsam to Mitchell.

8 The Colonial Asylum Regime

1. *JG*, 13 June 1872.
2. CO 137/359, Public Hospital and Lunatic Asylum Commission, 15 May 1861, p.42.
3. Bancroft, *A Letter to the Honourable Hector Mitchel*, pp.9–10; Scott, *A Reply to a Letter by Lewis Quier Bowerbank*, p.14; *Journal of Mental Science* VI (1859), p.162.
4. CO 7/71, 10 January 1842, Antigua Lunatic Asylum, Rules and Regulations, enclosed with 15 January 1842, McPhail to Stanley; CO 7/121, Antigua Lunatic Asylum, Returns, p.9, enclosed with 4 November 1863, Hill to Newcastle; CO 101/104, fo.164, Rules for the Government of the Grenada Lunatic Asylum, enclosed with fos.158–65, 15 May 1851, Colebrooke to Grey; *St George's Chronicle and Grenada Gazette*, 21 September 1872.
5. *Grenada: Medical Reports on the Charitable Institutions, &c for the Year 1899*, p.12.
6. *Post of Spain Gazette*, 4 February 1863.
7. *Trinidad: Report of Surgeon-General*, 1887, p.13.
8. *Report of Surgeon-General*, 1913/14, p.52.
9. CO 28/196, Barbados Lunatic Asylum, Returns, enclosed with 16 May 1863, Walker to Newcastle, fo.248, p.9.
10. CO 321/43, fo.646, Barbados Assembly, 19 April 1881, p.5.
11. CO 31/74, Barbados Assembly, 6 October 1885, HYR, January–June 1885, p.5.
12. MSJ, 23 August 1878–8 August 1890.
13. CO 31/81, Barbados Assembly, 10 April 1894, HYR, July–December 1893, p.20; CO 31/93, Barbados Assembly, 17 April 1906, HYR, July–December 1905, p.32.
14. CO 137/382, 12 December 1863, Lake to Trench, fos.87–8.

15. CO 137/382, 2 April 1864, Allen to Eyre, fos.377, 440–44.
16. CO 137/388, 21 October 1864, Allen to Austin, fos.133–36.
17. CO 140/194, Annual Report of Jamaica Lunatic Asylum, Year Ended 30 September 1886, p.47.
18. *JG*, 16 June 1891.
19. Smith, *'Cure, Comfort and Safe Custody': Public Lunatic Asylums in Early Nineteenth-Century England*, Chapter 8.
20. CO 137/359, Public Hospital and Lunatic Asylum Commission, 15 May 1861, evidence of Lewis Bowerbank, p.74.
21. CO 7/71, Rules and Regulations, in 9 March 1842, Stanley to Graham; CO 7/74, Rules and Regulations of the Lunatic Asylum, in 14 September 1842, Daniell to Fitzroy.
22. CO 7/122, 23 June 1864, 'Remarks' by Dr A. Nicholson.
23. CO 7/137, 27 September 1869, Price to Granville, appended note 15 October 1869.
24. CO 28/196, Returns enclosed with 16 May 1863, Walker to Newcastle, fo.251.
25. Co 28/209, 15 October 1869, Rawson to Granville, fos.160–61; CO 321/5, 29 November 1875, Pope Hennessy to Carnarvon, fo.590; Report of Thomas Allen to the Governor of Barbados, pp.8, 22–3.
26. CBW, p.63, no.831, Elizabeth G., 13 July 1878; p.79, no.885, Mercy Anne S., 25 January 1882; MSJ, 21, 24 April, 12 June, 7, 14 July, 14 August 1882, 13 January, 23 August, 7 October 1883, 9 July, 4 December 1888, 16 July, 5, 13, 22, 25, 30 September 1891.
27. CO 137/388, 21 October 1864, Allen to Austin, fo.93; CO 140/183, Report of Medical Superintendent for the Year 1878/79, p.24.
28. CO 140/181, Report of Medical Superintendent for the Year 1879/80, p.197.
29. CO 298/39, Report of Medical Superintendent for 1882, p.1.
30. *TAJ*, 1 March 1881, p.2.
31. *British Guiana: Report of Surgeon-General for the Year 1899–1900*, p.41.
32. CO 140/186, Report of Medical Superintendent for Year Ended 30 September 1883, p.166.
33. *Jamaica: Asylums and Hospitals (Returns)*, Lunatic Asylum Returns, 1897–98, p.6.
34. BPP 1894, Vol.LVI, Annual Series of Colonial Reports, 113, Barbados, Report for 1893, p.17; CO 31/89, Barbados Assembly, 14 May 1901, HYR, July–December 1900, p.22; CO 31/93, Barbados Assembly, 12 September 1905, HYR, January–June 1905, p.47.
35. *Grenada: Medical Reports*, Reports on the Lunatic and Poor Asylums, 1901, 1903, p.17; Report on the Lunatic Asylum, 1911, p.18; St Lucia Archives, Lunatic Asylum Return, 1913–14.
36. *TAJ*, 1 March 1881, p.2; 15 January 1883, p.91; 15 January 1886, p.149.
37. CO 321/21, 7 February 1878, Massiah to Harley, fos.180–81.
38. CO 321/21, Returns of Lunatic Asylum, 1877, fo.178.
39. *Grenada: Medical Reports*, Reports on the Lunatic Asylum – 1906; 1908, p.43; 1911, p.18; 1914, p.43.
40. CO 28/209, 15 October 1869, Rawson to Granville, fos.160–61; enclosed Lunatic Asylum Return, fo.172.

41. CO 31/73, Barbados Assembly, 22 April 1884, HYR, July–December 1883, p.4.
42. CO 31/74, Barbados Assembly, 6 October 1885, HYR, January–June 1885, p.5.
43. CO 31/95, Barbados Assembly, 26 May 1908, HYR, January–June 1907, p.7; CO 31/101, 19 January 1915, HYR, January–June 1912, p.6.
44. This was also the case in British colonial institutions elsewhere, e.g. in India, see Mills, *Madness, Cannabis and Colonialism*, pp.118–19.
45. CO 137/382, 2 April 1864, Allen to Eyre, fos.435–36.
46. *TAJ*, 1 April 1881, p.7.
47. *Journal of Mental Science XXX*, 1884, p.143
48. CO 298/40, Report of Medical Superintendent for 1883, p.1; CO 298/41, Report of Medical Superintendent for 1884, p.3; *Report of Surgeon-General, 1887*, p.13; CO 295/310, 13 March 1886, Seccombe to Crane, fo.298.
49. CO 295/310, 13 March 1886, Crane to Colonial Secretary, fo.296.
50. See annual reports of the Jamaica and Trinidad lunatic asylums, the *Asylum Journal* for British Guiana, and the post-1900 reports of the Poor Law Inspector for Barbados.
51. CO 295/310, 13 March 1886, Seccombe to Crane, fo.299; *Report of Surgeon-General, 1887*, p.14.
52. CO 137/388, 21 October 1864, fos.98–101.
53. *JG*, 13 June 1872.
54. CO 137/346, 22 September 1859, Trench to Darling, fo.305; CO 137/359, Public Hospital and Lunatic Asylum Commission, 15 May 1861, evidence of Lewis Bowerbank, pp.74–5; CO 137/366, 'Report on the Management of the Public Hospital', 20 November 1861, p.11.
55. *JG*, 22 September 1886.
56. CO 31/84, Barbados Assembly, 17 November 1896, HYR, January–June 1896, p.25.
57. CO 31/92, Barbados Assembly, 15 November 1904, HYR, January–June 1904, p.7.
58. CO 31/96, Barbados Assembly, 31 August 1909, HYR, July–December 1907, p.7; CO 31/101, 19 January 1915, HYR, January–June 1913, p.5.
59. *TAJ*, 15 July 1884, p.47.
60. CO 137/388, 21 October 1864, Allen to Austin, fo.97.
61. CO 298/40, Report of Medical Superintendent for 1883, p.3.
62. CO 298/40, pp.2–3; CO 298/42, Report of Medical Superintendent for 1885, p.17.
63. *TAJ*, 2 May 1881, p.21; 2 September 1881, p.54; 15 April 1882, p.22; 15 February 1883, p.6; 15 January 1884, p.91; 15 November 1884, p.61; 15 January 1886, p.150; Gramaglia, 'Colonial Psychiatry in British Guiana: Dr Robert Grieve', pp.194–95.
64. CO 140/194, Report of Medical Superintendent for Year Ended 30 September 1886, p.46; *JG*, 1 August 1890; *DG*, 18, 23, 29 December 1891, 29 September, 21 November 1893, 27 August 1914.
65. CO 28/218, 4 March 1873, Rawson to Kimberley, fo.170; CO 31/74, Barbados Assembly, 23 March 1885, HYR, July–December 1884, p.7; MSJ, 1876–1891.

66. CO 31/89, Barbados Assembly, 14 May 1901, HYR, July–December 1900, p.23; 19 November 1901, HYR, January–June 1901, p.28; CO 31/92, 11 April 1905, HYR, July–December 1904, p.10; CO 31/101, 16 June 1914, HYR, January–June 1909, p.7.
67. *Grenada: Medical Reports*, Report on the Lunatic Asylum, 1911; CO 23/223, fos.52–4, Bahamas, Lunatic Asylum Returns for 1882.
68. In Grenada and St Lucia, with their French colonial heritage, Catholic worship was periodically provided.
69. CO 137/ 382, 2 April 1864, Allen to Eyre, fo.439; *Jamaica: Asylums and Hospitals (Returns)*, Lunatic Asylum Returns, 1897–98, p.7.
70. *TAJ*, 15 December 1884, p.66. These numbers would suggest that many Hindus were attending Christian services.
71. CO 140/194, Report of Medical Superintendent for Year Ended 30 September 1886, p.47.
72. MSJ, 1876–91.
73. The Barbados Case Books indicate that, although a majority of patients were described as Church of England, there were significant numbers of Wesleyans and Moravians.
74. Moore and Johnson, *Neither Led Nor Driven*, Chapter 3.
75. CO 298/40, Report of Medical Superintendent for 1883, p.3.
76. CBW, p.77, no.881, 25 July 1879.
77. CBM, p.74, no.823, 9, 15 June, 15, 29 November, 18 December 1878.
78. Scull, *The Most Solitary of Afflictions*, p.290.
79. CO 321/21, 7 February 1878, Massiah to Harley, fo.181.
80. *TAJ*, 15 September 1882, Case VII, p.60.
81. *TAJ*, 15 April 1885, Case XXI, p.86; 15 July 1885, Case XXIII, p.110; 16 November 1885, Case XXVI, p.136.
82. MSJ, 8, 10 February 1891.
83. MSJ, 12, 13 September 1889.
84. Scull, *The Most Solitary of Afflictions*, pp.290–91.
85. *TAJ*, 15 September 1882, Case VII, p.60; 15 April 1885, Case XXI, p.86; 15 August 1885, Case XXIV, p.119; 16 November 1885, Case XXVI, p.136.
86. CBW, p.19, no.712, 28, 29 February, 2, 4 March 1876.
87. CBM, p.78, no.830, 8, 11, 30 June, 31 July, 30 August 1878; MSJ, 10, 11, 14, 15, 17, 21, 24 June 1878.
88. MSJ, 30 August 1883. One of these men appears to have been Samuel A.
89. *TAJ*, 15 February 1882, p.5; 15 September 1882, pp.60–1; 16 July 1883, p.44; 15 September 1883, pp.60–1; 15 July 1885, p.110; 16 November 1885, p.136; CBM, p.40, no.755, Thomas H., 30 November 1877; MSJ, 9 March, 27 May 1878; 14 November 1889; 10, 15 February 1891.
90. Smith, 'Cure, Comfort and Safe Custody', pp.202–5; Scull, *The Most Solitary of Afflictions*, pp.169, 290–91.
91. CO 137/361, Public Hospital and Lunatic Asylum Commission, 19 June 1861, pp.194–96; CO 137/366, 18 January 1862, Scott to Darling, fos.342, 350–51.
92. *Port of Spain Gazette*, 4 February 1863.
93. CO 7/122, 23 June 1864, 'Remarks'.
94. CBW, p.92, no.906, Mary M., 15 September–18 November 1879.

95. CBM, p.5, no.674, John Dixon E., 26 November, 2, 4 December 1875; MSJ, 5 June 1891.
96. MSJ, 29–31 December 1888 (William Henry B.); 15 January–15 February (Margaret S.), 13–21 March 1890 (Georgiana P.).
97. MSJ, 26 March–7 May 1883 (Mabel D.); 23 August 1888, 26 August–9 September (Anne T.), 5 October 1889.
98. MSJ, 20 February 1888–3 March 1890.
99. CBW, p.71, no.877, Mary Eliza C., 3 April 1882; MSJ, 18 February 1882, 18 October 1883, 3 September 1888.
100. MSJ, 21–24 January, 29–31 December 1888, 8 December 1890, 24, 29 June, 3 August 1891.
101. Asylum staff titles varied over time, according to gender and status. In Jamaica, for many years, some subordinate staff were called 'labourers'. Elsewhere, staff were referred to generically as 'servants'. In the mid-nineteenth century the term 'keepers' was in common usage, with 'warders' also used occasionally.
102. *TAJ*, 2 January 1882, p.85.
103. J. Conolly, *The Treatment of the Insane Without Mechanical Restraints* (London: Smith, Elder and Co., 1856), p.98; Smith, *'Cure, Comfort and Safe Custody'*, pp.131–32.
104. Monk, *Attending Madness*, pp.83–92, 147–50, 177–82; Ernst, *Mad Tales From the Raj*, pp.82, 106–7, 153; Mills, *Madness, Cannabis and Colonialism*, pp.150–61.
105. CO 7/74, 14 September 1842, Daniell to Fitzroy, enclosing 'Rules and Regulations of the Lunatic Asylum'.
106. See Chapter 3. The labourer Alexander Flemming was dismissed in 1858 for having allegedly fathered Elizabeth Green's child – Rouse, *New Lights on Dark Deeds*, pp.29–30, CO 137/360, Public Hospital and Lunatic Asylum Commission, 29 May 1861, evidence of Edward Hull, p.120; CO 137/363, Henrietta Dawson's statement, p.283.
107. CO 137/368, fos.497–500, 29 November 1862, Eyre to Newcastle.
108. CO 137/368, fos.524–44.
109. CO 137/368, fo.530.
110. See Chapter 4.
111. Barbados Archives, PAM A 398, 'Rules for the Management and Conduct of the Lunatic Asylum' (not dated: *circa* 1900).
112. 'Rules for the Management', pp.1–2.
113. Ibid., pp.2–3.
114. Ibid., pp.13–15.
115. CO 137/388, 21 October 1864, Allen to Austin, fo.122.
116. *TAJ*, 1881, 'Asylum Staff', listed after title page; 2 May 1881, p.21.
117. CO 298/40, Report of Medical Superintendent for 1883, pp.2–3.
118. CO 295/310, fos.296–97, 13 March 1886, Crane to Robinson; fos.298–301, 13 March 1886, Seccombe to Crane.
119. CO 298/43, Report of the Surgeon-General for 1886, p.2.
120. *TAJ*, 2 May 1881, pp.21–2. The women continued to wear brown uniform.
121. CO 298/40, Report of Medical Superintendent for 1883, p.2; Report of Surgeon-General, p.2.

122. CO 31/91, Barbados Assembly, 10 November 1903, HYR, January–June 1903, p.7.
123. CO 31/92, Barbados Assembly, 11 April 1905, HYR, July–December 1904, p.9. The Barbados Executive Council was unwilling to spend any additional money – see Chapter 5.
124. See Chapter 4.
125. CO 137/537, fos.71–3, 10 October 1888; *DG*, 14 September 1906, 22 June 1908.
126. *TAJ*, 1 March 1881, p.3.
127. CO 295/427, fos.228–34, 2 May 1904, Recruitment of Male Head Attendant; CO 298/36, Report of Medical Superintendent for 1880, p.2; *Trinidad: Reports of the Surgeon-General* – 1887, p.12; 1904/5, p.52; 1910/11, p.27; 1911/12, p.28.
128. CO 137/355, 20 June 1861, Darling to Newcastle, fo.210.
129. CO 137/361, Public Hospital and Lunatic Asylum Commission, 20 June 1861, p.205.
130. CO 137/366, 'Report on the Management of the Public Hospital', 20 November 1861, pp.5, 11.
131. CO 137/372, Asylum Returns, enclosed with 2 May 1863, Eyre to Newcastle, fo.40.
132. CO 137/382, 28 May 1864, Trench to Austin, fo.510. See Chapter 4.
133. CO 137/388, 21 October 1864, Allen to Austin, fos.92–3, 103–4.
134. CO 140/186, Report of Medical Superintendent for Year Ended 30 September 1883, pp.165–66.
135. CO 137/510, fos.225–29, 6 July 1883, 'Memorandum by the Medical Superintendent of the Lunatic Asylum'.
136. CO 140/199, Report of Medical Superintendent for Year Ended 30 September 1888, p.99.
137. *TAJ*, 1 March 1881, p.3; 1 April 1881, p.9.
138. *TAJ*, 1 July 1881, p.36.
139. *TAJ*, 15 June 1883, p.38.
140. *TAJ*, 15 April 1885, p.88.
141. *TAJ*, 15 June 1882, p.36.
142. *TAJ*, 15 April 1884, pp.19–20.
143. CO 298/41, Report of Medical Superintendent for 1884, p.4.
144. *Report of the Surgeon-General*, 1887, pp.12–13.
145. *Report of the Surgeon-General*, 1892, p.45.
146. *Reports of the Surgeon-General* – 1893, p.43; 1894, p.48; 1895, p.56.
147. *Report of the Surgeon-General*, 1896, p.59.
148. *Report of the Surgeon-General*, 1905/6, p.52
149. CO 31/78, Barbados Assembly, 29 October 1889, HYR, p.36.
150. MSJ, 24 April, 5, 8, 28, 31 May, 31 August 1883; 31 May, 30 June, 15 July, 9 October 1889.
151. CO 298/40, Report of Medical Superintendent for 1883, p.2; CO 298/41, Report for 1884, p.4; *Reports of the Surgeon-General* – 1892, p.45; 1894, p.48; *TAJ* – numerous entries from 1 March 1881–15 August 1885.
152. *Grenada: Medical Reports*, Reports on the Lunatic and Poor Asylums – 1899, p.13; 1900, p.6; 1902, p.20; Reports on the Lunatic Asylum – 1905, 1906, 1909, 1911, 1912, 1913.

153. *JG*, 24 January 1905.
154. CO 137/484, 31 May 1876, Allen to Colonial Secretary, fo.588.
155. *JG*, 6 September 1893.
156. MSJ, 15 June 1878; CBM, p.75, no.824, 15 June 1878.
157. MSJ, 24 April 1888. Julia Paul had worked at the asylum for about 35 years – MSJ, 24 December 1883.
158. Pratt, *Seven Months in the Kingston Lunatic Asylum*, p.17.
159. CO 137/363, Public Hospital and Lunatic Asylum Commission, Henrietta Dawson's Statement, p.256.
160. CO 137/388, 21 October 1864, Allen to Austin, fos.135–36.
161. CO 137/484, 31 May 1876, Allen to Colonial Secretary, fo.602.
162. *JG*, 4 December 1905.
163. BPP 1884–85, Vol. LII, Papers Relating to Her Majesty's Colonial Possessions, 1884, Grenada, p.147.
164. *TAJ*, 1 June 1881, p.29; *British Guiana: Report of the Surgeon-General for the Year 1896–97*, p.38.
165. CO 298/40, Report of Medical Superintendent for 1883, p.2.
166. CO 31/96, Barbados Assembly, 31 August 1909, HYR, July–December 1907, p.8; HYR, January–June 1908, p.49.

Conclusion

1. *The Lancet*, 23 October, 25 December 1858, 26 March, 6 August, 3 September 1859, 17 December 1860; *The Times*, 30 August, 1, 8 September 1859.
2. *The Anti-Slavery Reporter*, 2 May 1859, cited in *The Lancet*, 6 August 1859.
3. *The Lancet*, 3 September 1859, p.243.
4. Swartz, 'The Regulation of Colonial Lunatic Asylums and the Origins of Colonial Psychiatry, 1860–1864'.
5. Joseph Plaxton in Jamaica also succumbed. Robert Grieve was perhaps the exception, because of his elevation to the post of Colonial Surgeon in 1886 – Gramaglia, 'Dr Robert Grieve (1839–1906), An "Apostle of Science"', p.xxiii.
6. CO 885/3/4, 'Colonial Hospitals and Lunatic Asylums', pp.1–5, 24–7.
7. As recent studies have demonstrated, these developments were perpetuated in Jamaica in the inter-war period, as the plight of the asylum patients worsened – D.H. Heuring, '"In the Cheapest Way Possible..."': Responsibility and the Failure of Improvement at the Kingston Lunatic Asylum, 1914–1945', *Journal of Colonialism and Colonial History* 12, no.3 (Winter 2011), Johns Hopkins University, Baltimore (online); H. Altink, 'Modernity, Race and Mental Health Care in Jamaica, c1918–1944', *Journal of the Department of Behavioural Sciences* 2, no.1 (December 2012), 1–19, University of the West Indies, St Augustine's – Accessed at http://journals.sta.uwi.edu/jbs/index.asp
8. Beckwith, *Black Roadways: A Study of Jamaican Folk Life*, pp.168–71. Bedward remained in the asylum for several years until his death.

Bibliography

Primary sources

Acts of Parliament

48 Geo. III, Cap. 96, An Act for the Better Care and Maintenance of Lunatics, Being Paupers or Criminals in England, 1808.

8 & 9 Vic., Cap. 100, An Act for the Regulation of the Care and Treatment of Lunatics, 1845.

8 & 9 Vic., Cap. 126, An Act to Amend the Laws for the Provision and Regulation of Lunatic Asylums for Counties and Boroughs, and for the Maintenance and Care of Pauper Lunatics, in England, 1845.

Barbados – 3 Victoria, Cap. 28, 4 June 1840, An Act for the Better Care and Maintenance of Lunatics.

Barbados – 6 Victoria, Cap. 7, 20 December 1842, An Act to Amend an Act of this Island, Entitled 'An Act for the Better Care and Maintenance of Lunatics'.

British Parliamentary Papers

House of Commons Sessional Papers of the Eighteenth Century, Vols. 72–3, 1790, Select Committee on the Slave Trade.

1814–15, Vol. IV, Select Committee on the State of Madhouses.

1816, Vol. VI, Select Committee on the State of Madhouses.

1826, Vol. XXVI, Second Report of the Commissioner of Inquiry into the Administration of Civil and Criminal Justice in the West Indies.

1826–27, Vol. VI, Select Committee on Pauper Lunatics in the County of Middlesex and on Lunatic Asylums.

1826/27, Vol. XXIII, Report of His Majesty's Commissioners of Legal Inquiry on the Colony of Trinidad.

1826/27, Vol. XXIV, Third Report of the Commissioner of Inquiry into the Administration of Civil and Criminal Justice in the West Indies.

1830–31, Vol. XII, Gaols, West Indies: Copies of Correspondence Relative to the State of Gaols in the West Indies and the British Colonies in South America.

1837, Vol. LIII, Papers Presented to Parliament, In Explanation of the Measures Adopted by Her Majesty's Government, For Giving Effect to the Act for the Abolition of Slavery Throughout the British Colonies.

1837/38, Vol. XL, Report of Captain J.W. Pringle on Prisons in the West Indies.

1841, Vol. III, Papers Relative to the West Indies.

1843, Vol. XXXIII, Papers Relative to the Earthquake in the West Indies.

1846, Vol. XXVII: British Colonies (West Indies and Mauritius) – Papers Relating to the Labouring Population of the British Colonies.

1847, Vol. XXXVII; 1847/8, Vol. XLVI; 1849, Vol. XXXIV; 1850, Vol. XXXVI; 1851, Vol. XXXIV; 1852, Vol. XXXI; 1856, Vol. XLII; 1857, Vol. X; 1857–58,

Vol. XI; 1861, Vol. XL; 1865, Vol. XXXVII; 1867–68, Vol. XLVIII; 1868–69, Vol. XLIII; 1868–69, Vol. XLIII; 1870, Vol. XLIX, Reports to Secretary of State on Past and Present State of Her Majesty's Colonial Possessions.

1847–48, Vol. XXII, Seventh Report From the Select Committee on Sugar and Coffee Planting.

1852–53, Vol. LXVII, Copies of Extracts of Despatches Relative to the Condition of the Sugar Growing Colonies.

1854, Vol. XLIII, Cholera (Jamaica).

1856, Vol. XLIV, British Guiana. Copies or Extracts of Correspondence on the Subject of the Recent Disturbances in the Colony.

1873, Vol. XLVIII; 1875, Vol. LI; 1877, Vol. LIX; 1878, Vol. LVI; 1881, Vol. LXIV; 1882, Vol. XLIV; 1886, Vol. XLV; 1888, Vol. LXXII; 1889, Vol. LIV, Papers Relating to Her Majesty's Colonial Possessions.

1873, Vol. L, Report on Leprosy and Yaws in the West Indies, Addressed to Her Majesty's Secretary of State for the Colonies, by Gavin Milroy, M.D.

1876, Vol. LIII, Papers Relating to the Late Disturbances in Barbados.

1884, Vol. XLVI, Report of the Royal Commission to Enquire into the Public Revenues, Expenditure, Debts, and Liabilities of the Islands of Jamaica, Grenada, St Vincent, Tobago, and St Lucia, and the Leeward Islands.

1890–91, Vol. LV; 1892, Vol. LV; 1893–94, Vols. LV, LIX; 1894, Vol. LVI; 1896, Vol. LVII, Annual Series of Colonial Office Reports.

Official reports (other)

Metropolitan Commissioners in Lunacy, *Report of the Metropolitan Commissioners in Lunacy to the Lord Chancellor* (London: Bradbury and Evans, 1844).

1895–1900. *British Guiana: Reports of the Surgeon General* (Georgetown, Demerara: C.K. Jardine, Printer to the Government).

1896–1913. *British Honduras: Medical Reports* (London: Waterlow and Sons).

1899–1914. *Grenada: Medical Reports on the Charitable Institutions, &c* (St George: Government Printing Office).

1898–1907. *Jamaica: Annual Reports of the Superintending Medical Officers* (Kingston, Jamaica: General Printing Office).

1896–1909. *Jamaica: Asylums and Hospitals* (Returns).

1910–17. *St Lucia: Annual Reports of the Medical Officers* (Castries: Government Printing Office).

1888–1914. *Trinidad: Reports of the Surgeon General* (Port of Spain: Government Printing Office).

Archives

Antigua National Archives

Minutes of Legislative Council, 1839–63.

Barbados National Archives

Barbados Lunatic Asylum: Male and Female Casebooks, 1875–1910.
Barbados Lunatic Asylum: Medical Superintendent's Journals, 1876–1900.

Barbados Lunatic Asylum: Register of Admissions, 1846–88.
Report of the Commission on Poor Relief, 1875–77, Appendix, Report of Thomas Allen, M.D., Medical Superintendent and Director of the Jamaica Lunatic Asylum, to His Excellency the Governor of Barbados.
St Michael's Vestry, Minutes, 1823–48.

British National Archives (TNA)
CO 7. Antigua. Original Correspondence.
CO 28. Barbados. Original Correspondence.
CO 31. Barbados. Proceedings of the House of Assembly and Legislative Council.
CO 33. Barbados. Blue Books.
CO 101. Grenada. Original Correspondence.
CO 111. British Guiana. Original Correspondence.
CO 137. Jamaica. Original Correspondence.
CO 140. Jamaica. Votes of the Assembly.
CO 295. Trinidad. Original Correspondence.
CO 298. Trinidad. Legislative Council.
CO 321. Windward Islands. Original Correspondence.
CO 854/57 (1864), fos.447–67, 'Colonial Hospitals and Lunatic Asylums' – See also CO 885/3/4, 1–32.
CO 1069. Colonial Office Photographic Collection.
MH 50. Commissioners in Lunacy. Minutes.

Library of the Society of Friends
John Candler: West Indies Journal 1849, 1850.

Lincolnshire Archives
Minute Book of Visitors of Lincolnshire County Lunatic Asylum, 1852–68.

London Metropolitan Archives
Metropolitan Asylum District: Board Minutes, 1875–84.

St Lucia National Archives
Colonial Reports, Annual, 1890–1915.
Blue Books, Lunatic Asylum Returns, 1913–15.

Trinidad National Archives
Minutes of Proceedings of Legislative Council, 1877.

Newspapers and periodicals

The Asylum Journal (British Guiana), 1881–85.
Jamaica Gleaner/Daily Gleaner, 1867–1914.
Journal of Mental Science

Port of Spain Gazette, 1857–65.
Royal Gazette (Jamaica), 1815–21.
St George's Chronicle and Grenada Gazette, 1871–81.
The Lancet, 1858–60.
The Times, 1858–60.

Printed books, pamphlets and articles

Abuses in the Jamaica Lunatic Asylum. Magisterial Investigation into a Charge of Manslaughter, of Mrs Matilda Carey (Kingston, Jamaica: Geo. Henderson, Savage & Co., 1860).

Bancroft, E.N., *A Letter to the Hon. Hector Mitchel, Chairman of the Committee of Public Accounts, Representing the Total Unfitness of the Present Asylum for Lunatics, and the Urgent Necessity for Building a New Lunatic Asylum, in a Proper Situation* (Kingston, Jamaica: Jordon, Osborn & Co., 1840).

Battie, William, *A Treatise on Madness* (London: Whiston and White, 1758).

Beckwith, Martha Warren, *Black Roadways: A Study of Jamaican Folk Life* (Chapel Hill, North Carolina: University of North Carolina Press, 1929).

Bowerbank, Lewis Q., *A Circular Letter to the Individual Members of the Legislative Council and of the House of Assembly of Jamaica, Relative to the Public Hospital and Lunatic Asylum of Kingston* (Kingston, Jamaica, 1858a).

Bowerbank, Lewis Q., *A Third Letter to the Commissioners of the Public Hospital and Lunatic Asylum of Kingston Jamaica, Relative to Dr Scott's Reply* (Kingston, Jamaica: Ford and Gall, 1858b).

Caines, Clement, *Letters on the Cultivation of the Otaheite Cane ... and Also a Speech on the Slave Trade, the Most Important Feature in West Indian Cultivation* (London: Messrs Robinson, 1801).

Candler, John, *Extracts From the Journal of John Candler, Whilst Travelling in Jamaica, Part 1* (London: Harvey and Darton, 1840).

Collins, D. ('A Professional Planter'), *Practical Rules for the Management and Medical Treatment of Negro Slaves, in the Sugar Colonies* (London: J. Barfield, 1811).

Conolly, John, *The Treatment of the Insane without Mechanical Restraints* (London: Smith, Elder and Co., 1856).

Conolly, John, *The Construction and Government of Lunatic Asylums and Hospitals for the Insane* (London: John Churchill, 1847; reprinted London: Dawson, 1968).

Dancer, Thomas, *The Medical Assistant: or Jamaica Practice of Physic: Designed Chiefly for the Use of Families and Plantations* (Kingston, Jamaica: Alex Aikman, 1801).

Des Voeux, Sir G. William, *My Colonial Service in British Guiana, St Lucia, Trinidad, Fiji, Australia, Newfoundland and Hong Kong With Interludes* (London: John Murray, 1903).

Donald, James S., 'Notes on Lunacy in British Guiana', *Journal of Mental Science* XX (1876–77), 76–81.

Ellis, H. Warner, *'Our Doctor,' or Memorials of Sir William Charles Ellis* (London: Seeler, Jackson and Halliday, 1868).

Fifth Annual Report of the Lincolnshire County Lunatic Asylum, at Bracebridge, Near Lincoln (Lincoln: W. and B. Brooke, 1858).

Gardner, William J., *A History of Jamaica From the Discovery by Christopher Columbus to the Present Time* (London: Elliot Stock, 1873).

Grieve, Robert, *The Asylum Journal* (Berbice, British Guiana: The Asylum Press, 1881–85; republished in two volumes, Guyana: The Caribbean Press, 2010).

Halliday, Sir Andrew, *A General View of the Present State of Lunatics, and Lunatic Asylums, in Great Britain and Ireland, and in Some Other Kingdoms* (London: Thomas and George Underwood, 1828).

Hill, Robert Gardiner, *Total Abolition of Personal Restraint in the Treatment of the Insane: A Lecture on the Management of Lunatic Asylums and the Treatment of the Insane* (London: Simpkin, Marshall and Co, 1839).

'John Candler's Visit to Antigua', *Caribbean Studies* 5, no.3 (October 1965), 51–7.

'John Candler's Visit to British Guiana', *Caribbean Studies* 4, no.2 (July 1964), 52–61.

'John Candler's Visit to St Kitts', *Caribbean Studies* 5, no.4 (January 1966), 35–8.

'John Candler's Visit to Trinidad', *Caribbean Studies* 4, no.3 (October 1964), 66–71.

Lanaghan, Frances, *Antigua and the Antiguans: A Full Account of the Colony and Its Inhabitants From the Time of the Caribs to the Present Day* (London: Saunders and Otley, 1844).

Macfadyen, J., 'On Medical Topography as Connected with the Choice of a Site for a Lunatic Asylum in a Tropical Country', *Edinburgh Medical and Surgical Journal* LXXI (1849), 114–25.

Martineau, Harriet, 'The Hanwell Lunatic Asylum', *Tait's Edinburgh Magazine* (1834), 305–10.

Monro, John, *Remarks on Dr Battie's Treatise on Madness* (London: John Clarke, 1758).

Official Documents on the Case of Ann Pratt, the Reputed Authoress of a Certain Pamphlet, Entitled "Seven Months in the Kingston Lunatic Asylum, and What I Saw There" (Kingston and Spanish Town, Jamaica: Jordon and Osborn, 1860).

Pinder, Rev. Edward, *Letters on the Labouring Population of Barbados* (Barbados: Barbados, National Cultural Foundation, Heritage Reprint Series, 1858, reprinted 1990).

Pratt, Ann, *Seven Months in the Kingston Lunatic Asylum, and What I Saw There* (Kingston, Jamaica: George Henderson, Savage, & Co, 1860).

Rolph, Dr Thomas, *A Brief Account, Made During a Visit to the West Indies, and a Tour Through the Unites States of America, in Parts of the Years, 1832–33* (Dundas, Upper Canada: Heyworth, Hackstaff, 1836).

Rouse, Richard, *New Lights on Dark Deeds: Being Jottings From the Diary of Richard Rouse, Late Warden of the Lunatic Asylum of Kingston, Edited by His Son* (Kingston, Jamaica: Gall and Myers, 1860).

Schomburgk, Sir Robert, *The History of Barbados* (London: Longman, Brown, Green and Longmans, 1847).

Scott, James, *A Reply to a Letter by Lewis Quier Bowerbank, M.D. Edinburgh, to the Commissioners of the Public Hospital and Lunatic Asylum of Kingston, Jamaica, Relative to the Present State and Management of Those Institutions* (Kingston and Spanish Town, Jamaica: Jordon and Osborn, 1858).

Sturge, Joseph and Thomas Harvey, *The West Indies in 1837: Being the Journal of a Visit to Antigua, Montserrat, Dominica, St. Lucia, Barbados, and Jamaica, Undertaken for the Purpose of Ascertaining the Actual Condition of the Negro Population of Those Islands* (London: Hamilton, Adams & Co., 1838).

Tempany, H.A., *Antigua B.W.I.: A Handbook of General Information, 1911* (London: Waterlow & Sons, 1911).

Tenth Annual Report of the Lincolnshire County Lunatic Asylum, at Bracebridge, Near Lincoln (Lincoln: W. and B. Brooke, 1863).

'The Jamaica Lunatic Asylum', *Journal of Mental Science* VI (1859), 157–67.

Thomson, James, M.D., *A Treatise on the Diseases of Negroes, as They Occur in the Island of Jamaica with Observations on the Country Remedies* (Jamaica: Alex Aikman, 1820).

Treves, Sir Frederick, *The Cradle of the Deep: An Account of a Voyage to the West Indies* (London: Smith, Elder & Co., 1912).

Trollope, Anthony, *The West Indies and the Spanish Main* (London: Chapman and Hall, 4th Edition, 1860).

Secondary works

Books

Andrews, Jonathan, Asa Briggs, Roy Porter, Penny Tucker and Keir Waddington, *The History of Bethlem* (London and New York: Routledge, 1997).

Bartlett, Peter and David Wright (eds.), *Outside the Walls of the Asylum: The History of Care in the Community 1750–2000* (London and New Brunswick: Athlone, 1999).

Beckles, Hilary, *Great House Rules: Landless Emancipation and Workers' Protest in Barbados 1838–1938* (Oxford: James Currey, 2004).

Beckles, Hilary McD., *Britain's Black Debt: Reparations for Caribbean Slavery and Native Genocide* (Mona, Jamaica: University of West Indies Press, 2013).

Beckles, Hilary and Verene Shepherd (eds.), *Caribbean Freedom: Economy and Society From Emancipation to the Present* (London: James Currey, 1993).

Bell, Leland, *Social and Mental Disorder in Sub-Saharan Africa: The Case of Sierra Leone, 1787–1990* (New York, London: Greenwood Press, 1991).

Bolt, Christine, *Victorian Attitudes to Race* (London: Routledge Kegan Paul, 1971).

Brathwaite, Kamau, *The Development of Creole Society in Jamaica 1770–1820* (Oxford University Press, 1971; Miami: Ian Randle, 2005).

Brereton, Bridget, *Race Relations in Colonial Trinidad* (Cambridge: Cambridge University Press, 1979).

Brereton, Bridget, *A History of Modern Trinidad, 1783–1962* (Port of Spain, London: Heinemann, 1981).

Brereton, Bridget and Kelvin A. Yelvington (eds.), *The Colonial Caribbean in Transition: Essays on Post-Emancipation Social and Cultural History* (Jamaica: University of West Indies Press, 1999).

Brown, Vincent, *The Reaper's Garden: Death and Power in the World of Atlantic Slavery* (Cambridge, MA and London: Harvard University Press, 2008).

Bryan, Patrick, *The Jamaican People, 1880–1902* (London: Macmillan Caribbean, 1991).

Burnard, Trevor, *Mastery, Tyranny and Desire: Thomas Thistlewood and His Slaves in the Anglo-Jamaican World* (Chapel Hill and London: University of North Carolina Press, 2004).

Carter, Henderson, *Labour Pains: Resistance and Protest in Barbados 1838–1904* (Kingston, Jamaica: Ian Randle, 2012).

Cherry, Stephen, *Mental Health Care in Modern England: The Norfolk Lunatic Asylum/St Andrew's Hospital, 1810–1998* (Woodbridge and Rochester, NY: Boydell, 2003).

Coleborne, Catharine, *Madness in the Family: Insanity and Institutions in the Australasian Colonial World, 1860–1914* (Basingstoke: Palgrave Macmillan, 2010).

Cox, Catherine and Hilary Marland (eds.), *Migration, Health and Ethnicity in the Modern World* (London: Palgrave Macmillan, 2013).

Curtin, Philip D., *Two Jamaicas: The Role of Ideas in a Tropical Colony 1830–1865* (Cambridge, MA: Harvard University Press, 1955).

Curtin, Philip D., *Death by Migration: Europe's Encounter with the Tropical World in the Nineteenth Century* (Cambridge: Cambridge University Press, 1989).

De Barros, Juanita and Sean Stitwell (eds.), *Colonialism and Health in the Tropics*, in *Caribbean Quarterly* 49, no.4 (December 2003).

De Barros, Juanita, Sean Palmer and David Wright (eds.), *Health and Medicine in the Circum-Caribbean, 1800–1968* (London: Routledge, 2009).

Digby, Anne, *Madness, Morality and Medicine: A Study of the York Retreat, 1796–1914* (Cambridge: Cambridge University Press, 1985).

Ernst, Waltraud, *Mad Tales From the Raj: The European Insane in British India, 1800–1858* (London and New York: Routledge, 1991).

Ernst, Waltraud and Bernard Harris (eds.), *Race, Science and Medicine, 1700–1960* (London and New York, Routledge, 1999).

Evans, Julie, *Edward Eyre, Race and Colonial Governance* (Dunedin, New Zealand: University of Otago Press, 2005).

Fanon, Frantz, *The Wretched of the Earth* (Harmondsworth: Penguin, 1985 edition).

Foucault, Michel, *History of Madness*, ed. Jean Khalfa (London and New York: Routledge, 2006).

Garton, Stephen, *Medicine and Madness: A Social History of Insanity in New South Wales*, (Kensington, New South Wales: New South Wales University Press, 1988).

Green, William A., *British Slave Emancipation: The Sugar Colonies and the Great Experiment* (Oxford: Clarendon Press, 1976, reprinted 1981).

Hall, Catherine, *Civilising Subjects: Metropole and Colony in the English Imagination* (Cambridge: Polity Press, 2002).

Hall, Douglas, *Free Jamaica 1838–1865: An Economic History* (Aylesbury: Ginn and Company, 1981 edition – first published 1959).

Heuman, Gad, *Between Black and White: Race, Politics, and the Free Coloreds in Jamaica, 1792–1865* (Westwood, Connecticut: Greenwood Press, 1981).

Heuman, Gad, *'The Killing Time': The Morant Bay Rebellion in Jamaica* (London: Macmillan, 1994).

Heuman, Gad and David V. Trotman (eds.), *Contesting Freedom: Control and Resistance in the Post-Emancipation Caribbean* (Oxford: Macmillan, 2005).

Hickling, Frederick W. and Eliot Sorel (eds.), *Images of Psychiatry: The Caribbean* (Kingston, Jamaica: Stephenson's Litho Press, 2005).

Higman, Barry W., *Slave Populations of the British Caribbean, 1807–1834* (Baltimore: Johns Hopkins University Press, 1984).

Howells, John G. (ed.), *World History of Psychiatry* (New York: Brunner/Mazel, 1975).

Hunter, Richard and Ida Macalpine, *Three Hundred Years of Psychiatry 1535–1860* (London: Oxford University Press, 1963).

Jackson, Lynette A., *Surfacing Up: Psychiatry and Social Order in Colonial Zimbabwe, 1908–1968* (Ithaca and London: Cornell University Press, 2005).

Johnson, Ryan and Anna Khalid (eds.), *Public Health in the British Empire: Intermediaries, Subordinates, and the Practice of Public Health, 1850–1960* (New York and London: Routledge, 2012).

Jones, Kathleen, *A History of the Mental Health Services* (London: Routledge and Kegan Paul, 1972).

Jones, Margaret, *The Hospital System and Health Care: Sri Lanka, 1815–1960* (New Delhi: Orient Blackswan, 2009).

Jones, Margaret, *Public Health in Jamaica, 1850–1940* (Jamaica: University of West Indies Press, 2013).

Keller, Richard C., *Colonial Madness: Psychiatry in French North Africa* (Chicago and London: University of Chicago Press, 2007).

Kirk-Greene, Anthony, *Britain's Imperial Administrators, 1858–1966* (Basingstoke: Macmillan, 2000).

Lorimer, Douglas A., *Colour, Class and the Victorians: English Attitudes to the Negro in the Nineteenth Century* (Leicester: Leicester University Press, 1978).

Mahone, Sloan and Megan Vaughan (eds.), *Psychiatry and Empire* (London: Palgrave Macmillan, 2007).

McCarthy, Angela and Catharine Coleborne (eds.), *Migration, Ethnicity and Mental Health: International Perspectives, 1840–2010* (New York and London: Routledge, 2012).

McCulloch, Jock, *Colonial Psychiatry and the 'African Mind'* (Cambridge: Cambridge University Press, 1995).

Mellett, David J., *The Prerogative of Asylumdom: Social, Cultural and Administrative Aspects of the Institutional Treatment of the Insane in Nineteenth-Century Britain* (New York: Garland, 1982).

Melling, Joseph and Bill Forsythe (eds.), *Insanity, Institutions and Society: A Social History of Madness in Contemporary Perspective* (London and New York: Routledge, 1999).

Melling, Joseph and Bill Forsythe, *The Politics of Madness: The State, Insanity and Society in England, 1845–1914* (London and New York: Routledge, 2006).

Mills, James H., *Madness, Cannabis and Colonialism: The 'Native-Only' Lunatic Asylums of British India, 1857–1900* (Basingstoke: Palgrave Macmillan, 2000).

Mills, James H., *Cannabis Britannica: Empire, Trade and Prohibition* (Oxford: Oxford University Press, 2003).

Monk, Lee-Ann, *Attending Madness: At Work in the Australian Colonial Lunatic Asylum* (Amsterdam and New York: Rodopi, 2008).

Moore, Brian L., *Race, Power and Social Segmentation in Colonial Society: Guyana After Slavery 1838–1891* (Montreux, Switzerland: Gordon and Breach, 1987).

Moore, Brian L., *Cultural Power, Resistance and Pluralism: Colonial Guyana 1838–1900* (Mona, Jamaica: University of West Indies Press: 1995).

Moore, Brian L. and Michelle A. Johnson, *Neither Led Nor Driven: Contesting British Cultural Imperialism in Jamaica, 1865–1920* (Mona, Jamaica: University of West Indies Press, 2004).

Moran, James E., *Committed to the State Asylum: Insanity and Society in Nineteenth-Century Quebec* (McGill: Queen's University Press, 2000).

Morgan, Philip D. and Sean Hawkins (eds.), *Black Experience and the Empire* (Oxford: Oxford University Press, 2004).

Newton, Melanie J., *The Children of Africa in the Colonies: Free People of Colour in Barbados in the Age of Emancipation* (Baton Rouge: Louisiana State University Press, 2008).

Parle, Julie, *States of Mind: Searching for Mental Health in Natal and Zululand, 1868–1918* (Scottsville, South Africa: University of KwaZulu Natal Press, 2007).

Paton, Diana, *No Bond But the Law: Punishment, Race and Gender in Jamaican State Formation* (Durham and London: Duke University Press, 2004).

Pope-Hennessy, James, *Verandah: Some Episodes in the Crown Colonies, 1867–1889* (London: George Allen and Unwin, 1964).

Porter, Bernard, *The Absent-Minded Imperialists: Empire, Society and Culture in Britain* (Oxford: Oxford University Press, 2004).

Porter, Roy, *Mind Forg'd Manacles: A History of Madness in England From the Restoration to the Regency* (London: Athlone, 1987).

Porter, Roy and David Wright (eds.), *The Confinement of the Insane: International Perspectives, 1800–1965* (Cambridge: Cambridge University Press, 2003).

Reinarz, Jonathan and Leonard Schwarz (eds.), *Medicine and the Workhouse* (Rochester, New York and Woodbridge: 2013).

Roopnarine, Lomarsh, *Indo-Caribbean Indenture: Resistance and Accommodation, 1838–1920* (Kingston, Jamaica: University of West Indies Press, 2007).

Sadowsky, Jonathan, *Imperial Bedlam: Institutions of Madness in Colonial Southwest Nigeria* (Berkeley and London: University of California Press, 1999).

Scull, Andrew, *Social Order: Mental Disorder: Anglo-American Psychiatry in Historical Perspective* (London: Routledge, 1989).

Scull, Andrew, *The Most Solitary of Afflictions: Madness and Society in Britain, 1700–1900* (New Haven and London: Yale University Press, 1993).

Shepherd, Verene and Glen L. Richards, *Questioning Creole: Creolisation Discourses in Caribbean Culture* (Oxford: James Currey, 2002).

Sheridan, Richard B., *Doctors and Slaves: A Medical and Demographic History of Slavery in the British West Indies, 1680–1834* (Cambridge and New York: Cambridge University Press, 1985).

Smith, Leonard, *'Cure, Comfort and Safe Custody': Public Lunatic Asylums in Early Nineteenth-Century England* (London: Leicester University Press, 1999).

Smith, Leonard, *Lunatic Hospitals in Georgian England, 1750–1830* (London: Routledge, 2007).

Stevenson, Christine, *Medicine and Magnificence: British Hospital and Asylum Architecture, 1660–1815* (New Haven and London: Yale University Press, 2000).

Thomas, Hugh, *The Slave Trade: The History of the Atlantic Slave Trade 1440–1870* (London: Picador, 1997).

Thompson, Alvin O. (ed.), *In the Shadow of the Plantation: Caribbean History and Legacy* (Kingston, Jamaica: Ian Randle, 2002).

Topp, Leslie, James Moran and Jonathan Andrews (eds.), *Madness, Architecture and the Built Environment: Psychiatric Spaces in Historical Context* (London: Routledge, 2007).

Trotman, David V., *Crime in Trinidad: Conflict and Control in a Plantation Society, 1838–1900* (Knoxville: University of Tennessee Press, 1986).

Vaughan, Megan, *Curing Their Ills: Colonial Power and African Illness* (Cambridge: Polity Press, 1991).

Walvin, James, *Britain's Slave Empire* (Stroud: Tempus, 2007).

Wood, Donald, *Trinidad in Transition: The Years After Slavery* (Oxford: Oxford University Press, 1968).

Woodward, John, *To Do the Sick No Harm: A Study of the British Voluntary Hospital System to 1875* (London and Boston: Routledge and Kegan Paul, 1974).

Yeong, Ng Beng, *Till the Break of Day: A History of Mental Health Services in Singapore, 1841–1993* (Singapore: Singapore University Press, 2001).

Articles and chapters in edited volumes

Adair, Richard, Joseph Melling and Bill Forsythe, 'Migration, Family Structure and Pauper Lunacy in Victorian England: Admissions to the Devon County Pauper Lunatic Asylum, 1845–1900', *Continuity and Change* 12 (1997), 373–401.

Altink, Henrice, 'Modernity, Race and Mental Health Care in Jamaica, c1918–1944', *Journal of the Department of Behavioural Sciences* 2, no.1 (December 2012), 1–19, University of the West Indies, St Augustine's – Accessed at http:/journals.sta.uwi.edu/jbs/index.asp.

Baker, Melvin, 'Insanity and Politics: The Establishment of a Lunatic Asylum in St John's, Newfoundland, 1836–1855', *Newfoundland Quarterly* 77 (1981), 27–31.

Barry, Lorelle and Catharine Coleborne, 'Insanity and Ethnicity in New Zealand: Maori Encounters with the Auckland Mental Hospital, 1860–1900', *History of Psychiatry* 22, no.3 (2011), 285–301.

Beaubrun, Michael H., P. Bannister, L.F.E. Lewis, G. Mahy, K.C. Royes, P. Smith and Z. Wisinger, 'The West Indies', in John G. Howells (ed.), *World History of Psychiatry* (New York: Brunner/Mazel, 1975), 507–27.

Belle, George, 'The Abortive Revolution of 1876 in Barbados', in Beckles and Shepherd (eds.), *Caribbean Freedom: Economy and Society From Emancipation to the Present* (1993), 181–91.

Boa, Sheena, 'Discipline, Reform or Punish? Attitudes Towards Juvenile Crimes and Misdemeanours in the Post-Emancipation Caribbean, 1838–88', in Heuman and Trotman (eds.), *Contesting Freedom: Control and Resistance in the Post-Emancipation Caribbean* (2005), 65–86.

Bolland, O. Nigel, 'Systems of Domination After Slavery: The Control of Land and Labour in the British West Indies After 1838', in Beckles and Shepherd (eds.), *Caribbean Freedom: Economy and Society From Emancipation to the Present* (1993), 107–23.

Borome, Joseph A., 'John Candler's Visit to America, 1850', *The Bulletin of Friends Historical Association of Philadelphia* 48, no.1 (Spring 1959), 21–62.

Borome, Joseph A. (ed.), 'John Candler's Visit to Barbados, 1849', *Journal of the Barbados Museum and Historical Society* 28, no.4 (August 1961), 128–36.

Borome, Joseph, 'John Candler's Visit to the Virgin Islands', *Caribbean Studies* 4, no.1 (April 1964), 40–8.

Carter, Henderson, 'The Bridgetown Riot of 1872', in Thompson (ed.), *In the Shadow of the Plantation: Caribbean History and Legacy* (2002), 334–48.

Challenger, Denise, 'A Benign Place of Healing? The Contagious Diseases Hospital and Medical Discipline in Post-Slavery Barbados', in De Barros, Palmer and Wright (eds.), *Health and Medicine in the Circum-Caribbean, 1800–1968* (2009), 98–120.

Coleborne, Catharine, 'Making "Mad" Populations in Settler Colonies: The Work of Law and Medicine in the Creation of the Colonial Asylum', in D. Kirby and C. Coleborne (eds.), *Law, History, Colonialism: The Reach of Empire* (Manchester: Manchester University Press, 2001), 106–22.

Coleborne, Catharine, 'Passage to the Asylum: The Role of the Police in Committals of the Insane in Victoria, Australia, 1848–1900', in Porter and Wright (eds.), *The Confinement of the Insane: International Perspectives* (2003), 129–48.

Coleborne, Catharine,'Families, Patients and Emotions: Asylums for the Insane in Colonial Australia and New Zealand, c1880–1910', *Social History of Medicine* 19 (2006a), 425–42.

Coleborne, Catharine, ' "His Brain Was Wong, His Mind Astray": Families and the Language of Insanity in New South Wales, Queensland and New Zealand, 1880s–1910', *Journal of Family History* 31 (2006b), 45–65.

Craton, Michael, 'Continuity Not Change: The Incidence of Unrest Among Ex-Slaves in the British West Indies, 1838–1876', in Beckles and Shepherd (eds.), *Caribbean Freedom: Economy and Society From Emancipation to the Present* (1993), 192–206.

Deacon, Harriet, 'Madness, Race and Moral Treatment: Robben Island Lunatic Asylum, Cape Colony, 1846–90', *History of Psychiatry* 7 (1996), 287–97.

Deacon, Harriet, 'Racial Categories and Psychiatry in Africa: The Asylum on Robben Island in the Nineteenth Century', in Ernst and Harris (eds.), *Race, Science and Medicine, 1700–1960* (1999), 101–22.

Deacon, Harriet, 'Insanity, Institutions and Society: The Case of the Robben Island Lunatic Asylum, 1846–1910', in Porter and Wright (eds.), *The Confinement of the Insane: International Perspectives* (2003), 20–53.

De Barros, Juanita, 'Sanitation and Civilization in Georgetown, British Guiana', in De Barros and Stillwell (eds.), *Colonialism and Health in the Tropics* (2003), 65–86.

De Barros, Juanita, ' "Working Cutlass and Shovel": Labour and Redemption at the Onderneeming School in British Guiana', in Heuman and Trotman (eds.), *Contesting Freedom: Control and Resistance in the Post-Emancipation Caribbean* (2005), 39–64.

De Barros, Juanita, 'Dispensers, Obeah and Quackery: Medical Rivalries in Post-Slavery British Guiana', *Social History of Medicine* 25, no.2 (August 2007), 243–61.

De Barros, Juanita, ' "Improving the Standards of Motherhood": Infant Welfare in Post-Slavery British Guiana', in De Barros, Palmer and Wright (eds.), *Health and Medicine in the Circum-Caribbean, 1800–1968* (2009), 165–94.

De Barros, Juanita, ' "A Laudable Experiment": Infant Welfare Work and Medical Intermediaries in Early Twentieth-Century Barbados', in Johnson and Khalid (eds.), *Public Health in the British Empire: Intermediaries, Subordinates, and the Practice of Public Health, 1850–1960* (2012), 100–17.

De Barros, Juanita and Sean Stilwell, 'Introduction: Public Health and the Imperial Project', in De Barros and Stitwell (eds.), *Colonialism and Health in the Tropics* (2003), 1–11.

Ernst, Waltraud, 'Asylums in Alien Places: The Treatment of the European Insane in British India', in W. Bynum, R. Porter and M. Shepherd (eds.), *The Anatomy of Madness: Essays in the History of Psychiatry*, Vol. III, *The Asylum and Its Psychiatry* (London and New York: Routledge, 1988), 48–70.

Ernst, Waltraud, 'Colonial Policies, Racial Politics and the Development of Psychiatric Institutions in Early Nineteenth-Century British India', in Ernst and Harris (eds.), *Race, Science and Medicine, 1700–1960* (1999a), 80–100.

Ernst, Waltraud, 'Out of Sight and Out of Mind: Insanity in Early Nineteenth-Century British India', in Melling and Forsythe (eds.), *Insanity, Institutions and Society: A Social History of Madness in Contemporary Perspective* (1999b), 245–67.

Ernst, Waltraud, 'Madness and Colonial Spaces – British India, c.1800–1947', in Moran Topp and Andrews (eds.), *Madness, Architecture and the Built Environment: Psychiatric Spaces in Historical Context* (2007), 215–38.

Ernst, Waltraud, 'Institutions, People and Power: Lunatic Asylums in Bengal, c1800–1900', in Biswamoy Pati and Mark Harrison (eds.), *The Social History of Health and Medicine in Colonial India* (London: Routledge, 2009), 129–50.

Finnane, Mark, 'Asylums, Families and the State', *History Workshop Journal* 20 (Autumn 1995), 134–48.

Fryar, Christienna D., 'The Moral Politics of Cholera in Postemancipation Jamaica', *Slavery and Abolition* 34, no.4 (December 2013), 598–618.

Gramaglia, Letizia, 'Colonial Psychiatry in British Guiana: Dr Robert Grieve', in K. White (ed.), *Configuring Madness: Representation, Context and Meaning* (Oxford: Inter-Disciplinary Press, 2009), 191–206.

Gramaglia, Letizia, 'Dr Robert Grieve (1839–1906), An "Apostle of Science"' – 'Introduction', in R. Grieve (ed.), *The Asylum Journal* (Guyana: The Caribbean Press, 2010), Vol. 1, xiii–xxxii.

Gramaglia, Letizia, 'Migration and Mental Illness in the British West Indies 1838–1900: The Cases of Trinidad and British Guiana', in Cox and Marland (eds.), *Migration, Health and Ethnicity in the Modern World* (2013), 61–82.

Hall, Douglas, 'The Flight From the Estates Reconsidered: The British West Indies, 1838–1842', in Beckles and Shepherd (eds.), *Caribbean Freedom: Economy and Society From Emancipation to the Present* (1993), 55–63.

Heuman, Gad, 'The British West Indies', in A. Porter (ed.), *The Oxford History of the British Empire*, Vol. III, *The Nineteenth Century* (Oxford and New York: Oxford University Press, 1999), 470–93.

Heuman, Gad, 'From Slavery to Freedom: Blacks in the Nineteenth-Century British West Indies', in Morgan and Hawkins (eds.), *Black Experience and the Empire* (2004), 141–65.

Heuman, Gad, '"Is This What You Call Free": Riots and Resistance in the Anglophone Caribbean', in Heuman and Trotman (eds.), *Contesting Freedom: Control and Resistance in the Post-Emancipation Caribbean* (2005), 104–17.

Heuring, Darcy Hughes, '"In the Cheapest Way Possible...": Responsibility and the Failure of Improvement at the Kingston Lunatic Asylum, 1914–1945', *Journal of Colonialism and Colonial History* 12, no.3 (Winter 2011), Johns Hopkins University, Baltimore (online).

Hickling, Frederick W. and Roger C. Gibson, 'Philosophy and Epistemology of Caribbean Psychiatry', in Hickling and Sorel (eds.), *Images of Psychiatry: The Caribbean* (2005), 75–93.

Hickling, Frederick W. and Hari D. Maharajh, 'Mental Health Legislation', in Hickling and Sorel (eds.), *Images of Psychiatry: The Caribbean* (2005), 43–74.

Howell, Philip and David Lambert, 'Sir John Pope Hennessy and Imperial Government: Humanitarianism and the Translation of Slavery in the Imperial Network', in David Lambert and Alan Lester (eds.), *Colonial Lives Across the*

British Empire: Imperial Careering in the Long Nineteenth Century (Cambridge: Cambridge University Press, 2006), 228–56.

Johnson, Howard, 'The Black Experience in the British Caribbean in the Twentieth Century', in Morgan and Hawkins (eds.), *Black Experience and the Empire* (2004), 317–46.

Jones, Margaret, 'The Most Cruel and Revolting Crimes: The Treatment of the Mentally Ill in Mid-Nineteenth-Century Jamaica', *The Journal of Caribbean History* 42, no.2 (2008), 290–309.

Jones, Margaret, 'Surviving the Colonial Institution: Workers and Patients in the Government Hospitals of Mid-Nineteenth-Century Jamaica', in Johnson and Khalid (eds.), *Public Health in the British Empire: Intermediaries, Subordinates, and the Practice of Public Health, 1850–1960* (2012), 82–99,

Keller, Richard, 'Madness and Colonization: Psychiatry in the British and French Empires, 1800–1962', *Journal of Social History* 35 (Winter 2001), 295–322.

Laurence, Keith O., 'The Development of Medical Services in Trinidad and British Guiana, 1841–1873', in Beckles and Shepherd (eds.), *Caribbean Freedom: Economy and Society From Emancipation to the Present* (1993), 269–73.

Leckie, Jacqueline, 'Unsettled Minds: Gender and Madness in Fiji', in Mahone and Vaughan (eds.), *Psychiatry and Empire* (2007), 99–123.

Mahone, Sloane, 'East African Psychiatry and the Practical Problems of Empire', in Mahone and Vaughan (eds.), *Psychiatry and Empire* (2007), 41–66.

Malcolm, Elizabeth 'Mental Health and Migration: The Case of the Irish, 1850s–1950s', in McCarthy and Coleborne (eds.), *Migration, Ethnicity and Mental Health: International Perspectives, 1840–2010* (2012), 15–38.

Marks, Shula, ' "Every Facility that Modern Science and Enlightened Humanity have Devised": Race and Progress in a Colonial Hospital, Valkenberg Mental Asylum, Cape Colony, 1894–1910', in Melling and Forsythe (eds.), *Insanity, Institutions and Society: A Social History of Madness in Contemporary Perspective* (1999), 268–91.

Marks, Shula, 'The Microphysics of Power: Mental Nursing in South Africa in the First Half of the Twentieth Century', in Mahone and Vaughan (eds.), *Psychiatry and Empire* (2007), 67–98.

Marshall, Woodville, 'Peasant Development in the West Indies Since 1838', in Beckles and Shepherd (eds.), *Caribbean Freedom: Economy and Society From Emancipation to the Present* (1993a), 99–106.

Marshall, Woodville, ' "We Be Wise to Many Things": Blacks' Hopes and Expectations of Emancipation', in Beckles and Shepherd (eds.), *Caribbean Freedom: Economy and Society from Emancipation to the Present* (1993b), 12–20.

Martyr, Philippa, ' "Behaving Wildly": Diagnoses of Lunacy Among Indigenous Persons in Western Australia, 1870–1914', *Social History of Medicine* 24, no.2 (2011), 316–33.

McCandless, Peter, ' "Build! Build!" The Controversy Over the Care of the Chronically Insane in England, 1855–70', *Bulletin of the History of Medicine* 53 (1979), 553–74.

McCarthy, Angela, 'Ethnicity, Migration and the Lunatic Asylum in Early Twentieth-Century Auckland, New Zealand', *Social History of Medicine* 21, no.1 (2008), 47–65.

Mellett, David J., 'Bureaucracy and Mental Illness: The Commissioners in Lunacy 1845–90', *Medical History* 25 (1981), 221–50.

Millette, James, 'The Wage Problem in Trinidad and Tobago, 1838–1938', in Brereton and Yelvington (eds.), *The Colonial Caribbean in Transition: Essays on Post-Emancipation Social and Cultural History* (1999), 55–76.

Moore, Brian L., 'Leisure and Society in Postemancipation Guyana', in Brereton and Yelvington (eds.), *The Colonial Caribbean in Transition: Essays on Post-Emancipation Social and Cultural History* (1999), 108–25.

Moore, Brian L. and Michele A. Johnson, 'Married But Not Parsoned: Attitudes to Conjugality in Jamaica, 1865–1920', in Heuman and Trotman (eds.), *Contesting Freedom: Control and Resistance in the Post-Emancipation Caribbean* (2005), 197–214.

Moore, Robert J., 'Colonial Images of Blacks and Indians in Nineteenth-Century Guyana', in Brereton and Yelvington (eds.), *The Colonial Caribbean in Transition: Essays on Post-Emancipation Social and Cultural History* (1999), 126–58.

Morissey, Carol M., 'The Road to Bellevue: Conditions and Treatment of the Mentally Ill in Jamaica, 1776–1861', *Jamaica Journal* 22, no.3 (August–October 1989), 2–10.

Newton, Melanie, 'Race for Power: People of Colour and the Politics of Liberation in Barbados, 1816–c1850', in Heuman and Trotman (eds.), *Contesting Freedom: Control and Resistance in the Post-Emancipation Caribbean* (2005), 20–38.

Patterson Smith, James, 'The Liberals, Race and Political Reform in the British West Indies, 1866–1874', *The Journal of Negro History* 79 (Spring 1994), 131–46.

Roopnarine, Lomarsh, 'The Other Side of Indo-Caribbean Indenture: Landownership, Remittances and Remigration 1838–1920', *The Journal of Caribbean History* 42, no.2 (2008), 205–30.

Roopnarine, Lomarsh, 'The Indian Sea Voyage Between India and the Caribbean During the Second Half of the Nineteenth Century', *The Journal of Caribbean History* 44, no.1 (2010), 48–74.

Scull, Andrew, 'Psychiatry and Social Control in the Nineteenth and Twentieth Centuries', *History of Psychiatry* 2, no.2 (June 1991), 149–69.

Smith, Leonard, 'The County Asylum in the Mixed Economy of Care', in Melling and Forsythe (eds.), *Insanity, Institutions and Society: A Social History of Madness in Contemporary Perspective* (1999), 33–47.

Swartz, Sally, 'Changing Diagnoses in Valkenburg Asylum, Cape Colony, 1891–1920: A Longitudinal View', *History of Psychiatry* 6 (1995a), 431–51.

Swartz, Sally, 'The Black Insane in the Cape, 1891–1920', *Journal of Southern African Studies* 21, no.3 (September 1995b), 399–415.

Swartz, Sally, 'The Great Asylum Laundry: Space, Classification, and Imperialism in Cape Town', in Topp, Moran and Andrews (eds.), *Madness, Architecture and the Built Environment: Psychiatric Spaces in Historical Context* (2007), 193–213.

Swartz, Sally, 'The Regulation of Colonial Lunatic Asylums and the Origins of Colonial Psychiatry, 1860–1864', *History of Psychology* 13, no.2 (2010), 160–77.

Trotman, David, 'Capping the Volcano: Riots and Their Suppression in Post-Emancipation Trinidad', in Heuman and Trotman (eds.), *Contesting Freedom: Control and Resistance in the Post-Emancipation Caribbean* (2005), 118–41.

Vaughan, Megan, 'Idioms of Madness: Zomba Lunatic Asylum, Nyasaland in the Colonial Period', *Journal of Southern African Studies* 3, no.2 (April 1993), 218–38.

Walton, John K., 'Lunacy in the Industrial Revolution: A Study of Asylum Admissions in Lancashire, 1848–50', *Journal of Social History* 13 (1979), 1–22.

Welch, Pedro L.V., 'From Laissez Faire to Disinterested Benevolence: The Social and Economic Context of Mental Health Care in Barbados, 1870–1920', *The Journal of Caribbean History* 32 (1998), 121–42.

Welch, Pedro L.V., 'Gendered Health Care: Legacies of Slavery in Health Care Provision in Barbados Over the Period 1870–1920', *Caribbean Quarterly* 49, no.4 (December 2003), 104–29.

Wilkins, Nadine Joy, 'Doctors and Ex-Slaves in Jamaica 1834–1850', *Jamaican Historical Review* 17 (1991), *Health, Disease and Medicine in Jamaica*, 19–30.

Wilmot, Swithin, 'Emancipation in Action: Workers and Wage Conflict in Jamaica 1838–1848', in Beckles and Shepherd (eds.), *Caribbean Freedom: Economy and Society From Emancipation to the Present* (1993), 55–61.

Wright, David, 'Getting Out of the Asylum: Understanding the Confinement of the Insane in the Nineteenth Century', *Social History of Medicine* 10 (1997), 137–55.

Wright, David, 'The Certification of Insanity in Nineteenth-Century England and Wales', *History of Psychiatry* 9 (1998), 267–90.

Wright, David, James Moran and Sean Douglas, 'The Confinement of the Insane in Victorian Canada: The Hamilton and Toronto Asylums', in Porter and Wright (eds.), *The Confinement of the Insane: International Perspectives* (2003), 100–28.

Index

Note: Letter 'n' followed by the locators refer to notes.

Printed in the United States
By Bookmasters